Groundbreaking Scientific Experiments, Inventions and Discoveries of the 18th Century

Recent Titles in
Groundbreaking Scientific Experiments, Inventions and
Discoveries through the Ages

Groundbreaking Scientific Experiments, Inventions and Discoveries of the 17th Century
Michael Windelspecht

Groundbreaking Scientific Experiments, Inventions and Discoveries of the 18th Century
Jonathan Shectman

Groundbreaking Scientific Experiments, Inventions and Discoveries of the 19th Century
Michael Windelspecht

GROUNDBREAKING SCIENTIFIC EXPERIMENTS, INVENTIONS AND DISCOVERIES OF THE 18TH CENTURY ───────────

JONATHAN SHECTMAN

Groundbreaking Scientific Experiments, Inventions and
Discoveries through the Ages
ROBERT E. KREBS, SERIES ADVISER

GREENWOOD PRESS
Westport, Conn. • London

Library of Congress Cataloging-in-Publication Data

Shectman, Jonathan, 1972–
    Groundbreaking scientific experiments, inventions, and discoveries of the 18th
century / Jonathan Shectman.
        p.   cm. – (Groundbreaking scientific experiments, inventions and
    discoveries through the ages)
    Includes bibliographical references and indexes.
    ISBN 0-313-32015-2 (alk. paper)
    1. Science–History–18th century.   2. Technology–History–18th century.
I. Title.   II. Series.
Q125.S5178 2003
509.033–dc21   2002075306

British Library Cataloguing in Publication Data is available.

Library of Congress Catalog Card Number: 2002075306
ISBN: 0-313-32015-2

First published in 2003

Greenwood Press, 88 Post Road West, Westport, CT 06881
An imprint of Greenwood Publishing Group, Inc.
www.greenwood.com

Printed in the United States of America

The paper used in this book complies with the
Permanent Paper Standard issued by the National
Information Standards Organization (Z39.48-1984)

10 9 8 7 6 5 4 3 2 1

For the girl who loved penny candy

CONTENTS

ALPHABETICAL LIST OF ENTRIES

SERIES FOREWORD

The material contained in five volumes in this series of historical ground-breaking experiments, inventions and discoveries encompasses many centuries from the prehistoric period up to the 20th century. Topics are explored from the time of prehistoric humans, the age of classical Greek and Roman science, the Christian era, the Middle Ages, the Renaissance period from the years 1350 to 1600, the beginnings of modern science of the 17th century, and great experiments, inventions, and discoveries of the 18th and 19th centuries. This historical approach to science by Greenwood Press is intended to provide students with the materials needed to examine science as a specialized discipline. The authors present the topics for each historical period alphabetically and include information about the women and men responsible for specific experiments, inventions, and discoveries.

All volumes concentrate on the physical and life sciences and follow the same historical format that describes the scientific developments of that period. In addition to the science of each historical period, the authors explore the implications of how historical groundbreaking experiments, inventions and discoveries influenced the thoughts and theories of future scientists and how these developments affected peoples' lives.

As readers progress through the volumes, it will become obvious that the nature of science is cumulative. In other words, scientists of one historical period draw upon and add to the ideas and theories of earlier periods. This is evident in contrast to the recent irrationalist philosophy of the history and philosophy of science that views science, not as a unique, self-correcting human empirical inductive activity, but as just another social or cultural activity where scientific knowledge is conjectural, scientific laws are contrived, scientific theories are all false, scientific facts are fickle, and scientific truths are relative. These volumes belie postmodern deconstructionist assertions that no scientific idea has greater validity than any other idea and that all "truths" are a matter of opinion.

For example, in 1992 the plurality opinion by three jurists of the U.S. Supreme Court in *Planned Parenthood v. Case* restated the "right" to abortion by stating: "*at the heart of liberty is the right to define one's own concept of existence, of meaning of the universe, and of the mystery of human life.*" This is a remarkable deconstructionist statement, not because it supports the right to abortion, but because the Court supports the relativistic premise that anyone's concept of the universe is whatever that person wants it to be, and not what the universe actually is based on: what science has determined by experimentation, the use of statistical probabilities, and empirical inductive logic.

When scientists develop factual knowledge as to the nature of nature they understand that "rational assurance is not the same thing as perfect certainty." By applying statistical probability to new factual data this knowledge provides the basis for building scientific hypotheses, theories and laws over time. Thus, scientific knowledge becomes self-correcting as well as cumulative.

In addition, this series refutes the claim that each historical theory is based on a false paradigm (a methodological framework) that is discarded and later is just superseded by a new, more recent theory also based on a false paradigm. Scientific knowledge is the sequential nature that revises, adds to and builds on old ideas and theories as new theories are developed based on new knowledge.

Astronomy is a prime example of how science progressed over the centuries. Lives of people who lived in the prehistorical period were geared to the movement of the sun, moon and stars. Cultures in all countries developed many rituals based on observations of how nature affected the flow of life, including the female menstrual cycle and people's migrations to follow food supplies or adaptations to survive harsh winters. Later, after the discovery of agriculture around 8000 to 9000 B.C.E., people learned to relate climate and weather, the phases of the moon and the periodicity of the sun's apparent motion in relation to the Earth, because these astronomical phenomena seemed to determine the fate of their crops.

The invention of bronze by alloying first arsenic and later tin with copper occurred about 3000 B.C.E. Much later, after discovering how to use the iron found in celestial meteorites, and still later, in 1000 B.C.E. when people learned how to smelt iron from ore, civilization entered the Iron Age. The people of the Tigris-Euphrates region invented the first calendar based on the phases of the moon and seasons in about 2800 B.C.E. During the ancient and classical Greek and Roman periods (about 700 B.C.E. to A.D. 100) mythical gods were devised to explain what was viewed in the heavens or to justify their behavior. Myths based on astronomy, such as the sun and planet gods as well as Gaia the Earth mother, were part of their religions and affected their way of life. This period was the beginning of the philosophical thoughts of Aristotle and others concerning astronomy and nature in general that pre-dated modern science. In about 235 B.C.E. the Greeks first proposed a helio-centric relationship of the sun and planets. Ancient people in Asia, Egypt and India invented fantastic structures to assist the unaided eye in viewing the posi-tions and motions of the moon, stars and sun. These instruments were the forerunners of the modern telescopes and other devices that make modern astronomical discoveries possible. Ancient astrology was based on the belief that the positions of bodies in the heavens controlled one's life. Astrology is still confused with the science of astronomy and is still not based on any reli-able astronomical data.

The ancients knew that a dewdrop on a leaf seemed to magnify the leaf's surface. This led to the invention of a glass bead that could be used as a magnifying glass. In 1590 Zacharias Janssen, an eyeglass maker, dis-covered that two convex lenses, one at each end of a tube, increased the magnification. In 1608 Hans Lippershey's assistant turned the instrument end-to-end and discovered that distant objects appeared closer; thus the telescope was discovered. The telescope has been used for both navigation and astronomical observations since the 17th century. The invention of new instruments, such as the microscope and the telescope, led to further discov-eries such as of the cell by Robert Hooke and the four moons of Jupiter by Galileo, who made this important astronomical discovery that revolutionized astronomy with a telescope of his own design and construction. These inven-tions and discoveries enabled the expansion of astronomy from an ancient "eyeball" science to an ever-expanding series of experiments and discoveries leading to many new theories about the universe. Others invented and improved astronomical instruments, such as the reflecting telescope combined with photography, the spectroscope, and Earth-orbiting astronomical instru-ments that resulted in the discovery of new planets and galaxies as well as new theories related to astronomy and the universe in the 20th century. The age of "enlightenment" through the 18th and 19th centuries culminated in an explosion of new knowledge of the universe that continued through the 20th and into the 21st centuries. Scientific laws, theories, and facts we now know about astronomy and the universe are grounded in the experiments, discov-eries and inventions of the past centuries, just as they are in all areas of science.

The books in the series Groundbreaking Experiments, Inventions and Dis-coveries, are written in easy-to-understand language with a minimum of scientific jargon. They are appropriate references for middle and senior high school audiences as well as for the college-level nonscience major and for the general public interested in the development and progression of science over the ages.

Robert E. Krebs
Series Adviser

ACKNOWLEDGMENTS

First, I wish to thank Bob Krebs, the editorial adviser and project coordinator for this series of books. His thoughtful, reflective criticism was immensely helpful during the writing of this book. More than once, I have thought of Bob as a writer's best advocate. I also wish to thank this book's editors, Debby Adams and Emily Birch, as well as the many other folks at Greenwood. Thanks also to Michael Windelspecht for his helpful production advice. Special thanks to Jeff Reumann for making the superb illustrations for this book. Our fruitful conversations helped gel in my mind the structure and approach for several entries. Jeff's illustrations have also proven critical to effectively representing several abstract scientific concepts in this book. I want to thank my beloved family, especially my parents and grandparents, for their enthusiastic support during the writing of this book. Finally and most of all, I wish to thank my wonderful spouse, Katie—not only for being my research assistant, constructive critic, primary source of advice, chef, therapist, ad hoc rabbi, moral supporter, and late-night coffee brewer—but also for being my soul mate and best friend.

INTRODUCTION

Near the end of the eighteenth century, James Watt (1736–1819) wrote that "the very existence of Britain as a nation seems to me, in great measure, to depend upon her exertions in science. . . ." When he wrote this, Watt was already a legendary figure in England. His steam engines powered nearly every factory in nearly every industry, surpassing all forms of energy that the world had ever seen. His name was, in the minds of scientists and commoners alike, very nearly synonymous with the industrial revolution. Many people even thought that the steam engine might one day carry them all the way to the moon.

By the end of the century, scientific research and its practical applications had become two sides of a golden coin. Science could not only remake the world, it could very well carry ordinary people to the far-flung reaches of their imagination. However, unity of science and practical application was a relatively new concept during the late eighteenth century.

In ancient Greece, Archimedes (ca. 287–212 B.C.) used principles of geometry to build ingenious war machines that effectively repelled invading Roman armies. Centuries later, Plutarch (ca. 46–120) admonished his predecessor for using mathematics for meekly practical purposes. Even the greatest of Archimedes' contemporaries never conceived of science in the enlightenment terms of practical application. In fact, a cornerstone of ancient philosophy was the distinction between knowledge and usefulness.

A millennium and a half later, Renaissance scholar Francis Bacon (1561–1626) began the arduous task of conceptually removing "philosophy"—and science—from the ivory tower and placing it at the service of ordinary people. The seventeenth century "age of genius" began delineating applied science for collective benefit, but individual reactions of philosopher-scientists varied wildly. In the heavily chronicled life of Isaac Newton

(1642–1727), only a single laugh is officially recorded. Newton lent a copy of Euclid's *Elements* to a friend. After perusal, Newton's friend asked him about the useful benefits of such a study, thereby eliciting the historic chortle.

At end of the seventeenth century, few scientists could have imagined the world at the end of the following century. Conceptually, the world was a different place. Scientifically, the world was changing faster than it ever had before. Rarely, these events came together on a single day. On November 20, 1783, a crowd of 400,000 Parisians gathered to witness the first human flight in a hot air balloon. Most scholars believe that the crowd's scientists, royalty and ordinary people—all 400,000 of them—made up the largest human gathering to that point in history.

Just 10 days later, a smaller crowd gathered for the first launch of a hydrogen (lighter than air) balloon, which carried a human passenger to a height of 9,000 feet. On seeing the ascent, a spectator somehow failed to see the practical aspects of the new invention, turned to the man next to him, and asked what good the new invention was. The man next to him was Benjamin Franklin (1706–1790), minister to France from the newly formed United States. To this man's question, Franklin replied, "What good is a newborn baby?"

Franklin's hapless spectator represents an unwitting eighteenth-century counterpart to Newton's Euclid-reading friend from a century earlier. Franklin himself represents a conceptual shift on a level the world had never before witnessed, a link between empirical science and technical application in the person of the scientist. Franklin (or one of his contemporary colleagues) might have laughed at the spectator's absurd question, but never at its being asked. The very spirit of the century—its urgent, flourishing ethos impatiently straining against its own temporal yoke—required and even demanded an answer.

That this answer came in the form of Franklin's cheerful question is emblematic of the optimistic spirit of the times. Science must and will have a purpose. Science will make things better, easier, more efficient. Science will improve the lives of ordinary people. Science will do something. In the easy, scientific optimism of the late eighteenth century, one can almost hear the words of seventeenth-century mathematician Blaise Pascal (1623–1662), like a scientific prophet from a former era: "Joy, Joy, Joy, Tears of Joy."

When Franklin began his study of electricity, there existed no practical use for the substance. In a famous letter, Franklin apologized for this fact. (He later made amends when he invented the lightning rod, which is still in use today.) Franklin wrote that he was "Chagrin'd a little that We have hitherto been able to discover Nothing in the Way of Use to Mankind" from his many early electrical experiments. Similarly, when German mathematician Gottfried Wilhelm Leibniz (1646–1716) invented his binary system of numbers, eighteenth-century scientists and mathematicians accepted it as rationally sound. However, they paid no further attention because it offered no practical use. (It was entirely abdicated when binary numbers formed the basis of modern computing in the 1940s.) The point is that, without practical application, scientific knowledge was not enough. The eighteenth century opened all forms of esoteric scientific practice to public scrutiny.

In the space of little more than a century, science was completely transformed. In the same space, the world also was completely transformed. It took few more than 100 years to journey from Newton's time-honored laugh (a laugh heard since antiquity) to Franklin's easy optimism about the practice of science. At this point, a question presents itself: "How did this happen?" Or, more urgently, "How did it happen so *quickly*?"

In a very literal sense, these questions form the basis of this book. Each entry represents a groundbreaking experiment, invention or discovery on the road from science-in-transition of the seventeenth century "age of genius" to science-as-benefit of the eighteenth-century "age of reason" (and beyond). The seventeenth century claimed the very scientific precepts that the eighteenth century demonstrated through observation, experiment, and analysis; calculation, invention, and application; technique, classification, and nomenclature; and, especially, widespread dissemination, for eighteenth-century scientists realized their work would be nothing, if not readily available to as many people as possible.

For example, the Franklin heating stove would have made its inventor one of the wealthiest men in America. Instead of patenting it, Franklin printed its plans—along with (and this is key) an easily accessible, scientific explanation of thermal energy. A decade earlier, Pierre Fauchard (1678–1761) began founding modern dentistry with publication of *Le Chirurgien Dentiste* in 1728. Previously, dental knowledge was the closely guarded, esoteric knowledge of barber-surgeons and tooth-drawers. After Fauchard's publication, dentists began publishing their findings and freely sharing information for the benefit of their patients and fellow practitioners.

At the eighteenth century's end, mathematician and surveyor Benjamin Banneker (1731–1806) published impeccably precise information—complete with his scientific calculations—in his excellent *Farmer's Almanac*. During this period, scientists Denis Diderot (1713–1784) and Jean D'Alembert (1717–1783) began publishing the first modern (17-volume) encyclopedia. Never before had such a diversity of scientific topics appeared in one single compilation. Diderot and D'Alembert's *Encyclopédie* became emblematic of the free exchange of ideas that began taking place among scientists of the eighteenth century. For this reason, the eighteenth century is called not only the "age of reason," but also the "age of the encyclopedia."

So how did this revolution in applied science happen so quickly? One straightforward answer is that the age of reason had one more permutation: the age of revolution—French, American, chemical, industrial, and so forth. In other words, this answer says that the same century birthed democracy and also applied science quite literally from the same general milieu. A tougher answer says that it happened a little at a time, through the hard work of experimenters, inventors and discoverers engaged in a process much larger than their personal subjectivities and ending far beyond their own historical moments.

The eighteenth century opened with just such a revolutionary theory. In 1700, Georg Ernst Stahl (1660–1734) proposed the phlogiston theory of substances. This theory explained combustion, calcification (or oxidation), and animal respiration. It stated that these processes release the hypothetical substance phlogiston from an object to the surrounding air. Phlogiston was the first reasonable, rational theory of substances. It explained both everyday events (such as a burning candle) and scientific phenomena (such as calcification experiments). In terms of practical uses for both, phlogiston was wildly successful.

Virtually all discoveries of pneumatic chemistry were made by scientists subscribing to the theory of phlogiston. Discoveries of every one of the first independent gases to be identified—carbon dioxide, nitrogen, hydrogen, chlorine, and especially oxygen—were a direct result of scientists consciously conducting phlogiston experiments. Because of pneumatic discoveries, scientists realized the compound nature of water. For the first time, scientists had unshackled chemistry from the ancient four-element theory of Aristotle (384–322 B.C.), which had led them astray for nearly two millennia. Now, scientists quit trying to discover how many elements existed (be it 4, 8, or 177) and began trying to discover as many as they could. Science was quickly becoming free of the ancient, backward practice of adapting conclusions to fit abstract theories.

Indeed, phlogiston was so practically successful that it paved the way for its own replacement by a better theory. In the 1780s, the founder of modern oxygen chemistry, Antoine Lavoisier (1743–1794), began paying close attention to the concept of mass in quantitative chemical reactions. In many reactions, he found that reactants gained weight. If phlogiston was escaping, he wondered how this could be. Because phlogiston had led to the discovery of the first distinct gases, Lavoisier began to suspect that something from the air was joining with an object during combustion. When Lavoisier discovered that something was, in fact, joining with an object—and found that that something was oxygen—the phlogiston theory self-destructed. It had served its practical use. Lavoisier also drew up a seminal list of elements and compounds, and created a rational system of chemical nomenclature. The basic transformation of chemistry took place during the space of a few frenetic years in the 1770s and 1780s. In fact, the chemical revolution could be said to be conceptually completed with Lavoisier's publication of the magnificently important book *Traité Élémentaire de Chimie* in 1789. Since Lavoisier's day, scientists have adopted his oxygen theory of chemistry—most succinctly laid out in *Traité*—with near universality.

Phlogiston was certainly not the basis of all scientific research, but it was fruitful in many different areas. Henry Cavendish (1731–1810) was one of the last scientists to subscribe to phlogiston. This theory led him to the discovery that hydrogen is an independent gas. Cavendish quickly discovered that hydrogen was only about one-fifth the density of ordinary air. Fellow chemist Joseph Black (1728–1799) began experimenting with the uses of this gas, when he captured hydrogen in a small bag, which began to "float." Several yeas later, French chemist Jacques Alexandre Cesar Charles (1746–1823) began a series of fruitful inquires with the gas. These trials led to the formulation of Charles's law, which states that a gas expands by the same fraction of its original volume as its temperature rises. In turn, this law led to Charles's understanding of gas density and to his successful hydrogen balloon launch (the one on December 1, 1783).

During this period, Joseph Black was also conducting a series of inquiries into latent heat. Black noticed that an "extra" amount of heat was required to turn ice into water and to turn boiling water into steam. Black shared his scientific findings with his friend, James Watt. Watt had noted a similar finding in his work with outdated steam engines. He found that he needed a tremendous amount of fuel just to heat and reheat the engine's single chamber, which was cooled to create every power stroke. Out of the fruitful conversations between Black and Watt came Watt's most important innovation, the separate condenser. Watt realized that, in effect, he had to overcome latent heat for each and every power stroke. Once he took latent heat out of the picture (with the separate condenser), he could run his steam engines on about a quarter of the fuel. Later, dependable steam engines were adapted for the first self-propelled vessels in history, which were river-navigating steamboats. Eventually, steamboats ushered in a new, heralded age of modern transportation.

In biology, a new age was dawning, too. Stephen Hales (1677–1761), the founder of pneumatic chemistry, began a new study of plant physiology. In doing so, he ushered in a methodological shift in scientific practice so familiar it seems almost like "common sense" today. Since Aristotle, philosophers had studied plant physiology by projecting the features of animals onto plants. This practice was called the analogist approach because it draws comparisons between plants and animals. For example, many early enlightenment scientists thought that sap "circulated" similarly to blood in animals. Even today, one hears residual analogist phrases like the "heart" of wood and the "veins" of a leaf.

Hales not only rejected earlier findings. He also rejected this method. In earlier experiments, Hales had become the first person to measure blood pressure, when he did so on a horse. One day Hales was working with a "bleeding" grapevine when he suddenly realized he needed a similarly rigorous empirical approach to studying plant physiology. In a number of experiments to measure the "force" of sap in plants, Hales essentially replaced the analogist tradition with a new method of rational experimentation, decisive analysis and functional equilibrium. For this reason, Hales gets credit for founding the field of plant physiology. He also gets credit for founding many modern methods of scientific analysis.

Astronomy is often called the oldest science. It is also a field in which dedicated amateurs have made many of the most important discoveries. During the eighteenth century, an English sailor-turned-astronomer, Thomas Wright (1711–1786), spent many nights staring up at the Milky Way. He first suggested that the Milky Way is a slablike distribution of stars. This important discovery eventually led twentieth-century astronomers to one of that century's watershed discoveries: the classification of the Milky Way as a *spiral galaxy*.

A few years after Wright, Edmond Halley (1656–1742) demonstrated that comets follow a predictable orbit. Throughout the Middle Ages and well into the eighteenth century, the mere sighting of a comet routinely caused public panic and end-of-the-world scenarios. Halley demonstrated that, far from harbingers of earthly demise, the motion of comets is just as predictable as that of more conventional stellar objects. During the same decade, astronomer William Herschel (1738–1822) became the first person to discover a planet since the dawn of recorded history, when he realized that Uranus was indeed a planet (rather than a comet).

Throughout the history of astronomy, one sees time and again that astronomy is not just for professional astronomers. The practice of celestial navigation made amateur astronomers out of ancient sailors, who often had no formal education. When the age of exploration dawned in Europe, practical need propelled the need for astronomy to new and unforeseen levels. More than any other science, astronomy caught the attention (and funding) of Europe's royal governments. Even before the age of eighteenth-century scientific practice, astronomy had become a living practice of applied science on the high seas. Overwhelming need for practical instruments led to the revolutionary navigation inventions of the eighteenth century, navigational quadrants (strictly speaking, octants and sextants). Instruments for optically "lowering" a celestial body to the horizon for purposes of navigation were invented independently by John Hadley (1682–1744) in England and Thomas Godfrey (1704–1749) in the United States. In today's age of Global Positioning System (GPS) satellites, instruments like those of Hadley and Godfrey are still routinely found aboard vessels of varying sizes.

Not all eighteenth-century work followed the specific model of applied science. Pierre-Louis de Maupertuis (1698–1759) stumbled onto the principle of least action while trying to discern the scientific basis for the existence of a deity. An Enlightenment thinker to the core, Maupertuis thought he had found it with an eighteenth-century version of Ockham's razor. Voltaire (1694–1778) had a good laugh at the whole notion. Nonetheless, the principle of least action, stated broadly, is the idea that nature is thrifty in all its actions. This principle explains water running downhill as well as a ray of light "bending" when it enters water. The principle of least action turned out to be one of the most important generalizations in the history of science. This generalization was not specific to work done during the eighteenth century but continues in importance—along with many of the century's experiments, inventions, and discoveries.

In similar context, another of the eighteenth (and nineteenth) century's greatest chemists, Humphry Davy (1778–1829), wrote that "science, like that nature to which it belongs, is neither limited by time nor by space. It belongs to the world, and is of no country. . . ." Along with Count Benjamin Thompson Rumford (1753–1814), Davy made one of the century's final discoveries: that heat was synonymous neither with the substance phlogiston nor with the so-called heat fluid, caloric. Rumford and Davy made the lasting discovery during 1798 and 1799 that heat is a form of energy. At the end of the century, applied science had come so far since 1700 and the advent of phlogiston.

And here, at the eighteenth century's end, is Watt's statement of Britain's debt to science, from the beginning of this introduction. When he made the statement, Britain was on the threshold of a period of unprecedented growth. It had lost its crown colonial jewel and had lost recent political capital to the new French republic, but Britain was still the most technologically advanced country in the world. It stood on the threshold of unprecedented empire and unequaled riches, but Watt does not mention military prowess or economic supremacy. Science, he says, is the key to Britain's greatness—Watt hangs Britain's "very survival as a nation" on the single peg of applied science. At the end of the eighteenth century, this contested practice (of one nation and many, of all nations and none) had come so far since Plutarch's scold and Newton's laugh.

Selected Bibliography

Bowen, Catherine Drinker. *The Most Dangerous Man in America: Scenes from the Life of Benjamin Franklin*. Boston: Little, Brown, 1974.

Ketcham, Ralph L. *Benjamin Franklin*. New York: Washington Square Press, 1966.

Partington, J. R. *A Short History of Chemistry*. New York: St. Martin's Press, 1965.

TIMELINE OF
IMPORTANT EVENTS

1687   Isaac Newton (1642–1727) publishes his epoch-making book, *Principia*. This book contains Newton's laws of motion, his theory of universal gravitation and his initial findings on the nature of light. It also recapitulates many of the seventeenth century's most important scientific achievements. *Principia* provides much of the theoretical basis for the eighteenth century's revolution in applied science and its practical applications.

1700   Georg Ernst Stahl (1660–1734) proposes the first rational, reasonable theory of chemical substances. Stahl says that when substances combust or calcify, or when animals respire, they release the substance phlogiston (from the Greek for "burning"). When a candle burns, Stahl says it releases phlogiston. When the candle no longer will burn, it has rendered its store of phlogiston. This theory also explains independent gases. For example, objects burn brightly in "dephlogisticated air," or oxygen, because this air has lost all of its phlogiston and therefore hungrily grabs it back during combustion. This powerful theory leads to the eighteenth-century discoveries of the first independent gases, as well as many of the first applications of advanced chemical knowledge. Phlogiston is eventually replaced by a better theory, Antoine Lavoisier's (1743–1794) theory of oxygen chemistry (see 1779, 1789). Since the 1790s, the latter theory has been adopted with near universality.

1701   Gottfried Wilhelm Leibniz (1646–1716) first outlines his conception of a binary system of numbers. This is a base 2 system of mathematics that uses only the symbols 0 and 1 for the positional units of twos, fours, eights, sixteens and so on. This system takes its name from the Latin word for "two at a time." In the binary system, each number position has twice the value of the position to its right. For this reason, binary numbers tend to be much longer than their more familiar decimal counterparts. Scientists agree that Leibniz's system is valid. However, it holds little promise for practical application, so they universally ignore it. The binary system comes into widespread use during the mid-twentieth century when scientists invent the first computers, which process data in binary terms.

1704   Isaac Newton publishes *Opticks,* which explains his theories of the physical properties and contents of light. In the *experimentum cruces,* Newton demonstrates that colored (and not white) light is primary. Newton also reinvents the ancient idea of the corpuscular nature of light, which states that light comprises unbreakable particles called "corpuscles." This book is highly influential for the entire century, and well beyond its final years.

1705   Edmond Halley (1656–1742) predicts that the heavily chronicled comet of 1682, 1607, 1531, 1456, 1380 and 1305 will follow its predictable cycle of roughly 75-year returns. This prediction, which is the culmination of many years of mathematical calculations, has the comet returning to Earth in 1758. Halley believes that comets follow a predictable cycle of orbits, cycles and returns (see 1758).

1705   Francis Hauksbee (sometimes Hawksbee or Hawkesbee, ca. 1666–1713) determines that the "barometric light" produced when he shakes mercury in a glass vessel is caused by the action of electrical friction. This is the first realization that electricity can produce light. With this knowledge, Hauksbee also redesigns the electric friction machine, which is very important later in the century (see 1740s).

1709   Explorer John Lawson (ca. 1711) describes the fauna and flora found in the mid-Atlantic region of the New World.

1709   Johann Scheuchzer (1672–1733) introduces the idea that fossilized fish skeletons once belonged to living, organic beings.

1709   Guillaume Amontons (1663–1705) further develops the air thermometer after realizing that air volume decreases when air cools and increases when it is heated.

1710   Deeply influenced by Newton's corpuscular theory of light and color (see 1704), Jakob Christof Le Blon (1667–1741) proposes that three primary colors are sufficient to produce any color of the newly understood light spectrum. After many printing trials, Le Blon becomes convinced that cyan, magenta and yellow—or even blue, red and yellow—are the necessary "primary" colors for printing all other colors. He later publishes his theory of the fundamental nature of these three colors.

1712   Inventor Thomas Newcomen (1663–1729) designs and builds the first really usable steam engine. Though it is hampered by high fuel consumption and frequent breakdowns, Newcomen's engine becomes only the second self-acting machine in history (after clocks). More important, it becomes the first source of energy not known since antiquity. More than any other invention, steam engines power the late-century Industrial Revolution, first in England and then in every major European country.

1714   Daniel Fahrenheit (1686–1736) revolutionizes the design of thermometers when he begins using quicksilver (mercury) in place of alcohol. The use of this liquid metal eliminates the effects of atmospheric pressure and greatly expands the scale of freezing and boiling. Fahrenheit also proposes his temperature scale, based on approximately 30 degrees for the temperature of freezing water and approximately 90 degrees for the temperature of human blood (these are later adjusted). These are limited points, in terms of usefulness, and only the United States still uses the Fahrenheit scale for everyday (nonscientific) temperatures.

1715   Brook Taylor (1685–1731) publishes the watershed book *Methodus Incrementorum,* which most historians of mathematics credit as the founding work of differential equations. This work also contains the renowned Taylor series of expansion. With this book Taylor adds himself to the list of founders of calculus (see 1736, 1797) and also becomes one of the earliest mathematicians to use it for summation of series, as well as interpolation. *Methodus Incrementorum* also contains Taylor's proofs of a well-known theorem: $f(x + h) = f(x) + hf^1(x) + h^2f'''(x) + \ldots$.

1715   Led by Edmond Halley, astronomers throughout Europe carefully record a total solar eclipse for the first time. Previously, Halley correctly predicted the moon's path across the disk of the sun from an observational point in England. This allows him to organize the study and also allays superstitious fears of eclipses that have existed for many centuries.

1717	Lady Mary Montagu (1689–1762), the wife of the British ambassador to Turkey, becomes one of the first westerners to be inoculated against smallpox. Inoculation had been practiced for many centuries in the East. This procedure involves purposely infecting a healthy individual with a mild case of a disease to confer immunity. It is not always successful, and sometimes it proves disfiguring or deadly (see 1796).
1718	Edmond Halley establishes that the enormous differences between stellar positions recorded in antiquity and those recorded by his contemporaries are the result of the proper motion of the stars through space. In other words, Halley discovers that the stars are not fixed in place, as most astronomers accepted as fact, but are in motion (see 1782).
1718	Zabdiel Boylston (1679–1766) performs the first successful mastectomy on the diseased breast of a patient.
1724	John Floyer (1649–1734) writes a watershed book on geriatric medicine entitled *Medicina Gerocomica*.
1728	While trying to decipher stellar parallax, James Bradley (1693–1762) discovers the stellar aberration of light. Building on this discovery, Bradley prepares an accurate chart of the positions of more than 60,000 stars. He also determines the ratio of the speed of light to the speed at which Earth moves—a vastly more accurate method than previous ones. Bradley's figure for the speed of light is only about 1,000 miles per second shy of its actual speed. Also, Bradley provides the first empirical evidence of the Copernican hypothesis. Just as effectively as parallax would have done, aberration demonstrates that Earth is necessarily in orbit around the Sun.
1728	Pierre Fauchard (1678–1761) publishes *Le Chirurgien Dentiste*, which forms the basis of modern dentistry. This book both synthesizes more existing dental knowledge and disseminates Fauchard's revolutionary theories. Fauchard also disseminates plans for his bow-drill, the first really usable dental drill. After this year, modern dentists begin to quickly replace barber-surgeons and carnival quacks, just as drill-and-fill techniques gain popularity over extraction.
1729	Stephen Gray (1695–1736) draws the first empirical distinction between "electrics" and "non-electrics," or insulators and conductors.
1731	John Hadley (1682–1744) in England and Thomas Godfrey (1704–1749) in the United Sates independently invent the navigational quadrant (sextant and octant) within a few months

of one another. These instruments function according to the optical rule, which set forth the principles of light reflection and angles of incidence as outlined by Edmond Halley and Isaac Newton. To make measurements, a navigator holds the instrument vertical to the horizon and moves its arm until the image of a certain celestial body appears to "touch" the line of horizon. The double-reflecting instrument thereby "brings down" the celestial body to the level of the horizon, making it possible to record the vessel's relational position.

1731   René Réaumur (1683–1757) constructs a new temperature scale based on the freezing point of water, which he sets at 0 degrees, and the boiling point of water, which he sets at 80 degrees. One degree Réaumur is therefore one-eightieth the difference of the melting and boiling points of water. Scientists quickly find they can navigate Réaumur's 80 positive degrees with much greater efficiency than they previously could with Fahrenheit's 212 (see 1714). Use of the Réaumur scale becomes widespread (especially in France), though far from universal.

1733   Noting that a rubbed glass repels a piece of metal leaf, but rubbed amber, wax or gum attracts, Charles Dufay (1698–1739) concludes that there must be two kinds of electric "fluid." Later, Benjamin Franklin names these "positive" and "negative" (though not in the modern sense in which the terms are used today). Franklin also disproves the two-fluid theory of Dufay (see 1746–1752).

1733   Stephen Hales (1677–1761) publishes *Haemastaticks* and essentially shatters the analogist approach to plant and animal physiology. From this point forward, plant and animal physiology are considered independent of one another and studied as such.

1733   John Kay (1704–1764) invents the flying shuttle loom for weaving. With a little practice, experienced weavers can send the shuttle racing to and fro across the loom at much greater speed than either one or two weavers could previously manage with their hands. Nearly instantly, Kay's shuttle doubles the productivity of any given loom at very little cost to its owner. Once fitted with Kay's flying shuttle apparatus, the loom requires only one weaver, even for wide cloth. Moreover, one weaver can produce much more cloth than before (see 1785). Now, weavers can meet soaring demand for finished cotton products. This, in turn, drives the need for cotton yarn.

1734–1742   René Réaumur publishes his six-volume *History of the Insects*. In these volumes, Réaumur gives detailed descriptions of the insects and their heretofore esoteric lives. He also publishes many drawings (made by an uncredited artist he hired). His many descriptions are often lavish in the intricate knowledge of tiny creatures. In one striking study, Réaumur details the recently hatched, minute larvae of blister beetles, which he dubs "bee lice" because he observes them attached to the tiny hairs of bees.

1735 Carl von Linné (1707–1778), better known as Carolus Linnaeus, publishes the first edition of *Systema Naturae* (a later edition is published in 1758). This volume first proposes Linnaeus's revolutionary system of binomial classification (from the Latin for "two-name naming"). This is the modern system of naming organisms by two names, one for their genus and one for their species. Many of Linnaeus's groupings are still valid (see 1753).

1736 Leonhard Euler (1707–1783) publishes *Mechanica*. In this text, he systematically applies "the calculus" to mechanics and, in doing so, develops a range of methods that use power series to solve differential equations (see 1715).

1737 *Biblia Naturae*, by microscopist Jan Swammerdam (1637–1680), is published posthumously for the first time. This text contains excellent and previously unknown accounts of a wide variety of insect dissections.

1737 Carolus Linnaeus publishes *Flora Lapponica*, his descriptive work of the customs of the Lapp people and the flora of Lapland.

1738 Daniel Bernoulli (1700–1782) publishes his book *Hydrodynamica*, which contains Bernoulli's principle. For a steady, nonviscous, incompressible liquid flow, Bernoulli's principle is written: $p + \rho v^2/2 + \rho gy = A$, where p is pressure, ρ is density, v is velocity, g is acceleration of gravity, y is height and A is constant. This principle explains that energy is conserved in a moving fluid (liquid or gas). Moving in a horizontal direction, pressure decreases as the fluid's speed increases. Conversely, if the speed decreases, pressure increases. Bernoulli's law explains why water moves more quickly through a narrow part of a horizontal pipe than through its wider part. It also establishes that the pressure is lowest where the speed is greatest. This is the first correct analysis of water flowing in a container.

1740s Building on the work of Francis Hauksbee (see 1705), Jean Antoine Nollet (sometimes known as the Abbé Nollet, 1700–1770), successfully creates a series of "electric lights" in a small room. Using a Hauksbee-type electric machine, Nollet is able to power a series of glowing evacuated flasks through the use of a conductor. This marks one of the first practical applications of electricity as well as electrical conveyance. Years later, electric arcs and, later still, electric light bulbs grow out of the body of knowledge created by Hauksbee and Nollet.

1740   Benjamin Franklin (1706–1790) creates the Franklin stove, the first stove to achieve thermal efficiency. Using this stove, Franklin is able to make his room twice as warm with about a quarter of the fuel. This invention could have made Franklin very wealthy. However, he gives away his plans, as well as a free lesson in thermodynamics, in a pamphlet he prints and freely distributes. Franklin stoves are used with near universality in England and the United States until the mid- to late nineteenth century.

1742   Anders Celsius (1701–1744) proposes a new temperature scale based on two degrees, 0 and 100. However he designates 100 as the freezing point of water and 0 as the boiling point. For modern people habituated to things "heating up" and "cooling down," this thermometer may seem reversed and awkward (it was later changed). Many scientists and most lay people outside the United States use the Celsius temperature scale today.

1744   Building on the medieval principle know as Ockham's razor, Pierre-Louis de Maupertuis (1698–1759) states the principle of least action. This groundbreaking generalization explains that nature is thrifty in all its actions. Therefore, a ray of light "bends" when it enters water and water flows only downhill, because both minimize the product: action. Maupertuis further defines action as the product of mass (m), velocity (v) and distance (s) ($m \times v \times s =$ action).

1744   Jean Le Rond D'Alembert (1717–1783) performs a series of complex experiments involving the drag on a spherical object immersed in flowing liquid. Out of them comes his formation known as D'Alembert's paradox, a new generalization of Newton's third law of motion, which states that for every action, there is an equal and opposite reaction. This finding demonstrates that Newton's law holds not only for fixed bodies, but also for bodies that are free to move, such as flowing fluids. Suddenly, D'Alembert has shown that Newton's law explains a host of additional phenomena.

1745–1746   Ewald Jurgen von Kleist (1700–1748) and Peter van Musschenbroek (1692–1761) independently invent Leyden jars, the first devices capable of storing and discharging electricity. In modern terms, a Leyden jar acts as a simple capacitor.

1746	Andreas Marggraf (1709–1782) first isolates zinc.
1746–1752	Benjamin Franklin conducts his most fruitful electrical inquiries, discoveries and inventions. During this period, Franklin undertakes his famous kite experiment, which proves that lightning is a powerful form of electrical discharge. He also coins the terms "positive" and "negative" electricity (though his understanding of the terms differed from the modern notion) and demonstrates effectively that there is only one kind of electric "fluid." During this period, Franklin also designs the lightning rod, which is still in use, essentially unchanged, 250 years later. This is the first use of electrical knowledge for practical purposes (see 1733).
1747	Andreas Marggraf discovers sugar in beets, which later becomes an important source for this food.
1747	Benjamin Robins (1707–1751) becomes the first true experimentalist of classical physics when he studies the physics of spinning projectiles. In his book *New Principles of Gunnery* (published in 1742), Robins forms the basis for all subsequent theories of projectiles and artillery. He further experimentally explains the enormous forces acting on high-speed projectiles and lays the groundwork for the sophisticated aerodynamics of the Magnus effect. Robins is widely regarded as the founder of the science of ballistics.
1747	By conducting a series of controlled experiments of sick sailors aboard the *HMS Salisbury,* James Lind (1716–1794) determines that lime juice successfully treats the disease scurvy. However, Lind has no idea why this treatment works. It is not until 1928 that Albert von Szent-Györgyi (1893–1986) discovers vitamin C, the substance responsible for the cure.
1748	Julien Offray de LaMettrie (1709–1751) delineates the theory in his book *L'Homme Machine* that animal bodies, including those of humans, function as complex machines. This theory becomes the basis for explaining all bodily processes as primarily physical movements. (This explanation is later proven useful in only certain instances.)
1748	Jean Antoine Nollet discovers the processes of osmosis while conducting water experiments with an animal membrane.
1748	Antonio de Ulloa (1716–1795) discovers the element platinum in South American gold mines. He concludes that the metal is extremely resistant and will cause great problems for mining contiguous gold in the region.

1748	Jean Antoine Nollet invents the electroscope, an instrument used to detect electrical charges.
1748	Mikhail Lomonosov (1711–1765) provides strong empirical evidence of the law of conservation of mass when he carefully weighs the results of calcification experiments in his laboratory. During calcification, Lomonosov demonstrates, elements such as tin gain mass. These experiments form the first empirical challenge to the theory of phlogiston, though no one (including Lomonosov) fully realizes it for several decades.
1750	Thomas Wright (1711–1786) proposes that the Milky Way (Earth's galaxy) is a lens-shaped distribution of stars, which is thicker in the middle than at its thin edges. Building on Wright's work, one of the most important achievements of twentieth-century astronomy is the classification of the Milky Way as a spiral galaxy. This classification comes from the realization that dust, gases and even stars unfold from the Milky Way's bulge in a pinwheel pattern.
1751–1772	Denis Diderot (1713–1784) and Jean D'Alembert publish the *Encyclopédie, ou Dictionnaire Raisonné des Sciences, des Arts, et des Métiers (Encyclopedia, or Critical Dictionary of the Sciences, Arts, and Trades)*. This is the first encyclopedia in the modern sense. From the start, Diderot considers the *Encyclopédie*'s central purpose to be the spread of scientific knowledge for the specific purpose of improving the pursuits of ordinary people. He intends it to provide a thorough and complete treatment of rational knowledge. No similarly vast collection of general information has even been assembled before. Both Diderot and D'Alembert therefore see their project as a record of their civilization's accomplishments to that time—not just distribution, but also preservation of knowledge gained through scientific rationalism. They also understand that this is an acutely rhetorical exercise, meant to cause a revolution in the minds of its readers. Most historians credit the *Encyclopédie* as a major factor in the French Revolution.
1751	Axel Cronstedt (1722–1765) isolates nickel.
1752	René Réaumur proves that animal digestion is a primarily chemical process. In a decisive experiment, Réaumur induces a pet hawk to swallow small metal cylinders with mesh wire over each end. Inside each cylinder is a piece of meat. Because the hawk cannot digest the metal cylinders, it regurgitates each one. Réaumur finds that the meat inside is partially digested. Because the metal cylinders are intact, the hawk cannot have digested the meat mechanically. Réaumur correctly concludes

that the meat must have been digested chemically when the hawk's stomach juices filled the cylinders.

1752   In his dangerous and famous kite experiment, Benjamin Franklin charges a Leyden jar (see 1745–1746) through static electricity generated during a storm. This experiment demonstrates that lightning is a powerful kind of electric discharge.

1752   Albrecht von Haller (1708–1777) publishes *On the Irritable and Sensible Parts of the Body*. This book expresses von Haller's concept that certain parts of the body are "sensitive" to pain, whereas other parts react "irritably" by contracting to stimulations. For the first time in history, vital organ function is freed from abstract principle and therefore open to empirical observation. Von Haller further suggests that the heart is the body's most "irritable" organ because it beats continuously. He finds that such "irritable" organs can be made to contract post-mortem through the application of electricity (see 1780).

1753   Carolus Linnaeus publishes *Species Plantarum*, in which he classifies plants according to a "sexual" system of stamens and pistils. In this book's 1,200 pages, Linnaeus classifies 5,900 species in 1,098 genera—every plant he has ever known. Today, botanists generally accept *Species Plantarum* as the starting point of modern botanical nomenclature (see 1735).

1755   William Cullen (1710–1790) discovers that evaporating water registers lower on a bulb thermometer than does dry air. From this information, he concludes that evaporation causes cooling.

1756   Joseph Black (1728–1799) discovers the first independent gas in history when he identifies "fixed air," or carbon dioxide, in magnesia alba. For many years, this is the only known independent gas. In his inquiries, Black comes very close to theoretically dismantling the phlogiston theory many years before the end-of-century chemical revolution in France.

1757   Alexis Clairaut (1713–1765) deciphers the mass of Venus from a series of elaborate mathematical calculations.

1758   Andreas Marggraf establishes the validity of flame test as a method of chemical analysis. He finds that sodium salts consistently burn with a yellow flame, whereas potassium salts burn with a pale purple one.

1758   German amateur astronomer Johann Georg Palitzsch (1723–1788) spots a comet on Christmas Day. This comet's return was famously predicted in 1705 by Edmond Halley, for whom it is named. Since 1758, Halley's comet has made three regular, predicted visits in approximately 75-year intervals: in 1835, 1910 and 1986. Scholars have also uncovered records of sightings of this comet going back to 240 B.C. in China. Halley's comet is scheduled to make its next visit in 2061.

1760s–1770s   Beginning in the early 1760s, Joseph Priestley initiates a period of unprecedented chemical inquiry. During these years, Priestley identifies carbon dioxide in brewery vats near his house; invents soda water; demonstrates that ordinary air is a collection of many different gases; invents or perfects many of the pneumatic techniques for freeing "airs," or gases; perfects flame test analysis as a tool of chemical inquiry; discovers hydrogen chloride, nitric oxide, ammonia, oxygen, ammonium chloride and sulfur dioxide; and presents his findings to the founder of modern chemistry, Antoine Lavoisier, in Paris.

1761   Joseph Black discovers that it takes an "additional" amount of heat to turn water into steam. Black calls this quantity "latent heat" and explains it to his friend, inventor James Watt (1736–1819). Watt realizes that latent heat accounts for the problem of heating and cooling the same chamber of his steam engine—a realization that eventually leads to the separate condenser, a revolutionary innovation. Black also coins the term "specific heat," the amount of heat needed to raise the temperature of 1 gram of a specific substance 1 degree Celsius.

1761   Mikhail Lomonosov observes a transit of Venus and suggests that the well-documented, infuriating "ring" around the planet is caused by intensified sunlight created by something akin to a lens. This "something," Lomonosov correctly suggests, is an atmosphere. However, because few scientists outside Russia can speak Russian (and because Russia has not yet produced many scientists of international stature), few scientists credit Lomonosov with the discovery until 1910.

1761   Joseph Koelreuter (1733–1806) conducts experiments leading him to believe that insects are primarily responsible for pollination.

1762   Scottish ironworkers begin successfully converting cast iron into malleable iron.

1763–1799   James Watt perfects his steam engine with ongoing, multiple improvements. More than any other individual, Watt is credited with creating the means to power the Industrial Revolution. Without doubt, Watt's greatest improvement over the Newcomen steam engine is the separate condenser, which frees his engine from the fuel and time requirements of heating and cooling in the same chamber. Another major innovation is the rotary flywheel, which uses the principle of inertia to transfer the piston's up-and-down motion to a uniform, circular motion. After these improvements, steam engine use becomes nearly universal in industry.

1764   James Hargreaves (ca. 1722–1778) invents the spinning jenny, the first machine to spin multiple threads at the same time. To use the machine, a spinner turns a large wheel that turns a row of spindles—at first 8, then 16 and then as many as 120 per machine. The effect of the invention of spinning machines (see 1769 and 1779) is huge, creating and fulfilling enormous demand for cotton thread.

1765   Building on his own groundbreaking experiments that disproved spontaneous generation of microorganisms, Lazaro Spallanzani (1729–1799) successfully preserves several food samples by a process of thoroughly boiling them, drawing out the air by vacuum and sealing them in glass laboratory vials (see 1795).

1766   Henry Cavendish (1731–1810) establishes that hydrogen is, beyond doubt, an element. He names the gas "inflammable air" and believes, for a time, that he may have isolated the elusive phlogiston itself. Cavendish's discovery shatters the ancient belief that both air and water are elements. It also establishes Cavendish as the scientific benefactor of gas balloon flight, because he demonstrates that hydrogen is about one-fourteenth the density of ordinary air.

1766   Building on the knowledge of Le Blon's "primary" colors (see 1710), Moses Harris (1731–1785) publishes *The Natural System of Colours.* Written in the tradition of English naturalism, Harris's book examines the relationship between colors. It also examines how three primary colors can mechanically produce many others. This book also contains a printed color wheel. For printers and other non-scientists, this is the first opportunity to study a visual representation of an abstract color theory.

1767   Richard Edgeworth (1744–1817) constructs the first line of mechanical telegraphs so he can report the outcome of a horserace in Newmarket to his friends in London. Using a series of signals, these machines, which resemble large windmills, form the first major improvement in distance communication since antiquity (see 1791).

1767	Joseph Black performs his famous sack trick, in which he fills a sack with "inflammable air," or hydrogen, and releases it. When the sack floats to the ceiling, several prominent on-lookers accuse Black of an elaborate deception involving thin black thread. Black's sack trick forms the basis for large-scale hydrogen balloon trials by J. A. C. Charles (1746–1823) a decade and a half later (see 1783).
1768	Joseph Priestley dissolves his "fixed air" (carbon dioxide) in water, thereby creating soda water. Within a few years, Priestley has tipped off a European craze for this drink, which is erroneously believed to possess healing properties.
1769	Sir Richard Arkwright (1732–1792) invents the water frame spinning machine. This machine is capable of spinning multiple threads of rough, coarse yarn at the same time. The effect of spinning machines (see 1764 and 1779) is huge, creating and fulfilling enormous demand for cotton thread.
1770	Joseph Black proposes that his discovery, "fixed air" (carbon dioxide), is responsible for extinguishing a candle's flame in a closed container. He believes that the "fixed air" results when the phlogiston in a quantity of air is used up by combustion. Several years later, French scientists such as Lavoisier begin discovering that oxygen (and not phlogiston) is necessary for combustion.
1771–1772	Carl Scheele (1742–1786) discovers oxygen gas as a by-product of burning many different chemicals in his laboratory. He also identifies that gas in the respiration of plants and fish and names it "fire air." Modern historians have established that Scheele's discovery of oxygen occurred several years earlier than Joseph Priestley's independent discovery of the same gas. However, a delay in publishing Scheele's discovery allowed Priestley to announce first, and thus gain credit for many years. Today, Scheele and Priestley generally share credit (see 1774).
1772	Daniel Rutherford (1749–1819) discovers nitrogen as the "left-over" air in combustion and respiration experiments.
1773	British sea captain James Cook (1728–1779) becomes the first person to cross the Antarctic Circle when his Pacific voyage takes him to the extreme south. Because no inhabited land mass is near the Antarctic Circle (Tierra del Fuego is more than 600 miles north), Cook is very likely the first person in history to cross it.
1774	In a series of thirty-four brilliant copper engravings, William Hunter (1718–1783) first depicts the uterus as an independent organ during a study of late-term pregnancy form.

1774	Joseph Priestley discovers oxygen gas in combustion experiments and realizes that it is the same gas he discovered 2 years earlier in plant respiration. He names his gas "dephlogisticated air" and quickly publishes his results. He never becomes aware that Carl Scheele independently discovered the gas 3 years previously. Today, both scientists share credit for this discovery (see 1771–1772).
1774	Joseph Priestley travels to Paris, where French chemists at the forefront of the chemical revolution greet him warmly. During this trip, Priestley introduced his "dephlogisticated air" to Antoine Lavoisier, who would later name it "oxygen."
1774	On treating a piece of pyrolusite with hydrochloric acid, Carl Scheele isolates a greenish-yellow gas he calls "dephlogisticated acid of salt." Scheele also discovers many of the gas's special properties and uses. In 1810, Sir Humphry Davy (1778–1829) names the gas "chlorine."
1775	Joseph Priestley discovers gaseous hydrogen chloride and sulfur dioxide gas. When he dissolves these gases in water, they respectively form hydrochloric acid and sulfurous acid.
1777	Joseph Huddart (1741–1811) writes the first reliable account of color blindness in a letter to Joseph Priestley.
1778	Jean Baptiste Lamarck (1744–1829) classifies the wild plants native to France in his three-volume work, *Flore Française*.
1778	Huber Höfer first detects boric acid in a Tuscan natural hot spring.
1778	John Hunter (1728–1793) classifies teeth and reverses centuries-old beliefs about tooth decay. He conclusively demonstrates that decay begins on the outside (and not the inside) of a tooth, most especially on surfaces where particles of food repeatedly lodge.
1779	Building on the work of Mikhail Lomonosov (see 1761), Johann Schröter (1745–1816) begins a series of observations of Venus. Years later Schröter notes a pronounced "filmy" streak on the otherwise uniformly bright surface of the planet. Subsequent close examinations reveal several other markings. Schröter concludes that each of the markings is wandering and imprecise. He draws the correct conclusion that the markings stem from Venus's atmosphere. This is the first empirical observation that Venus indeed has an atmosphere.
1779	Antoine Lavoisier publishes the magnificently important *Traité Élémentaire de Chimie*. Single-handedly and almost immediately, this book disproves the ancient four-element idea and the pesky

notion of transmutation; obliterates the notion that air and water are elements by demonstrating their compound nature; shatters the phlogiston theory of substances; and provides modern chemical names to elements and compounds (including the most important element during this time—oxygen). The fundamental transformation of chemistry may be said to be completed with this publication.

1779   Samuel Crompton (1753–1827) introduces his spinning mule. Just as a mule is a cross between two animals (a horse and a donkey), the spinning mule is a cross between two machines, the spinning jenny and the water frame spinning machine (see 1764 and 1769). Like its animal namesake, the spinning mule takes the best of both worlds. It can produce whichever yarn is desired, strong or fine. A secondary advantage is that the quality of yarn produced by the spinning mule is much easier to control than with the jenny or water frame. The effect of spinning machines is huge, creating and fulfilling enormous demand for cotton thread.

1779   Lazzaro Spallanzani establishes the role of spermatozoa in fertilization. In one difficult experiment, Spallanzani fits a male frog with a pair of tight-fitting taffeta pants. He finds that this frog, on mating, is unable to fertilize its partner's eggs because of the effective segregation of spermatozoa.

1779   Jan Ingenhousz (1730–1779) publishes *Experiments upon Vegetables*. This book confirms Joseph Priestley's experiments on green plant respiration (see 1760s–1770s). More importantly, it details Ingenhousz's experimental discovery of the process now known as photosynthesis (one of the most important processes in nature). This is the process by which plants make carbohydrates from carbon dioxide and water, given the presence of chlorophyll and light. During this process, they also release oxygen as a by-product, which is essential for animal survival.

1780   Luigi Galvani (1737–1798) performs his famous experiment on electrical "fluid" and animal tissue. When he sees a dead frog's leg twitch during a powerful electrical storm, Galvani arrives at the erroneous conclusion that electricity is somehow "produced" in the moist tissues of animals.

1780   James Watt invents the wet-transfer copy process (a.k.a. the copying press). For the first time, a person can copy a letter or memo without having to write multiple copies by hand. By "pressing" a letter written with special ink, a person can create a "negative" copy—read from the back—for his or her records.

Virtually all business offices have at least one copying press until the combination of typewriter and carbon paper finally begins to replace these machines in the 1920s. Advertisements for copying presses are routinely found in periodicals as recently as the 1950s.

1780s

John Hunter collects an enormous archive of anatomical specimens—13,00 of which still exist. He also dissects more than 500 different species. During this period, Hunter publishes outstanding books on several topics, including dentistry, communicable disease and blood inflammation.

1781

William Herschel (1738–1822) discovers the first planet to be discovered since the advent of history. This is the seventh planet from the sun, which Herschel names Georgium Sidus, after King George III. The French refuse this name for political reasons and begin calling the planet Herschelium. Sixty years later, the planet is finally named Uranus, after the Greek god of the sky. The discovery of Uranus alters the entire understanding of the solar system. Instead of six planets, there are now seven. What's more, the known size of the planetary system roughly doubles. All at once, scientists realize that they have much to discover about the workings of the universe. Since Herschel's discovery, two more planets, Neptune and Pluto, have been discovered scientifically. Herschel's discovery also allows scientists to broadly accept a formula for approximating planetary distances from the Sun, first published in 1772 by Johann Bode (1747–1826).

1782

Building on the work of Edmond Halley (see 1718), William Herschel discovers that the Sun is moving in the direction of the constellation Hercules. Where Copernicus (1473–1543) had set the Earth in motion on its axis and around the Sun, Herschel sends the Sun and its relatively tiny solar system on a gigantic journey toward the star Lambda in the constellation Hercules.

1782

Building on the work of René Réaumur (see 1752), Lazzaro Spallanzani demonstrates that human digestion is a primarily chemical process. Always one to put science first, he constructs a number of small, hollow wooden blocks. Then Spallanzani inserts various foods into the blocks and swallows them. After a set number of minutes, which he varies for each experiment, Spallanzani induces himself to vomit so that he can study the changes in the food. Spallanzani also coins the term "gastric juice" during the inquiry.

1783	Caroline Herschel (1750–1848) becomes one of the most important female astronomers of enlightenment Europe when she discovers three nebulae. She later discovers eight comets and publishes two volumes of her other findings. She also improves the famous star catalog of John Flamsteed (1646–1719), England's first Astronomer Royal.
1783	Paper makers-turned-inventors Joseph (1740–1810) and Étienne (1745–1799) Montgolfier conduct the first trials of hot-air balloons. They are the result of many experiments the brothers performed on the physical properties of hot air. In November, two human passengers soar to an altitude of approximately 3,000 feet. Ballooning quickly becomes a craze in France and then throughout Europe.
1783	Fueled by his sophisticated trials with lighter-than-air hydrogen gas (which result in Charles's law of gases), J. A. C. Charles conducts the first gas balloon trials. These culminate in a launch in which Charles and an assistant drift high above France for 2 hours and travel a distance of 27 miles.
1783	John Michell first proposes the existence of black holes in purely Newtonian terms (see 1704). He contends that a black hole is an area of space with such a powerful gravitational force that even corpuscles of light cannot escape (like cannonballs shot from the surface of Earth). Today, quantum theory conceives of black holes differently from Michell's purely Newtonian terms.
1783	Using a eudiometer of his own making, Henry Cavendish establishes that the composition of the lower atmosphere is constant regardless of altitude (several thousand feet up) or geography. For this finding, Cavendish pioneers the use of lighter-than-air balloons for scientific research. He also anticipates the discovery of argon gas by more than 100 years.
1783	On July 15, French nobleman the Marquee Claude de Jouffroy d'Abbans (1751–1832) debuts the steamboat *Pyroscaphe*. Driven by a Watt steam engine, this boat steams against the current of the river Saone for a full 15 minutes. This event marks the first time in history that a vessel moves under its own power.
1784	Deeply influenced by Joseph Priestley's "fixed air" experiments with birds, Antoine Lavoisier and Simon de Laplace (1749–1827) establish that animal respiration gives off carbon dioxide and consumes oxygen. They find that the levels of both gases vary with activity. They also theorize the existence of "caloric," which is later explained as just a form of energy.

1784   Henry Cavendish shatters the ancient belief that water is an element when he demonstrates its compound nature. When he ignites oxygen and ordinary air, Cavendish carefully records a measurable amount of "dew" formed inside his glass vials. A staunch phlogistonist, Cavendish never shares the modern interpretation of water as a compound of oxygen and hydrogen, but he does realize that water is, in essence, "made" as a by-product of his experiments.

1784   To see objects at a variety of distances, especially while traveling in France, Benjamin Franklin invents bifocal lenses. Along with the lightning rod, bifocals are one of Franklin's inventions still widely used today.

1785   Claude Berthollet (1748–1822) establishes that ammonia consists primarily of two substances: nitrogen and hydrogen.

1785   Building on John Kay's flying shuttle loom (see 1733), Edmund Cartwright (1743–1823) designs the first usable steam-powered loom. Though only minimally successful at harnessing reciprocating steam power, Cartwright's power loom provides the conceptual framework for more successful ones during the next century. These looms are generally credited with founding the modern Scottish textile industry.

1785   Charles Coulomb (1736–1806) states the rule of electrical forces, also called Coulomb's law. This law states that the force between two electric or magnetic charges varies inversely to the square of the distance between them.

1787   On August 22, inventor John Fitch (1743–1798) and several others become the first steamboat passengers in history. On this day, Fitch's vessel makes a long, successful trip up the Delaware River. Three years later, Fitch establishes the first reliable, regular transportation provided by a vessel operating under its own power. Steamboats have a catastrophic effect on transportation during the next century, particularly in the vast wilderness of the western United States.

1787   Jacques Alexandre Cesar Charles states the law of gases that bears his name. While pioneering lighter-than-air flight in a gas balloon (see 1783), Charles realizes that a gas expands by the same fraction of its original volume as its temperature rises. Charles's law is written as V/T = constant. Scientists now generally accept this law for all "ideal" gases.

1787   A task force of French chemists, led by Antoine Lavoisier and Louis-Bernard Guyton de Morveau (1737–1816), publishes the magnificently influential *Méthode de Nomenclature Chimique*.

Almost single-handedly, this book establishes a rational system of chemical nomenclature. The effect on chemistry is international and catastrophic; now chemists everywhere begin to use the same chemical names.

1789   William Herschel discovers two satellites of Saturn: Enceladus and Mimas.

1789   Building on the work of Lomonosov and others, Antoine Lavoisier begins paying strict attention to the concept of mass during calcification. Phlogiston theory states that phlogiston *separates from* a substance during combustion (including calcification). However, in a series of experiments, Lavoisier demonstrates that certain substances actually *gain* mass. Lavoisier concludes that something from the air is *joining with* the substance during calcification and that this "something" is the newly discovered gas, oxygen. This finding establishes the law of conservation of mass and virtually completes the transformation of chemistry.

1790   A commission of twenty-eight scientists appointed by the French Academy (and unofficially led by Antoine Lavoisier) outlines the metric system of weights and measures. French scientists quickly adopt the new standards, which give them enormous international advantage.

1791   Building on the mechanical telegraphs of Richard Edgeworth in England (see 1767), Claude Chappe (1763–1805) and his brothers construct the first semaphore line in France. Mechanical telegraphs relay a message through a theoretically transparent medium. This proves very cumbersome. The success of Chappe's semaphore is that it only has to signify a predetermined sign, analogous to a modern stoplight (where red means stop and green means go). Napoleon quickly makes use of the semaphore to relay military instructions a great distance.

1792–1796   Benjamin Banneker (1731–1806) publishes the *Farmer's Almanac*, which many historians regard as the finest almanac ever published. Banneker publishes the results of his scientific inquiries, as well as the mathematical and scientific methods behind his results. This wins both the almanac and Banneker the deep respect of many important people in high places (including President Thomas Jefferson) during a time that generally afforded few opportunities for African American scientists.

1793   The French Revolutionary Convention disbands the French Academy of Sciences.

1793	Christian Sprengel (1750–1816) discovers a direct relationship between flower form and pollinating insects. Building on this knowledge, Sprengel discovers that the "male" and "female" sexual parts of many flowers do not mature at the same time. He thereby concludes that cross-pollination between individual flowers is much more crucial than scientists have previously thought.
1793	Eli Whitney (1765–1825) invents the cotton gin, a machine capable of cleaning short-grain cotton, the variety most widely grown in the southern United States. Its effect on the U.S. economy is catastrophic, as the South proclaims its "king" crop, cotton. On the other hand, exploding cotton profits immediately revive the ailing plantation system and expand the Atlantic trade in slaves. As a direct result, the slave population of many southern states grows tenfold during the next quarter century.
1793	George Washington's favorite dentist, John Greenwood (1760–1819), is deeply influenced by the 1728 publication of Fauchard's *Le Chirurgien Dentiste*. Accordingly, he invents the foot-powered dental drill by adapting his mother's foot-treadle spinning wheel to rotate his drill. Though it is successful in Greenwood's practice, this ingenious drill never catches on.
1793	The Revolutionary government of France appoints Philippe Pinel (1745–1826) head of a large insane asylum. Pinel believes that the patients' mental illnesses are broadly analogous to now-familiar bodily ailments. He therefore abolishes the use of cruel chain restraints and makes the first case studies of patient progress.
1794	Antoine Lavoisier, widely regarded as the founder of the modern chemical revolution, is guillotined during the French Reign of Terror for his partnership in a tax-collecting company many years earlier. Seconds after Lavoisier's execution, Lagrange comments, "Only a moment to cut off his head, and perhaps a century before we shall have another like it.
1795	The new French republic officially adopts the metric system as the official system of weights and measures. Today, all major countries except the United States have adopted the Systeme International d'Unites (International System of Units), known commonly as SI.
1795	More than forty years after James Lind's discovery that lime juice cures scurvy, Sir Gilbert Blaine (1749–1834) makes it a mandatory ration in the British navy. This earns British sailors the familiar nickname "limey" (see 1747).

1795   Nicolas-François Appert (1750–1841) studies Spallanzani's ingenious demonstrations of small-scale food preservation (see 1765). In the service of Napoleon's army, Appert designs early methods of large-scale "canning" by sealing food into sterile glass and metal containers.

1796   Smithson Tennant (1761–1815) conclusively demonstrates that diamonds are exclusively made of carbon.

1796   Edward Jenner (1749–1823) discovers vaccination as a safe and effective prevention against the smallpox virus. This procedure involves the natural cross-immunity transferred to a human through infection with the relatively benign cowpox virus. This is a much better treatment than active-virus inoculation (see 1717). Today, owing to highly successful vaccination, smallpox exists only in rural "pockets" of developing countries.

1796   Deeply influenced by Galvani's work on animal electricity, Count Alessandro Volta (1745–1827) begins to believe that electrical charges are produced by a combination of different metals, not moist animal tissue (see 1780). To investigate his hypothesis, Volta layers pairs of zinc and silver disks (in between are layers of cloth saturated with saltwater). This voltaic pile contains the basic elements of modern batteries— several cells connected together with conducting materials that carry electrical current. These piles are widely used during the first half of the nineteenth century. Much later, Michael Faraday (1791–1867) finds that the source of electricity is chemical action, as when zinc is eaten away and continuous energy is liberated in the form of electricity. This finding does not violate the law of conservation of energy.

1797   French Balloonist André-Jacques Garnerin (1769–1823) first uses a parachute to safely descend from a balloon. The parachute works on the principle of a light object presenting a large surface, so that air resistance slows its descent.

1797   Henry Cavendish "weighs" the Earth by figuring out the gravitational constant. Isaac Newton's famous equation for gravitational attraction contains symbols for gravitational constant, the masses of the two bodies and the acceleration of their movement toward one another. The acceleration was clearly known for an object falling to Earth. This leaves the two unknowns, and if either could be determined, then the other could be mathematically calculated. Using a torsion balance of his own making, Cavendish calculates the gravitational constant, which in turn gives him the mass of the Earth:

	6,600,000,000,000,000,000,000 tons—very close to today's accepted value.
1797	Joseph-Louis Lagrange (1736–1813) extends the work of Euler (see 1736) in the theory and application of "the calculus." Lagrange seeks to place calculus on a rigorous basis algebraically, without reference to geometry of any kind. He establishes many of the fundamentals found in modern calculus volumes. In his *Theory of Analytic Functions*, Lagrange attempts a purely algebraic form for the derivative, with no recourse to geometric intuition of numeric graphs. This results in the primer notation used for derivatives today but does not, in the end, provide a firm foundation of "pure" calculus.
1798	Nicola Vauquelin (1763–1829) discovers beryllium.
1798	Eli Whitney manufactures 10,000 muskets for the U.S. government. He machines his musket parts with such precision that they are easily replaceable, because any given part is not specific to any given musket. This is the beginning of interchangeable parts, a key manufacturing concept in the Industrial Revolution.
1798–1799	Count Benjamin Thomson Rumford (1753–1814) and his student Sir Humphry Davy demonstrate that heat is a form of energy, not the so-called "element" caloric. Rumford says, "It is hardly necessary to add that anything which any insulated body, or system of bodies, can continue to furnish without limitation cannot possibly be a material substance: and it appears to me to be extremely difficult, if not quite impossible, to form any distinct idea of anything, capable of being excited and communicated, in the manner the heat was excited and communication in these, except it be MOTION." In other words, when objects are rubbed together, or cut, or bored, heat is produced by friction as a form of energy.
1799	After a long, public debate with Claude L. Berthollet, Joseph Proust (1754–1826) proves that compound composition is always constant (and never "fickle"). He states these findings in the law of constant proportions, which is sometimes called Proust's law. This law states that an "invisible hand" holds the composition of compounds so that it is always the same proportion, regardless of how it is made in nature or a laboratory. A modern statement of this law is: a chemical compound always contains the same proportion (mass) of its constituent elements, regardless of its source or preparation. A simple, distilled version of Proust's law can be stated as: compounds always contain fixed proportions of elements.

EXPERIMENTS, INVENTIONS AND DISCOVERIES

A

Animal Physiology and Experimental Pathology
Stephen Hales, 1733
Albrecht von Haller, 1752
John Hunter, 1780s

As in many other fields, the Greeks conducted active investigations in the areas of animal structure and disease. In animal structure, the natural philosopher Alcmaeon of Croton (ca. 450 B.C.) made some very early studies in medicine and was possibly the first to attempt animal dissection. His finding that the brain was the body's sensory center became significant to several of the important physicians of latter antiquity. In medicine, Hippocrates (ca. 460–380 B.C.) challenged many unsound inherited practices. Many years later, Hippocrates became the most famous ancient physician, largely due to the writings of Galen (ca. 130–200). Hippocrates taught that diseases had natural rather than supernatural causes and that physicians could therefore cure diseases by studying the workings of nature. Hippocrates perceived illness as an imbalance in the four bodily **humors**, or liquids (phlegm, black bile, yellow bile, and blood). Versions of this theory lasted until the eighteenth century.

Among natural philosophers, animal studies were common by the time Aristotle (384–322 B.C.) founded his school. Aristotle performed both vivisections and dissections on many kinds of animals and published influential zoological texts. A major treatise, *About the Movement of Animals*, focused on animal movement and locomotion. Aristotle was not a scientist, and he never employed rigorous empirical testing or experimentation. In this treatise, Aristotle postulated that locomotion resulted from "breath" passing through an animal's heart. He also reverted to the much older idea that the heart was the center of sensory experience and intelligence. Aristotle set forth his metaphysical scale of being, which organized all organisms in order of increasing functional and structural complexity. This theoretical ordering of creatures was authoritative among European scientists and philosophers during the next millennium.

During the following century, Herophilus (ca. 300 B.C.) took an important step toward scientific anatomy when he established the systematic approach for identifying organs during dissection. Herophilus organized an animal's organs according to each one's purpose in relation to the brain. Centuries later, Galen made some of antiquity's greatest contributions to animal structure, while he followed the classical model of medicine outlined in his reading of Hippocrates. Galen recognized the general idea of the nervous system when he dissected an ox's brain. He also dissected the internal organs of the Barbary ape, though he drew errant anatomical comparisons between the brain of this species and that of human beings. Galen became the leading anatomist of his day and beyond, as his authority lasted virtually unchallenged until the Italian Renaissance.

The next anatomist on anything close to Galen's level was Leonardo Da Vinci (1452–1519), who lived a millennium and a half later. Da Vinci actually discovered many of the mistakes evident in Galen's work. He also made outstanding studies of the human form and detailed drawings of organs. Da Vinci especially demonstrated the mechanical action of many human joints, such as the ball and socket of the shoulder and hip. A near contemporary, Andreas Vesalius (1514–1564), became perhaps the greatest anatomist of the era. For the first time since antiquity, political and religious authorities allowed Vesalius and his contemporaries to dissect human cadavers (of executed criminals). His text, *De Humani Corporis Fabrica Libri Septem*, revolutionized the study of human anatomy as independent of the ancient animal studies of Galen.

After Vesalius, other important scientists turned to the study of the human and animal form. Galileo Galilei (1564–1642) adapted his mechanical theories to an analysis of animal motion. His treatise, *The Movement of Animals*, investigated human biomechanics and the elegant muscular movements of horses. After Galileo, scientists such as Jan Swammerdam (1637–1680) and William Croone (1633–1684) made important studies of muscle composition and contraction, touching off a fruitful area of study.

These were important steps for the study of anatomy and, eventually, of pathology, but the founding work of modern animal **physiology** took place a full century after William Harvey (1528–1657) discovered the circulation of blood in animals in 1628. When Harvey lacerated the arteries of certain animals, he observed that the blood spurted out forcefully. He quickly surmised that blood must be under pressure, due to the mechanical activity of the heart. Harvey also proposed the idea of blood flow, because his theory required blood to travel back to the heart.

Harvey's concepts attracted the attention of Stephen Hales (1677–1761), the founder of **pneumatic chemistry**. Hales had done many imperative tests on the phenomenon of rising sap in plants. He knew of theories dating from antiquity that drew physiological **analogies** between plants and animals, often even between sap and blood. Early in his work, Hales became skeptical of such analogies and began performing separate, categorically objective experiments on plants and animals. These challenged the analogist tradition methodologically. On publication of two books, *Vegetable Staticks* (1727) and *Haemastaticks* (1733), Hales experimentally shattered the analogist approach. (For more information, please see the entry on Plant Physiology.)

In the former book, Hales demonstrated that plants do not "circulate" sap. In the latter book, he published proof of animal blood circulation. In a key series of experiments, Hales became the first person to measure blood pressure when he inserted the end of an 11-foot glass tube into a horse's artery. Though it was many years before scientists developed a calibrated scale, Hales marked the height to which the blood rose in the tube as a measure of the force behind it. He later adapted his experiments to other species, including canines. In these studies, Hales demonstrated that the blood of different creatures flows at very different pressures.

The publication of *Haemastaticks* marked a new period of how the bodies of humans and animals were perceived. More than as simply machines, they now also consisted of the still largely mysterious workings of chemical processes. At this point, the exact role of blood as regards **oxygen** gas was very much unclear. However, in the 50 years that followed *Haemastaticks*, scientists such as René Réaumur (1683–1757) and Lazaro Spallanzani (1729–1799) demonstrated that digestion is a chemical process. After that, Antoine Lavoisier (1743–1794) showed that animals respire oxygen from ordinary air. The chemistry of the body was quickly coming into focus.

During this period, scientists were also delving ever more deeply into the dizzying combination of pumps, pulleys, levers and valves that they observed in the inner workings of animal bodies. The leading physiologist of the century, Albrecht von Haller (1708–1777), synthesized the work of others, especially his teacher Hermann Boerhaave (1668–1738). He also conducted many animal experiments and made two vital proposals. First, he proposed a system for organizing tissue into three separate groups based on the structure of the tissue in accordance with its function. Much later, von Haller's concept of the relationship between form and function would become important to cell theory.

Second, in 1752, von Haller published *On the Irritable and Sensible Parts of the Body*. This book expressed von Haller's concept that certain parts of the body are "sensitive" to pain, while other parts react "irritably" by contracting to stimulations. For the first time in history, vital organ function was freed from abstract principle and therefore opened to empirical observation. Von Haller further suggested that the heart was the body's most "irritable" organ because it beat continuously. He found that such irritable organs could be made to contract postmortem through the application of electricity. This finding became important to both animal physiology and to the study of electricity, especially the later electrical inquiries of Luigi Galvani (1737–1798).

Rounding out the tripartite of eighteenth-century animal experimenters is John Hunter (1728–1793). After assisting his brother William Hunter (1718–1783) as a dissectionist, John Hunter went on to collect an enormous archive of anatomical specimens—13,000 of which still exist. He also dissected over 500 different species. In the 1780s and beyond, Hunter published outstanding books on a number of topics, including dentistry, communicable disease and blood inflammation.

In his books, Hunter embraced both the established metaphor of the human body as a machine and the emergent idea that it hosts a series of chemical reactions. For many years, anatomists and dissectionists had been confused by lesions found postmortem on the stomach walls. Hunter knew of the conclusive work by Réaumur and Spallanzani that had established **digestion** as primarily a chemical process. Taking it a step further, Hunter demonstrated that the lesions formed only after some time had passed between a person's death and the surgical opening of the stomach. Hunter correctly concluded that the stomach tissue was protected from gastric juice only while it was living and that digestion continued after death, thus damaging the stomach walls. Previously, Spallanzani had demonstrated that digestion could occur outside the stomach. Now, Hunter had shown it could also occur after an organism had died. Digestion was more of a chemical process than anyone had ever imagined. (For more information, please see the entry on Digestion.)

Though a full-blown microbe theory of disease was many years away, Hunter began to conceive a start. An early proponent of **smallpox inoculation**, he deliberately infected himself with pus derived from a patient suffering from venereal disease. Of course, this lent no immunity. Though the diseases were not yet medically distinguishable, Hunter probably contracted both gonorrhea and syphilis. He knew how to treat venereal disease with mercury, but he delayed treatment for a long time in order to record observations of the diseases' natural courses and publish them in a book in 1786. These became the first medically reliable descriptions of venereal disease and began the long task of delineating the differences in terms of symptom, cause, and eventual treatment.

Hunter published several other important books, especially *Observations on Certain Parts of the Animal Economy* (1786) and *Treatise on the Blood, Inflammation and Gunshot Wounds* (1794). Many later surgeons were deeply influenced by Hunter's books, and most scientists consider him a founder of experimental pathology.

SEE ALSO Animal Respiration; Digestion; Oxygen; Plant Physiology; Immunization Prevents Smallpox.

Selected Bibliography

Allan, D. G. C. and R. E. Schofield. *Stephen Hales: Scientist and Philanthropist.* London: Scholar Press, 1980.

Krebs, Robert E. *Scientific Laws, Principles, and Theories: A Reference Guide.* Westport, CT: Greenwood Press, 2001.

McGrew, Roderick E. *Encyclopedia of Medical History.* New York: McGraw-Hill, 1985.

Porter, Roy. *The Cambridge Illustrated History of Medicine.* New York: Cambridge University Press, 1996.

Rhodes, Philip. *An Outline History of Medicine.* Boston: Butterworths, 1985.

Animal Respiration Obtains Oxygen and Eliminates Carbon Dioxide
Antoine Lavoisier and Simon de Laplace, 1784

The processes of respiratory breathing in animals vary greatly, as determined by animals' adaptation to their environments. Many aquatic organisms, such as fish and tadpoles, obtain **oxygen** that is dissolved in water. Many insects breathe through tiny tubes called trachea. Many amphibians use neither lungs nor gills, but breathe directly through their moist skin. This method is well suited to damp environments.

These processes may seem simple enough today, but little was known about animal **respiration** until the late eighteenth century. Early on, the English pneumatic chemist Joseph Priestley (1733–1804) discovered that his **dephlogisticated air** (or oxygen) made a candle burn longer and with greater intensity than ordinary air. He found it supported the life of enclosed mice much longer, too. In the prevailing **phlogiston** theory, the hypothetical substance phlogiston is released during **combustion**. Priestley thought that "dephlogisticated air" had lost all of its phlogiston and therefore hungrily grabbed it back during combustion (hence the dazzling candle). Priestley began wondering about the relationship between combustion and respiration. If his gas supported both a candle and mouse, perhaps there was some direct connection between combustion and respiration. Many scientists had studied the former process, but most had not yet investigated the latter.

In October 1774, the famous Priestley traveled to Paris, where he was well received by French chemists, especially Antoine Lavoisier (1743–1794). During their meetings, Priestley first introduced Lavoisier to his "dephlogisticated air."

Over the next two years, Lavoisier conducted many new experiments on the gas. These foundational studies in oxidation and mass formed a large part of the chemical knowledge with which Lavoisier shattered the phlogiston theory of substances. Through a careful study of mass, Lavoisier demonstrated that phlogiston was not *separating from* substances during combustion. Rather, oxygen was *joining with* them. Lavoisier published these results in one of the most important books in history, *Traité Élémentaire de Chimie*. This book was concerned primarily with the chemistry of gases rather than of animals.

However, along the way, Lavoisier had recorded some puzzling results from working with animals. In one experiment, he put a bird inside a bell jar until it died. When he tried to light a candle in the bell jar, it would not burn. Much of the knowledge of animal respiration had been more or less unchanged since antiquity. The ancient Greeks had proposed that the role of respiration was to inflate the lungs so that blood could be "cooled." Lavoisier rejected this theory.

In fact, Lavoisier began to believe privately that the same gas that supported combustion also supported animal respiration. In *Traité*, Lavoisier had already demonstrated that ordinary air is not a single substance with elastic properties. Rather, he had shown that air consists of a variety of gases—many of which were not yet discovered. Lavoisier began designing a series of experiments on respiration. From the start, he began to believe that air was comprised of at least two gases—one that supports respiration and one that does not. Questions raced through his mind: Could these be the same gases responsible for supporting and quelling combustion? He had already identified oxygen. In the 1750s, Joseph Black (1728–1799) had shown that **fixed air**, or **carbon dioxide**, quells combustion. Could animal respiration involve these same gases? The experiments on combustion had finally led Lavoisier to respiration—from the chemistry of burning to the chemistry of respiration, from the candle to the bird. Lavoisier's physiological inquiries, which had grown out of his work in chemistry, began taking on a separate reality of their own.

In several experiments beginning where Priestley left off, Lavoisier had animals breathe in a confined space. He found that "fixed air" was formed during respiration. Lavoisier was beginning to understand respiration as primarily a chemical process in which an animal converts "pure air" (or oxygen) to "fixed air" (or carbon dioxide). From here, Lavoisier reasoned that respiration consumes and produces exactly the same gases as does burning. This discovery resulted in an exceedingly fruitful metaphor. The process of respiration, explained Lavoisier, is "similar in every way to that which takes place in a lamp or lighted candle" (quoted in Poirier 1996, 301). From now on, this metaphor would guide Lavoisier's respiration inquiries.

In dismantling phlogiston and building modern oxygen chemistry, Lavoisier employed his concept of **caloric**. Caloric was a vague kind of "fire air" that Lavoisier listed as an actual substance and even an **element**. He did not yet realize that heat was a form of energy, rather than the "element" caloric. (For more on this topic, please see the entry on Heat Is a Form of Energy and also the one on Substances: Elements and Compounds.) Naturally, Lavoisier sought a method for measuring amounts of both "fixed air" and caloric produced during animal respiration or chemical combustion.

In 1784, Lavoisier and his friend, astronomer–mathematician Simon de Laplace (1749–1827), invented an apparatus to measure both amounts. Appropriately, they named their invention the **ice calorimeter** (see Figure 1). This machine consisted of three concentric compartments. The inside compartment held the source of heat—say, a live guinea pig or a piece of burning charcoal. The middle compartment held a specific amount of ice for the heat source to melt. The outside compartment held packed snow for insulation.

In a series of tests and calibrations, Lavoisier and Laplace found that it took one pound of 167° **Fahrenheit** (F) water to melt 1 pound of ice at its

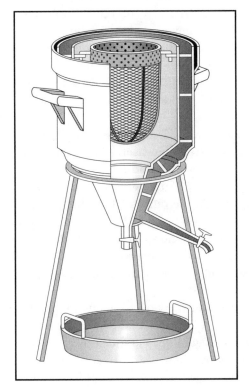

Figure 1. Lavoisier's ice calorimeter. Note the three concentric chambers of Lavoisier and Laplace's ice calorimeter. The inside chamber held a respiring animal or burning object. The middle chamber held a specific quantity of ice. The outside chamber held packed snow for insulation. By measuring the quantity of melted ice from the middle chamber, the scientists could measure heat produced in the innermost one.

freezing point. They also found that 7.7 pounds of iron shavings heated to 207.5°F melted 1.1 pounds of ice in 11 hours. In this time, they found that the iron had cooled by 174.4°F. The former measurements formed the basis of Lavoisier and Laplace's caloric scale. By carefully collecting the water as the ice melted, Lavoisier and Laplace found that a piece of burning charcoal melted a total of 10.5 ounces of ice. On the other hand, a live guinea pig melted 13 ounces of ice during a period of 10 hours. These were astonishing results. Given enough time, a respiring guinea pig actually produced more caloric (or heat) than a small piece of charcoal. Respiration was more like combustion than even Lavoisier had previously imagined with his original metaphor.

This was a significant discovery, but it was still only half of the experiment. Laplace designed a sleeve for the ice calorimeter. When he connected the sleeve to a separate container, it allowed Lavoisier and Laplace to collect the "fixed air" being given off through combustion or respiration. They found that charcoal released 224 grains (1 grain equals 1/7,000 pound). The guinea pig released approximately 236 grains. Respiration produced more "fixed air," as well as more caloric. No longer was the metaphor a matter of words only. Now, Lavoisier and Laplace could measure quantifiable and specific amounts of caloric and "fixed air" produced during respiration. In other words, they could demonstrate—in terms accessible to anyone—that combustion and respiration involved the same gases. As a guinea pig loses heat to its environment, Lavoisier explained, it maintains its body temperature by a complex kind of "burning" in its tissues, which gives off caloric (or generates heat). This theory was not exactly accurate, but it set down a basic statement for later scientists to refine.

Thus far, Lavoisier and Laplace had measured two important processes, which left one major hurdle—how to accurately measure the oxygen consumed during respiration. For this inquiry, Lavoisier and Laplace turned to a human volunteer. They had a friend, Armand Séguin, wear an airtight mask of leather with a tube in its mouthpiece. At the other end of the tube was a **pneumatic trough** of oxygen. In this way, Séguin breathed only pure oxygen from the trough. Lavoisier and Laplace hoped to measure the exact amount of oxygen that the man consumed over the course of an hour. From there, they thought it would be a matter of simple mathematics to compute oxygen consumption over long periods.

Their findings surprised them. Respiration was no easy matter, especially as far as oxygen was concerned. The level of oxygen that Séguin consumed varied wildly with each of his activities. Séguin needed much more oxygen when exercising than when resting, more when eating than fasting, and more when sitting quietly in a cold room than in a warm one. Right away, Lavoisier and Laplace realized that the production of caloric (or heat) required oxygen in direct proportion. More heat means more oxygen. The metaphor had taken them further than they ever dreamed. Just as combustion requires "fuel," so does animal respiration. The two were not identical processes, but they helped explain one another. And, just as Lavoisier and Laplace had suspected from the start, respiration was primarily a chemical process.

Modern scientists have shown that animal respiration produces energy, which is necessary for the maintenance of living organisms. Energy comes from the oxidation of three classes of organic substances: carbohydrates, lipids, and proteins. This process takes place in the mitochondrion—the microscopic center of an animal's cell. This process also consumes oxygen from the lungs.

Today, the study of chemical processes in living things is known as **biochemistry**. Biochemistry has become one of the most important fields, especially in its study of the complex processes of deoxyribonucleic acid (DNA). Through the study of DNA, biochemists have explained the molecular basis of genetic variance. Many scientists believe that biochemistry could become even more important in the twenty-first century and hope to decode the chemical instructions controlling heredity in all animals—humans and redoubtable guinea pigs alike.

SEE ALSO Carbon Dioxide; Heat Is a Form of Energy; Oxygen; Phlogiston; Substances: Elements and Compounds.

Selected Bibliography

Greenberg, Arthur. *A Chemical History Tour: Picturing Chemistry from Alchemy to Modern Molecular Science.* New York: John Wiley & Sons, 2000.

Guerlac, Henry. *Antoine-Laurent Lavoisier: Chemist and Revolutionary.* New York: Charles Scribner's Sons, 1975.

Holmyard, Eric John. *Makers of Chemistry.* Oxford: Clarendon, 1931.

Krebs, Robert E. *Scientific Laws, Principles, and Theories: A Reference Guide.* Westport, CT: Greenwood Press, 2001.

Plambeck, James. "Chemical Sciences: Chemical Thermodynamics, From Heat to Enthalpy" (website). Copyright 1995 James A. Plambeck, updated July 4, 1996. http://www.chem.ualberta.ca/~plambeck/che/p101/p01073.htm.

Poirier, Jean-Pierre. *Lavoisier: Chemist, Biologist, Economist.* Translated by Rebecca Balinski. Philadelphia: University of Pennsylvania Press, 1996.

Riedman, Sarah R. *Antoine Lavoisier, Scientist and Citizen.* New York: Abelard-Schuman, 1967.

Atmospheric Composition Is Constant
Henry Cavendish, 1783

Before the first manned balloon flights of the late eighteenth century, most people in England and France believed that breathing the air just a few thousand feet above the ground would be fatal. Of course, experience had already proven otherwise, as more than a few intrepid explorers had long since climbed mountains to very high altitudes. Nonetheless, the misconception endured until the first manned balloon flights shattered it.

On November 20, 1783, Joseph (1740–1810) and Étienne **Montgolfier** (1745–1799) launched the first manned **hot-air balloon** flight to a height of about 3,000 feet. Ten days later, on December 1, 1783, J.A.C. Charles (1746–1823) and his assistant made the first manned **hydrogen** balloon flight to a height of about 2,000 feet. The flight went smoothly until, on landing, Charles's assistant immediately stepped out of the balloon's gondola. Lightened of his load, the balloon suddenly soared to a height of 9,000 feet with Charles still aboard. The accidental ascent's only consequence came when Charles became extremely cold at 9,000 feet. Eventually, he landed safely, though elected never to fly again. This proved—once and for all—that the air several thousand feet up was indeed not fatal.

Modern scientists know that Charles's unplanned ascent was well within the **troposphere**, the atmospheric layer in which terrestrial weather occurs. The fatal effects of extreme temperature and lack of **oxygen** would have occurred only near the upper reaches of the troposphere. This layer extends about 10 miles over the equator and about 6 miles over the north and south poles. But even after Charles's relatively safe ascent to 9,000 feet, questions about the atmosphere's composition continued to puzzle scientists of the late eighteenth century.

One such scientist was the English aristocrat Henry Cavendish (1731–1810). Cavendish is often called the father of the **gas balloon** because of his discovery of **inflammable air**, or hydrogen. Cavendish also conducted many other important studies of gases and was especially influential on the topic of the composition of the atmosphere. Another scientist and a contemporary of Cavendish, Tiberius Cavallo (1749–1809), explored what he often called the "purity" of the air—the quantity of oxygen contained in it. Cavallo found that oxygen levels varied widely in different parts of the world. Cavendish, however, was less than convinced by this claim. He was joined in his skepticism by other prominent scientists. Studies of air collected at different locations were worse than inconclusive—they directly contradicted one another. What was the actual composition of this misleadingly complex "common air" in the atmosphere?

Cavendish set off to find out. Previously, he had improved an instrument known as the **eudiometer**, which measured the quantity of **dephlogisticated air**—or oxygen—in a sample of ordinary air. Cavendish and many of his contemporaries did not understand it in those modern terms. Nonetheless, Cavendish designed this ingenious device to estimate the combustible portion of a particular air sample. The eudiometer achieved this by measuring the loudness of the explosion when the air was detonated with hydrogen. Cavendish created both a uniform procedure and a standard of measure—the number 5—which he assigned to pure oxygen. From the start, Cavendish's experiment encountered a problem. He did not know that nitric oxide and oxygen often combine in ordinary air. This fact alone accounted for many skewed studies that measured different compositions for air.

Despite this obstacle, Cavendish successfully tested air collected on sixty different trial days in London and Kensington. Cavendish meticulously studied the results from his comprehensive tests and found that there was no significant difference in composition. Cavendish found neither temporal nor spatial variances. This was the most meticulous scientific study of the atmosphere ever undertaken. At the end of his study, Cavendish estimated the concentration of oxygen in the atmosphere at 20.83%. This estimate is astonishingly close to today's accepted value of 20.95%. In making this estimate, Cavendish lacked modern instruments of precision, and he also had no accurate body of research in the area of atmospheric meteorology on which to draw. Cavendish's accuracy continues to amaze scientists.

After the famous balloon rides of 1783, Cavendish became keenly interested in aeronautics. But unlike many others, Cavendish was interested in ballooning for the sole purpose of lifting his scientific experiments (quite literally) to new and higher levels. The following year, several aeronauts made a high balloon ascent over London. On this trip, they became the first people in history to collect scientific, in-flight data. Cavendish gave the aeronauts stoppered bottles filled with distilled water. At various heights, the aeronauts emptied the water, thereby allowing the bottles to fill with the air from that specific altitude.

Back on the ground, Cavendish tested these samples with his eudiometer. Then he compared the results with air he took from outside a window of his house. He found that the concentration of oxygen in these samples was no different than the concentration on the ground. The air a few thousand feet up was no more unique than it was fatal to breathe. Cavendish had scientifically and authoritatively established that little to no detectable variation existed between air on the ground and air a few thousand feet up. However, the eccentric (some say indifferent) Cavendish did not publish these important results. For many years, credit for this discovery erroneously went to Joseph Louis Gay-Lussac (1778–1850). Today, it is clear that Cavendish's findings predate those of Gay-Lussac by some 20 years.

Cavendish also anticipated the work of Lord John Rayleigh (1842–1919) and William Ramsay (1852–1916) some 100 years later. In 1784 and 1785,

Cavendish wrote two papers, each named "Experiments on Air." In the first paper, he established that the atmosphere was not composed of a common, unified air, but rather of two distinct substances—"dephlogisticated air" (or oxygen) and a substance he called "phlogisticated air." In the second paper, Cavendish established that "phlogisticated air" was the substance discovered by Daniel Rutherford (1749–1819) in 1772. (For more on this topic, please see the entry on Discovery of Nitrogen.) Rutherford named this gas **mephitic air** (or **nitrogen**). Cavendish demonstrated that this air made up the overwhelming part of the atmosphere. However, while he was experimenting with his eudiometer, Cavendish recorded a tiny residue, which he said was no more than 1/120 of the atmosphere's composition.

More than a century later in 1894, Ramsay and Rayleigh came across this paper by Cavendish. Based on Cavendish's observation, they began an inquiry into the question in context of their work on nitrogen. After extensive tests, Ramsay and Rayleigh determined that this residue noted so meticulously by Cavendish was actually a new atmospheric gas, which they named **argon**. Entirely inert, argon ushered in a new chapter in the study of the atmosphere. It also eventually became an important industrial product. After it was extracted from the atmosphere, argon was used in a new invention—electric lighting.

Today, scientists have shown that the atmosphere (just as Cavendish suggested) is about 78% nitrogen, 21% oxygen and about 1% argon. It also contains trace amounts of other gases (such as carbon dioxide, neon, helium, krypton, hydrogen, xenon, and ozone). Air in the atmosphere also contains water vapor and a significant amount of dust.

Knowledge of constant atmospheric composition has brought many practical benefits as well as scientific advances. The science of meteorology deals, in large part, with the chemical composition of the atmosphere. Using instruments to detect small changes of atmospheric gases, scientists can measure current weather conditions, forecast changes, and conduct research into long-term shifts. Modern instruments include satellites, radar devices, and (as in Cavendish's day) gas balloons, which gather a wide variety of data. Scientists also use knowledge of the atmosphere's composition to monitor public health dangers, such as air pollution, chemical contamination, industry waste, and smoke.

Atmospheric studies eventually aided balloon exploration as well. In 1932, Swiss physicist Auguste Piccard (1884–1962) invented an airtight balloon cabin and ascended in a hydrogen balloon 51,775 feet into the **stratosphere**, which is the atmospheric layer above the troposphere (which Cavendish studied). Without the airtight cabin, Piccard's flight would have been fatal for a number of reasons. At −55°C (−67°F), the air of the stratosphere is much colder than that which Charles encountered on his accidental ascent.

SEE ALSO Balloon Flight; Charles's Law; Hydrogen; Discovery of Nitrogen; Oxygen.

Selected Bibliography

Berry, A. J. *Henry Cavendish: His Life and Scientific Work*. New York: Hutchinson of London, 1960.

Crowther, J. G. *Scientists of the Industrial Revolution: Joseph Black, James Watt, Joseph Priestly, Henry Cavendish*. London: Cresset Press, 1962.

Darrow, Floyd L. *Masters of Science and Invention*. New York: Harcourt, Brace and Company, 1923.

Holmyard, Eric John. *Makers of Chemistry*. Oxford: Clarendon Press, 1931.

Jungnickel, Christa, and Russel McCormmach. Cavendish. *The Experimental Life*. Lewisville: Bucknell University Press, 1999.

B

Ballistic Pendulum and Physics of Spinning Projectiles
Benjamin Robins, 1747

Historians generally believe that the first small arms—forerunners of muskets and rifles—appeared in North Africa sometime during the fourteenth century. These weapons fired a variety of projectiles (including small lead balls) a short distance. For several centuries, small arms made up only a tiny part of the arsenal of military weapons, as they were extremely inefficient and ineffective at all but the closest ranges. During this time, the only people studying the flight of spinning projectiles were the gunners who made up a tiny part of each nation's army. What little **ballistics** information they had was closely guarded and purely esoteric.

Around the turn of the eighteenth century, riflemen began to emerge as a distinct group in many standing armies. During this century, the application of scientific principles to the problem of spinning projectiles began to expand outward from its modest beginnings during previous centuries. The earliest mathematical exercises regarding projectiles had begun in the sixteenth century. In 1522, a Bavarian philosopher, Herman Moritz, claimed that, because gunpowder is a "devilish" product, there must be small devils sitting upon all projectiles to make them stray from a straight course. Of course, scientists who lacked specific ballistics information still regarded this viewpoint as ludicrous. However, Moritz also made the observation that the practice known as rifling (cutting grooves inside the barrel of a firearm) spun the projectile and resulted in a significantly straightened trajectory. Though wildly misconstrued (Moritz thought spinning shook loose the tiny devil), Mortiz's observation about the efficacy of spinning was important to later studies.

In 1537, Niccolò Franco Tartaglia (1499–1557), an Italian mathematician, conducted the earliest scientific studies of the physics of projectiles. Tartaglia is most famous for first solving cubed roots. He was also fascinated with the mathematics of spinning projectiles. Tartaglia became the first to draw firing tables after he found what he thought was the perfect angle at which to aim a gun for maximum distance: 45 degrees. Mathematically, this angle was correct, but Tartaglia did not account for the effects of air resistance on the projectile. Tartaglia kept meticulous notes, but lacked the scientific theories of motion necessary to correct his finding.

Another Italian, the great Galileo Galilei (1564–1642), studied the effects of gravity on a trajectory. In fact, he came very close to explaining gravity as a universal force. Galileo pioneered many advanced equations and determined that a projectile follows a curved trajectory called a *parabola*. His equations were fundamentally sound, but (like Tartaglia) he did not arrive at the concept of air resistance or drag. Galileo's findings were nonetheless often accurate, due to the low velocities of projectiles during the sixteenth century. However, at times Galileo's explorations led to some incorrect conclusions. The mathematics of his vacuum trajectory theory told him that a 24-pound cannonball should travel a maximum of 16 miles. However, he knew its maximum range in the field was barely 3 miles.

During the late sixteenth century, Sir Isaac Newton (1642–1727) was pioneering the theory of classical physics. Newton's famous theories of motion were the first to deal quantitatively with the issue of air resistance. Newton surmised that the slowing of a projectile from air resistance must vary proportionally with the square of the velocity (assuming friction might be neglected). By this time, Newton had also explained his theory of gravity. He knew with reasonable certainty that gravity was a constant, universal force pulling a projectile toward the center of the earth (at 32 feet per second [fps]). By dropping round balls of different densities from the dome of a cathedral, Newton found that the theory held, but this was true only for the low velocities he attained in this manner, not for much faster projectile velocities being attained with newer muskets. (For more on Newton's laws of motion and effects of gravity, please see the entry on Newtonian Black Holes.)

As it turned out, even a sophisticated scientific theory like Newton's quadratic law could not explain the effects of air resistance on the flight of projectiles. Where Newton had led the way for the classical theory of physics, Benjamin Robins (1707–1751) would become its first true experimentalist. Robins knew of Newton's laws of motion and gravity as well as the 1731 mathematical text *A Treatise of Gunnery* by John Gray (died 1769). Though Gray did not conduct empirical tests, he speculated that air resistance played a much more vital role than any scientist had previously understood.

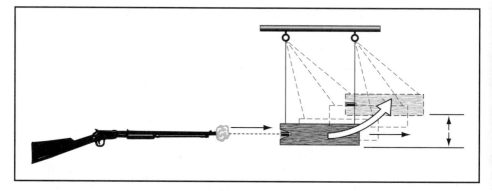

Figure 2. Ballistic pendulum. When a scientist fires a projectile at this instrument's wooden block, the block moves up and back in the arc of a pendulum. With careful measurements, the scientist can mathematically determine the projectile's velocity.

In 1742, Robins surprised the worlds of both physics and the military with publication of his book *New Principles of Gunnery*, which formed the basis for all subsequent theories of projectiles and artillery. In this book, Robins detailed his invention of the **ballistic pendulum** (see Figure 2). This instrument allowed Robins to conduct the first ballistics tests on the motion and behavior of projectiles. It consisted of a heavy wooden block suspended on a bar, allowing it to pivot around a fixed point on a tripod. A projectile striking the wooden block caused the block to move back and up in the arc of the pendulum. Because he knew the mass of the wooden block, Robins could mathematically determine the projectile's velocity and direction by carefully recording the distance of the arc, its period of oscillation, and its amplitude.

In subsequent experiments, Robins investigated the enormous forces acting on high-speed projectiles. He was the first to accurately determine the velocities of shots fired at a wide variety of ranges. By moving a weapon's muzzle progressively farther from the pendulum, Robins recorded a marked slowing of the projectile in ratio to the distance it had to travel. Robins suspected (like Gray) that this effect was due to air resistance. After many trials, he announced that the drag caused by air was 85 times more influential than the effects of gravity on any given projectile. This was an astonishing and even unbelievable discovery, but it was one that would hold. He also found that the force of air resistance was as high as 120 times the actual weight of the projectile. Using his ballistic pendulum, Robins measured velocities up to 1,700 fps and determined the reasonable loss of velocity for flights as long as 250 feet. Finally, Robins found that Newton's quadratic law satisfactorily explained velocities up to about 900 fps but failed to account for drag at higher velocities.

Robins also was the first to scientifically understand the practice of rifling when he explained that lateral deflection was due to a projectile's spin. In a sequence of ingenious experiments, Robins set up a series of evenly spaced paper "curtains" across a level field. When he fired a musket ball over 760 yards, the holes it ripped in the paper showed an enormous—and steadily predictable—deflection to the left. This previously mysterious problem of enormous deflection had long plagued riflemen. Robins correctly identified that the improper spin of the ball (in the barrel) caused the deflection.

In a subsequent experiment, Robins bent the barrel of his musket a few degrees to the left—*into* the deflection. As he expected, the projectile initially moved farther to the left. However, it later reversed its direction and crossed to the *right* of the firearm's barrel. This phenomenon is often observed in tennis, golf and baseball and is known widely as the **Magnus** (or sometimes Robins) **effect** (see Figure 3). Though no one knows for certain, many linguists believe this phenomenon (first observed by Robins, a British scientist) may have influenced the sports expression, putting "English" on a ball (meaning to give it a spinning motion).

Robins did not fully grasp the sophisticated aerodynamics of the Magnus effect. In the example of an improperly spun lead ball, the ball carries some of the air around with it due to the action of friction (a bit like a whirlpool). This creates a meeting of two airstreams traveling in opposite directions and

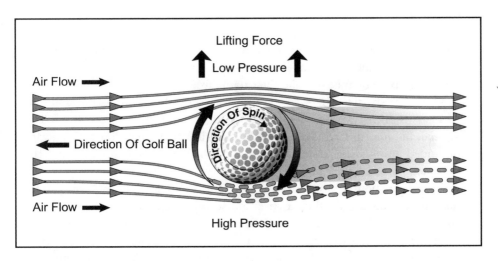

Figure 3. Magnus effect. Uneven spinning causes a ball (or projectile) to carry some air around with it. When the two air streams meet, they create a pressure differential, resulting in a difference in velocity between the ball's sides. This sends the ball askew. *See also:* **Hydrodynamics of lift** illustration, page 114.

an increased pressure on one side of the ball. The increased pressure results in enhanced velocity. On the opposite side, the motion has been opposed and the velocity is decreased. The ball is thereby carried askew by the unequal effects of different air pressures.

Robins drew up the first reasonably accurate ballistics tables, which greatly simplified the work of artillery. When his work was questioned, Robins simply compared his tables with the actual firing data collected in the field. When his tables were incorrect, Robins amended them to fit his experiments—not the other way around. Swiss mathematician Leonhard Euler (1707–1783) greatly enhanced Robins's tables. Later, Napoleon made extensive use of data from Robins and Euler in building superior artillery units for his conquest of Europe. Tables based on Robins's and Euler's work were still in use for high-angle and low-velocity mortar calculations during World War II.

Today, Robins is widely regarded as the founder of the science of ballistics. Modern ballistics is broken into four main categories. *Interior* ballistics studies the motion of a projectile as it travels down the barrel of a weapon. *Exterior* ballistics studies the behavior of a projectile after it leaves the weapon but before it strikes an object—while it is in flight. *Terminal* ballistics studies a projectile's effect on its target and surrounding area. This category especially (though by no means exclusively) deals with exploding weapons, such as missiles and torpedoes. *Forensic* ballistics is especially useful in law enforcement. Because every firearm leaves special markings (in large part due to its spin) on each projectile it fires, experts in forensic ballistics can determine whether a specific gun fired a specific bullet.

Research in ballistics is also conducted by all branches of the U.S. armed forces and comprises a significant branch of engineering worldwide. Large missile-type weapons are classified in one of two basic categories: *ballistic* or *nonballistic*. Ballistic missiles take their name from their trajectory—an arcing, "ballistic" path, like that of a basketball shot. Intercontinental Ballistic Missiles (ICBMs), which form a large part of the U.S. Air Force's nuclear arsenal, fall into this category. All other missiles, including the vast majority of guided missiles (such as the U.S. Navy's Tomahawk cruise missiles), fall into the category of nonballistic missiles.

This distinction is based on two different methods of propulsion. Many nonballistic missiles fly their entire course under engine power within the control of sophisticated guidance systems. On the other hand, the flight of ballistic missiles has two parts. First, a rocket engine blasts the missile into its long course. Once it reaches the desired speed, the engine simply shuts off and the missile coasts until it drops on its desired target. The ballistic missile is therefore so named because it is propelled for only a short time and at the very beginning its journey. Strictly in this sense as a projectile, it shares more in common with traditional ballistics (such as musket balls and bullets) than it does with guided and other types of nonballistic missiles.

SEE ALSO Newtonian Black Holes.

Selected Bibliography

Farrar, C. L. and D. W. Leeming. *Military Ballistics: A Basic Manual*. Elmsford, NY: Pergamon, 1983.

Halliday, David, and Robert Resnick. *Fundamentals of Physics*. Third edition. New York: John Wiley & Sons, 1988.

Herrmann, Ernest E. *Exterior Ballistics 1935*. Annapolis, MD: U.S. Naval Institute, 1935.

Lockett, Keith. *Physics in the Real World*. New York: Cambridge University Press, 1990.

Balloon Flight
Joseph and Etienne Montgolfier, hot air balloons, 1783
Jacques Alexandre Cesar Charles, hydrogen gas balloons, 1783

Among the more distinguished spectators to witness the world's first hydrogen balloon launch on December 1, 1783, was Benjamin Franklin (1706–1790), minister to France from the newly created United States. On seeing the balloon ascent, a fellow spectator somehow failed to see the practical aspects of the new invention, turned to Franklin and asked him what good it was. To this, the venerable old statesman replied, "What good is a newborn baby?"

Indeed, people have been dreaming about flying for thousands of years. Predictably (and often with macabre results), attempts at flight prior to the late eighteenth century ended in total disaster, because the attempts were made by people who did not approach the question of flight scientifically. Nonetheless, a few impressive attempts had been made. Two-thousand-year-old Peruvian legends clearly tell of successful **hot-air balloon** flights. The Peruvian desert also has many ancient area markings that are only discernible from the air. Chinese generals used large kites to approximate the distance of enemy fortifications as early as 200 B.C. In the following centuries, they sent intrepid observers in kites to float high above enemy encampments. Medieval European history is full of mostly unsubstantiated stories in which occupants of besieged castles donned billowing cloaks to help slow their descent after a terrified jump. Historians doubt the legitimacy of many of these claims. More impressively, as early as the eleventh century, several groups of monks are said to have made successful (read: non-deadly) winged glides after jumping from high atop cathedral towers. Historians believe at least a few such glide-jumps may have actually taken place.

These cases notwithstanding, the first people to approach the question of flight scientifically lived during the late eighteenth century. One such person was Joseph **Montgolfier** (1740–1810), a paper maker-turned-inventor. Joseph had a creative, scientific mind. He was a kind of free spirit with an aptitude for mechanics—the perfect disposition for an inventor. Throughout the history of science, big discoveries have often been made by several individuals working together to augment one another's talents. This was the case with Joseph and his youngest brother, Étienne Montgolfier (1745–1799), who was very much his opposite. Where Joseph was romantic, Etienne was passionate. Where Joseph was terrible with responsibility and organization, Étienne was dependable to the point of esteem. For all their differences, however, the Montgolfier brothers shared one very important trait: They both loved mechanics.

As early as 1777, Joseph was doing a lot of musing about the properties of heat and density—important topics for eighteenth-century scientists. Joseph had become perplexed by the question of why sparks from a hot fire often rose high into the air. Why didn't they just fall straight to the ground? After Joseph had been wondering about this question for several months, he was surprised by a staggering but simple occurrence. One night, he made a fire and proceeded, with his wife, to dry their laundry. When the hot gases caught his wife's chemise the right way, Joseph was amazed to see it billow high into the air. This effect practically made him seethe with curiosity.

By 1783, Joseph was experimenting with the properties of hot air. After many trials, he made his first major discovery, which led him to the idea of hot-air balloon flight. In his Paris apartment, Joseph built a miniature 3-foot by 4-foot balloon made of taffeta. When he ignited shards of paper beneath the balloon, it quickly rose to the ceiling. Joseph was so excited that he immediately shared his experimental findings with Étienne.

Joseph and Étienne conducted several further experiments in which they approached their topic more or less scientifically. Nonetheless, they never understood that a physical change (heat) is responsible for a balloon's ascent. Since 1700, the **phlogiston** theory of substances had dominated chemical thought. This theory claimed that a hypothetical "heat fluid" substance, phlogiston, was released during **combustion**. The Montgolfier brothers thought they had simply discovered one of phlogiston's properties—a much lower density than ordinary air. Throughout their trials, Joseph and Étienne actively tried to discover just the right mix of fuel to produce the phlogiston for flight. They arrived at damp straw with a few old leather shoes thrown in.

On June 5, 1783, Joseph and Étienne made the first hot-air balloon launch when they sent a tethered, unmanned 33-foot linen balloon about 6,000 feet into the air. Next, on Royal command, Joseph and Étienne traveled to Versailles for a second, larger launch before King Louis XVI. This time, they became the first to launch living creatures in a balloon. In the Montgolfier

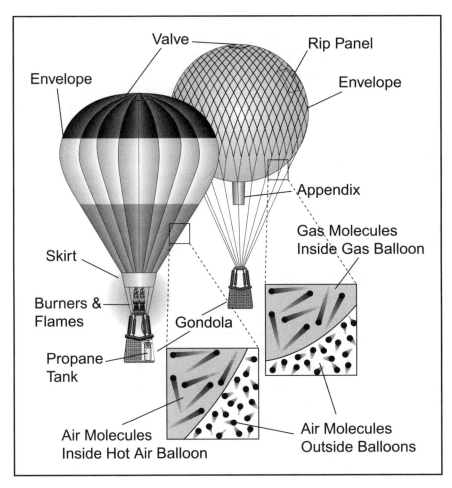

Figure 4. Montgolfiers and Charlieres, or hot air and gas balloons. Both hot air and gas balloons work on the property of lower density, which is a temporary physical property of hot air and a chemical property of hydrogen (and helium).

brothers' balloon, a rooster, a duck, and a sheep ascended to a height of about 1,500 feet. Eight minutes later, all three creatures landed safely 2 miles away. The only injury was to the rooster, who had been kicked in mid-flight by the frightened sheep (see Figure 4).

On November 20 of the same year, Joseph and Étienne Montgolfier's trials reached a successful pinnacle. On this day, two men, a science teacher-turned-balloon pilot named Jean-Francois Pilatre de Rozier (1756–1785) and an infantry officer-turned-passenger named the Marquis d'Arlandes (ca. 1742–1809) made the first manned balloon ascent in history. The two men floated across Paris at approximately 3,000 feet. Twenty-five minutes and 5 miles later, they safely landed, having floated over the river Seine. Many scholars believe that the 400,000 people who gathered for this launch comprised the largest assembly of people to this point in history.

While the Montgolfier brothers were teaching themselves about hot-air balloons, a single competitor was working independently. The most famous physics lecturer in Paris, Jacques Alexandre Cesar Charles (1746–1823), had just completed a revolutionary inquiry into the properties of gases. Charles built on the revolutionary discoveries of English pneumatic chemist Henry Cavendish (1731–1810). In 1766, Cavendish had isolated an explosive gas he called **inflammable air** (or **hydrogen**). This gas was so explosive, in fact, that for a while Cavendish thought he had isolated the elusive phlogiston itself. Cavendish soon realized that this gas had some unique properties. Most especially, hydrogen is far less dense than ordinary air. Cavendish measured accurately that a sample of hydrogen is about fourteen times lighter than an equal volume of ordinary air. In other words, Cavendish's special gas was only about one-fourteenth as dense as ordinary air. (For more on this topic, please see the entry on Hydrogen.)

The following year, one of Cavendish's fellow pneumatic chemists, Joseph Black (1728–1799), provided a simple but surprising demonstration. In front of an audience, Black filled a sack with hydrogen gas, sealed it, and—with much fanfare—released it into the air. Like Joseph Montgolfier's miniature balloon, Black's sack rose quickly to the ceiling. Several guests accused Black of an elaborate deception involving thin black suspension wire. Nonetheless, Black had demonstrated that Cavendish's gas was very suitable for lighter-than-air balloon flight.

After Black's sack trick, Charles began his important inquiries into the specific area of lighter-than-air flight. Charles understood better than anyone else how the hot-air balloons were able to fly. The Montgolfiers' balloon began to take flight only when its quantity of trapped air became very hot. From the start, they also implicitly accepted that the most efficient method of descent was to simply let the trapped air cool. During the next several years, Charles conducted a series of tests demonstrating that the air had become less dense when it heated because heat caused it to expand, decrease its density, and increase its volume. On cooling, the volume of air decreased and the balloon became much heavier. Hot-air balloons flew because of a temporary physical change.

Charles demonstrated conclusively that this was not the case with hydrogen. When Cavendish and others produced hydrogen in the laboratory, they also produced a lot of heat. However, hydrogen's lighter-than-air quality did not need any form of heat. It did not lose this quality after it cooled substantially. These observations led Charles to formulate a special law, which bears his name. **Charles's law** of gases states that a gas expands by the same fraction of its original volume as its temperature rises. If the pressure remains constant, its volume (V) and the temperature (T) do not change. Therefore, Charles's law is written as V/T = constant. Assuming no change in pressure, Charles's law states that the volume of a gas is in direct proportion to the temperature of the gas. (For more on this topic, please see the entry on Charles's Law of Gases.)

Charles's formulation of his law proved invaluable for his work with lighter-than-air balloons. Every one of Charles's tests convinced him that hydrogen **gas balloon** flight was possible. His law explained how. The biggest challenge left was how he would trap enough hydrogen for a balloon ascent. Charles quickly realized he needed to figure a way to collect a previously unimaginable amount of gas. Even his relatively small 12-foot-diameter balloon meant that he had to painstakingly pour sulfuric acid over iron filings until he had produced some 900 cubic feet of hydrogen. Previously, scientists had only ever thought to measure hydrogen gas in cubic inches for use inside laboratories. Charles calculated that producing the 900 cubic feet of hydrogen would require a staggering 498 pounds of acid and approximately 1,000 pounds of iron. He also foresaw another problem: Might the precious gas leak through ordinary material, especially permeable taffeta? Charles was especially concerned that this might happen during flight.

It just so happened that a prominent artisan had recently developed a new material. It was a kind of impenetrable India rubber that, when combined with taffeta, would hold the gas for a very long period of time. However, producing the gas would prove more of a challenge for several of Charles's assistants and a whole host of volunteers. Despite Charles's ingenious method using barrels of iron shavings and a "hose" of sulfuric acid, one writer describes the production trials this way:

> In a small, enclosed courtyard in a densely populated section of the city, a handful of largely inexperienced people were collecting an unprecedented quantity of the most inflammable gas known through a tube too hot to touch into the confinement of a rubberized bag that was close to catching fire if it was not first chewed through by sulfuric acid (Gillispie 1983, 30).

Just 10 days after the Montgolfiers' first flight, Charles and his assistant rose from the Tulieries Gardens in Paris. Charles's flight was much longer than the Montgolfiers', as it was a tranquil 2-hour, 27-mile drift. Indisputably, this was the first gas balloon flight in history.

Today, hot-air balloons are the most popular balloons, in large part because of their relatively inexpensive operating costs. These balloons are often called *montgolfiers,* in honor of their inventors. Gas balloons are still widely used, though their high cost limits their accessibility. These balloons are sometimes called **charlieres**. Gas balloons are used by serious private and sports enthusiasts; the U.S. Navy; and many scientific agencies, organizations, and companies for research purposes. They also are used extensively to monitor weather conditions. For many years, passenger gas balloons employed trapped hydrogen gas. However, this changed after May 6, 1937. On this date, the enormous German hydrogen airship *Hindenburg* burst into flames while approaching its docking in Lakehurst, New Jersey. Thirty-five people on board and many others on the ground were killed. To this day, no one is entirely certain how the ship's hydrogen gas ignited. Nonetheless, this disaster marked the end of rigid airships, as well as hydrogen passenger balloons. Today's passenger gas balloons use expensive but inflammable helium gas, which has about 92% of the lifting ability of hydrogen. Helium was discovered over a century after Charles's first gas balloon.

A third major kind of balloon also is used today, though less widely than the other two. A **roziere** is actually a combination of a gas balloon and a hot-air balloon. Primarily, a roziere is flown as a gas balloon, but it also employs a "cone" of hot air around its gas cell, which helps it compensate for the different temperatures encountered during day and night on long trips. By firing burners and indirectly warming its helium, a roziere balloon can more efficiently maintain altitude when the temperature drops at night. Rozieres are named for their inventor and the Montgolfiers' first pilot, Pilatre de Rozier. He was killed in 1785 when he tried to fly his roziere across the English Channel. Nonetheless, Pilatre de Rozier is remembered countless times each day, as his first name formed the basis of the word *pilot.*

SEE ALSO Atmospheric Composition Is Constant; Charles's Law; Hydrogen; Phlogiston.

Selected Bibliography

Baker, David. *Flight and Flying: A Chronology.* New York: Facts On File, 1994.
Bridgett Travers (ed.). *World of Invention.* Washington, DC: Gale Research Group, 1994.
Crouch, Tom D. *The Eagle Aloft: Two Centuries of the Balloon in America.* Washington, DC: Smithsonian Institution Press, 1983.
Denniston, George. *The Joy of Ballooning.* Philadelphia: Courage Books, 1999.
Gillispie, Charles Coulston. *The Montgolfier Brothers and the Invention of Aviation: 1783–1784.* Princeton, NJ: Princeton University Press, 1983.
Rolt, L. T. C. *The Aeronauts: A History of Ballooning, 1783–1903.* Gloucester, UK: Alan Sutton, 1985.
Waligunda, Bob, and Larry Sheehad. *The Great American Balloon Book: An Introduction to Hot Air Ballooning.* Englewood Cliffs, NJ: Prentice-Hall, 1981.

Binary System
Gottfried Wilhelm Leibniz, 1701 and 1730

In mathematics, the process of designating numbers according to a particular system is called numeration. Today, by far the most widely used numeration system is the **decimal system**, which uses the number 10 as its base. Decimal counting, therefore, tends to group objects to be counted in groups of 10. From 1 to 9, this system uses one-digit numbers, and from 10 to 99 it uses two-digit numbers. Many major currencies are decimal based. Ten pennies make one dime, ten dimes make one dollar and so forth. Many currencies also include non-decimal holdover exceptions, such as quarters and nickels in U.S. currency. Other holdovers, such as base 60 in time and angular degrees, are also evident. In the decimal system, the value of each digit depends on its place in a given numeral, and each position is ten times greater than the position directly to its right. For example, in the numeral 1,475, 1 is in the thousands place, 4 is in the hundreds place, 7 is in the tens place and 5 is in the ones place.

The first prehistoric counting systems emerged when people used objects (such as stones or sticks) to symbolize objects they wanted to keep track of (such as sheep or potatoes). Later, they developed symbols to stand for numbers. For example, the word for "wings" might also mean "two" and the word for "hand" might be used for "five." Finally, people began to arrange number-names in certain order (one, then two, then three and so on).

About 3000 B.C., the Egyptians began using hieroglyphics, or pictures, to write numerals. This system lacked both place value and a symbol for zero. Around 2100 B.C., the Babylonians developed a base 60 system of cuneiform (or wedge-shaped) numbers. In this system, one group stood for 1s, then the next group stood for 60s, the next for (60 × 60), and so on. By 1500 B.C., the Babylonian system was base 10 (1, then 10, then 100, then 1,000 and so on).

Around 500 B.C., the ancient Greeks developed a 10 base system on the basis of the twenty-seven-letter alphabet. The first nine letters stood for numbers 1 to 9, the next nine stood for 10 to 90, and the final nine stood for 100 to 900. The familiar Roman numeral system also used letters for numbers. The numerals VI, XV, and LX demonstrate the additive principle of mathematics, in which the lesser first number is added to the second. The numerals IV and IX demonstrate the subtractive principle, in which the value of the smaller first number is subtracted from the second. By the second century B.C., Hindu mathematicians developed the sophisticated mathematics that would become the modern decimal system. This system was superior to many others because it followed the principle of place value. Along these lines, by the year 600, the Hindus eliminated place names by inserting the "empty" placeholder, a figure for zero. During the next century, Arab scholars learned Hindu arithmetic. Four hundred years later, the Hindu system swept across Europe, when a book detailing it was translated from Persian to Latin.

There is nothing magical about using the base number 10. Most scientists believe that base 10 systems took hold independently in several different cultures because humans have 10 fingers for convenient grouping. There is nothing that mathematically prevents a person from using base 8, base 37 or base 1,845.5 just as effectively as base 10.

Gottfried Wilhelm Leibniz (1646–1716) pointed out the arbitrary nature of base 10 when he published his **binary system** in a 1701 paper, "Essay d'une Nouvelle Science des Nombres." Leibniz's binary system is base 2, as it requires only the symbols 0 and 1 for positional units 2s, 4s, 8s, 16s, and so on. This system groups numbers by 2s and powers of 2. In fact, the word *binary* comes from the Latin for "two at a time." The number 10_{two} stands for two, the system's base, meaning 1 two plus no ones.

In binary system place value, each position has twice the value of the position to its right. For this reason, binary numbers are generally much longer than the decimal equivalent. For example, the decimal number 90 appears as binary number 1011010—or right to left:

$$0 \times 2, 1 \times 2 \times 2, 0 \times 2 \times 2 \times 2, 1 \times 2 \times 2 \times 2 \times 2,$$
$$1 \times 2 \times 2 \times 2 \times 2 \times 2, 0 \times 2 \times 2 \times 2 \times 2 \times 2 \times 2,$$
$$1 \times 2 \times 2 \times 2 \times 2 \times 2 \times 2 \times 2$$

Figuring the reverse, the decimal equivalent of a binary number is calculated by adding together each digit multiplied by its power of 2. In the preceding example, this appears as:

$$(1 \times 2^6) + (0 \times 2^5) + (1 \times 2^4) + (1 \times 2^3) + (0 \times 2^2) + (1 \times 2^1) + (0 \times 2^0)$$
$$= 64 + 0 + 16 + 8 + 0 + 2 + 0 = 90$$

Leibniz's system was both philosophical and scientific, as he conceived of it as an attempt to move "thought" from a verbal to a universal condition of mathematical perfection. An enlightenment thinker, Leibniz believed that if he could reduce the noise of decimal numbers to a language of only 0s and 1s, then a beautifully ordered universe would result. Leibniz carried this idea to nearly utopian lengths, arguing that the binary system might one day reduce arbitrating disagreements to a matter of simple calculation. He saw evidence of a binary universe everywhere and anywhere in which opposites might be said to occur naturally.

Ultimately, Leibniz refined his philosophy into a weighty religious discourse, outlining the number 1 as evidence of higher power and the number 0 as paradoxical evidence of a void, or the universe before it was created.

Most scientists paid no attention to Leibniz's binary system when he first proposed it. Most found it impenetrable, abstract and overtly religious. They also found it totally devoid of practical application—the litmus test for many eighteenth-century scientific proposals.

However, through a friend, Leibniz began to learn about the binary system's ancient history. The general practice of using two symbols to encode all information is very old among African bush tribes, which sent complicated messages using two-toned drums, and among Australian aborigines, who counted only by 2 s. But the most sophisticated binary system—and the one Leibniz would discover—came from the ancient Chinese text, *I Ching, the Book of Changes*. This revered philosophical text dates from the Chou dynasty (ca. 1122–256 B.C.). It rests on the fundamental idea of natural binary distinction in the yin and yang, or opposites that contain one another in perpetual contradiction. On reading this, Leibniz found a confirmation of his binary system in the depiction of the entire universe as a series of contradictory dualities such as yes–no, experiencer–experience, thinker–thought, on–off and so forth. He became convinced more than ever that his system could reduce reasoned thought to a series of straightforward proposals.

After many years of studying ancient Chinese philosophy in the context of enlightenment mathematics, Leibniz refined his binary system. One important version was published posthumously in 1730 as the widely read treatise "On Binary Arithmetic." Contemporary mathematicians and scientists accepted Leibniz's refined binary system as fundamentally sound and rationally valid. However, they found no direct practical use for the binary system for many years. In the mid-nineteenth century, English mathematician George Boole (1815–1864) developed a new type of mathematics based on the binary system. Boolean algebra was able to solve complex operations with only the digits 0 and 1. It would become an important field of study to a number of disciplines and is still highly influential in the field of modern logic.

Direct application of the binary system would not occur until the 1940s when scientists developed the first generation of computers. Just as Leibniz described, computers "think" only in two states. In circuitry terms, they are open or closed. In electrical terms, they are on or off. And in binary terms, they are 0 or 1. This is the language of modern computing, with emphasis on "either–or" statements familiar to most computer programmers. In a computer's binary system, an individual digit, a 0 or a 1, is known as a *bit*, which is a contraction of "binary digit." A computer's microprocessor processes bits by switching them between millions of tiny electronic circuits. These circuits operate analogously to a light switch. When an individual circuit is off, it corresponds to the 0 bit. When it is on, it corresponds to the 1 bit. A computer performs basic mathematical functions by adding, subtracting, multiplying or dividing bits of information.

Currently, scientists are developing the so-called fifth generation of computers. In the infancy stage of development, these "artificial intelligence" computers are still mostly theoretical. In fact, some scientists publicly doubt whether computers will ever be made to truly "think," and others wonder what computer "thought" would actually entail. Yet others—echoing Leibniz—believe that, one day, computers will effectively reduce actual, complex thoughts to series of calculations involving only the binary digits 0 and 1.

SEE ALSO Calculus; Metric System.

Selected Bibliography

Asimov, Isaac. *Asimov's Chronology of Science and Discovery*. New York: Harper & Row, 1989.
Bruck, Richard Hubert. *A Survey of Binary Systems*. Berlin: Springer, 1958.
Flores, Ivan. *The Logic of Computer Arithmetic*. Englewood Cliffs, NJ: Prentice-Hall, 1963.
Parkinson, Claire L. *Breakthroughs: A Chronology of Great Achievements in Science and Mathematics, 1200–1930*. Boston: G. K. Hall, 1985.

C

Calculus of Finite Differences and Differential Equations
Brooke Taylor, *Methodus Incrementorum*, 1715
Leonhard Euler, *Mechanica*, 1736
Joseph-Louis Lagrange, *Theory of Analytic Functions*, 1797

The main ideas that underpin **differential equations** developed over a very long period of time. Indeed, they depended on **calculus**, which came about during the seventeenth century and matured in the eighteenth, after a very long history of finite differences. The ancient mathematician Archimedes (ca. 287–212 B.C.) greatly extended a way of determining areas of figures with curved boundaries, known as the *method of exhaustion*. Thousands of years later, mathematicians would rework this method into the mathematical field of calculus. Archimedes used it and similar methods to figure curved surface areas, like those of spheres and cones. In the seventh century, Indian and Arab mathematicians successfully used finite differences to calculate astronomical sinus tables.

Little further significant work was done in this area until the explosion of mathematical knowledge during the seventeenth century. In 1624, English mathematician Henry Briggs (1561–1630) used differential equations to calculate tables of logarithms. Briggs was first to introduce logarithms to base 10, and he began constructing a 14-base table of common logarithms. Later in the century, Dutch mathematician Adriaen Vlacq (1600–1667) finished Briggs's work. During this period, Blaise Pascal (1623–1662) undertook important work toward the understanding of infinite series in relation to the mathematics that would later become important to both differential and integral calculus. The prior revolutions in physics and astronomy led to the seventeenth-century birth of analytic geometry, number theory, probability theory and calculus.

Prior to the middle of the seventeenth century, mathematicians had developed highly sophisticated number series. Having each developed several series representations for functions, the great Isaac Newton (1642–1727) and the renowned mathematician–philosopher Gottfried Leibniz (1646–1716) independently invented mathematical systems for "the calculus," the field of mathematics that deals with infinitesimally tiny changes. Newton's calculus, which he referred to as the **method of fluxions**, stemmed from the mathematical relationship between differentiation and integration of small changes in events. It also was based on his crucial insight that the integration of a function is simply the inverse procedure to differentiating it. Once he realized differentiation was the basic operation, Newton produced many simple analytical methods that served to unify disparate techniques previously used to solve areas, tangents, curve lengths, and maxima and minima of functions. Motion became a fundamental basis for curves, tangents and related calculus phenomena.

Newton first wrote about his calculus in *De Methodis Serierum et Fluxionum*, and this manuscript enjoyed wide circulation among Newton's scientific colleagues. However, it was not published until 1736, when it appeared as an English translation. In this text, Newton first detailed his calculation of the series expansion for sine x and cosine x and the expansion for what would later be established as their actual exponential function. Earlier, in 1676, Newton had written the *Methodus Differentialis*, which was published in 1711. This method used detailed difference equations for interpolation and quadrature. In hindsight, *Methodus Differentialis* is now considered the first major work toward the calculus of finite differences.

During this same period, Leibniz was also developing his version of "the calculus." When he studied the ancient problem of determining the area and volume of curved surfaces, Leibniz realized he needed a method of notation, in which symbols could represent quantities. Leibniz first proposed the familiar notation ? $ydy = y^2/2$ and published his theory of calculus in 1684. Earlier, in 1676, Leibniz discovered the familiar notation $d(x^n) = nx^{n-1}dx$ for both fractional and integral n. Finally, in 1715, Leibniz revealed that differential equations formed the basis of his calculus, when he published *Historia et Origo*. On mastery of "the calculus" of Leibniz's formulas, a person could easily solve tangent problems without mastery of geometry.

Newton had developed his calculus many years before Leibniz. However, his long delay in publishing gave rise to a major, prolonged dispute over priority between the two mathematicians and their followers. Calculus was marred for many years. The fissure became so great that it led, at times, to nationalistic tendencies between Newtonians in England and the continental followers of Leibniz. This dispute was wholly irrelevant to the actual mathematics and utility of calculus. In fact, by the early part of the eighteenth century, other mathematicians were using calculus to solve many long-vexing mathematical problems. Solutions to several challenging problems led to development of differential equations during the early part of the century.

In 1715, English mathematician Brook Taylor (1685–1731) published the watershed book *Methodus Incrementorum*, which most historians of mathematics credit as the founding work of differential equations. This work also contained the renown Taylor series of expansion. With this book, Taylor added himself to the list of founders of calculus and also became one of the earliest mathematicians to use it for summation of series, as well as interpolation. *Methodus Incrementorum* also contained Taylor's proof of a well-known theorem:

$$f(x+h) = f(x) + hf^1(x) + \frac{h^2}{2!} f''(x) + \cdots$$

In **Taylor's theorem**, a function of a single variable can be expanded in its own powers. It does not consider series convergence. The importance of Taylor's theorem was recognized, however, as the foundational principle of differential calculus, and the term *Taylor series* was in use soon after.

Methodus Incrementorum also included several theorems on interpolation. It contained a change of variables formula and a way of relating function derivative to the derivative of the inverse function. Taylor demonstrated the possibility of a so-called calculus of operation. In *Methodus Incrementorum*, he also denoted the *n*th differential coefficient of *y* by *yn* and used *y*-1 for the integral of *y*.

In this book, Taylor set forth an additional theory of the transverse vibrations of strings (as used in musical instruments)—an understanding of which had eluded investigators for years. Taylor found that the number of half-vibrations each second is Π√ (DP/LN), in which L is the string's length, N its weight, P the weight at which it stretches, and D the length of a seconds pendulum. Essentially, Taylor's theory was on the right track. Later, it was rejoined after scientists demonstrated that every point on the string does not pass through equilibrium at the exact same instant. Regardless, Taylor not only originated the earliest mathematics of differential equations, but also investigated a wide range of mathematical problems as related to the new and expansive field of calculus.

Another early investigator of differential equations from calculus was Swiss mathematician Leonhard Euler (1707–1783). Euler created an enormous amount of analysis, revised most of the known branches of pure mathematics (by adding proofs and details), and arranged his own work in an overall consistent, accessible form. He also frequently used infinite series to develop new methods or to model problems of application. Euler published the book *Mechanica* in 1736. In this text, he systematically applied calculus to mechanics and, in doing so, developed a range of methods that used power series to solve differential equations.

Euler also developed many modern notation summations that are still in use today. Most notably, he developed notations such as e^x sine x, cosine x and the relationship of $e^{ix} = $ cosine $x + $ sine x. The notation e became the base of the natural logarithm, often referred to as "the calculus number." He also demonstrated the differential equation named for him, as well as the formula relating the number of faces, edges, and vortices of a polyhedron ($F + V = E + 2$). Euler's invention of the calculus of variations led to the general method for solving maximum and minimum value problems. Finally, Euler connected five fundamental mathematical numbers with the famous equation $e^{ipi} + 1 = 0$.

At the end of the eighteenth century, French mathematician Joseph-Louis Lagrange (1736–1813) extended the work of Euler in the theory and application of calculus. Lagrange sought to place calculus on a rigorous basis algebraically, without references to geometry of any kind. His textbooks of 1797 and 1801 established many of the fundamentals found in modern calculus volumes. In his 1797 *Theory of Analytic Functions*, Lagrange attempted a purely algebraic form for the derivative, with no recourse to geometric intuition or numeric graphs. This resulted in the primer notation used for derivatives today, but did not (in the end) provide a firm foundation of pure calculus.

In this enterprise, Lagrange approached the study of functions almost wholly through their Taylor series, which eventually made the study of functions possible in strictly algebraic terms. Therefore, he additionally demonstrated that the Taylor series played a fundamental role in the underpinnings of calculus.

Today, scientists use calculus in nearly all areas of physics, as well as many other fields of study. Broadly speaking, calculus is the scientific language that engineers and physicists use to address practical problems. Scientists also use differential equations in a wide variety of areas. Students generally learn calculus in advanced high school mathematics classes or in college, after they have successfully studied algebra, geometry, and trigonometry.

SEE ALSO Binary System.

Selected Bibliography

Baron, Margaret E. *The Origins of the Infinitesimal Calculus*. New York: Pergamon, 1969.

Edwards, C. H. *The Historical Development of the Calculus*. New York: Springer, 1979.

Hahn, Alexander. *Basic Calculus: From Archimedes to Newton to Its Role in Science*. New York: Springer, 1998.

Simmons, George Finlay. *Calculus Gems: Brief Lives and Memorable Mathematics*. New York: McGraw-Hill, 1992.

Carbon Dioxide Is a Distinct Gas
Joseph Black, 1756

When the important chemist Joseph Priestley (1733–1804) began his late-eighteenth-century work with gases, it was partially in response to a gas that escaped a brewery's vats very close to his house. In his experiments, Priestley discovered this was the gas **fixed air**. Later, Priestley forced this gas into water, touching off a European craze for soda water. Sparked by this early gas research, Priestley would later discover **dephlogisticated air** (**oxyge**n), which today is sometimes conventionally paired with **carbon dioxide**, due to the role of both gases in animal **respiration**. Priestley was one of two independent discoverers of oxygen. (For more on the discovery of oxygen, please see the entry on this gas.) However, carbon dioxide—or "fixed air"—had been identified as a distinct gas by Joseph Black (1728–1799) in 1756.

In the previous century, Stephen Hales (1677–1761) had experimented widely on gases and published two important books, *Vegetable Staticks* and *Haemastaticks*. He recorded the different amounts of "air" he could extract from various substances. Hales also catalogued the different properties (such as odor, color and combustibility) of each "air." Hales did not recognize that he had identified unique gases. Rather, he thought he had manipulated the elastic qualities of a single, unified "air."

Following Hales, Black began to experiment on the so-called mild alkalis (potassium and sodium carbonates) and the "caustic" alkalis (potassium and sodium hydroxides). At this time, the **phlogiston** theory of substances was dominant. This theory stated that a hypothetical substance, phlogiston, was released and absorbed in all reactions involving **combustion**, calcification or respiration. The phlogiston theory explained that the "mild" alkalis combined with phlogiston and yielded the "caustic" alkalis.

This alkali misconception was demolished in 1754 when Black published his celebrated inaugural dissertation, *Experiments upon Magnesia Alba, Quick-lime, and some other Alkaline Substances*. A physician as well as chemist, Black was originally trying to find a way to dissolve some of his patients' kidney stones. He began experimenting with the mild alkali "white magnesia." When he combined limestone and the white magnesia through combustion, he found that both substances lost weight. Though he did not bother collecting it, he assumed that the substances lost some kind of "air" (or gas) during combustion. He therefore postulated that the "air" had been "fixed" in the mild alkalis. Black did not yet understand that this air was the distinct gas carbon dioxide, which escaped during many reactions.

In 1756, Black began a series of important experiments with carbon dioxide produced in other ways, including burning charcoal and fermenting grapes.

His work with "magnesia alba" (or magnesium carbonate, $MgCO_3$) comprises his most important experiments. From previous work, Black knew that carbon dioxide escaped during combustion. He also had noted many times that the resultant magnesia oxide (MgO) was lighter than the original magnesium carbonate. However, he was surprised to learn that only a small proportion of the volatile parts of the powder consisted of water. He concluded that the volatile matter lost during combustion was mostly air. In the strictest sense, this is not at all what the phlogiston theory would have led him to believe.

Black's burgeoning discovery could thus far be written as this basic formulation: magnesia alba = magnesia oxide + water + fixed air. By adding acid to the magnesia alba, Black found another formulation: magnesia alba + acid = magnesia sale + (a large quantity of) fixed air.

Black continued his inquires by turning to limestone and quicklime (in place of magnesium alba). Here he found that, when limestone effervesces with acids, it immediately gives off carbon dioxide. Black was most intrigued with his discovery that heating limestone produced quicklime. In contrast to the magnesia alba, this reaction gave off only a trace of water. Yet, the odd thing was that the end product—time and again—demonstrated a considerable loss in weight. If water was not responsible for this loss in weight, then it must be a gas. But what gas was making its escape? After repeating the experiment many times with many variables, Black correctly concluded that the loss of weight was due to the escape of his "fixed air," carbon dioxide. Therefore, the only conclusion to acceptably account for his findings was this formula: limestone = quicklime + fixed air.

This conclusion was astonishing. To begin with, phlogiston was not central to Black's conception. Additionally, Black arrived at an accurate conception of limestone and magnesia alba. This was many years before the theoretical framework of oxygen chemistry made this conception widely available. Here, Black was supposing that carbon dioxide escaped in many reactions, thereby accounting for their loss in weight. Black was therefore paying close attention to the concept of mass in chemical reactions decades before Antoine Lavoisier (1743–1794) conceived of the **law of conservation of mass**. (For more on this topic, please see the entry on Law of Conservation of Mass.)

Black therefore came very close to dismantling the phlogiston theory many years before Lavoisier led the end-of-century chemical revolution. At the same time, Black began to wonder about the properties of his "fixed air" as unique from those of ordinary air. This enterprise also nearly led him to a frontal assault on phlogiston. In one near-hypothesis, Black broadly noted that "fixed air" was not the same substance as ordinary air. Specifically, he said that quicklime "does not attract air when in its most ordinary form, but is capable of being joined to one particular species only. . . . To this I have given the name of *fixed air*" (quoted in Stillman 1924, 466). Many chemical historians trace a theoretical change in Black's work to statements like this. This change put Black far in front of even his greatest contemporaries, who were thinking of the elastic properties of a phlogiston-centered, unified air. Black's realization of the specific makeup of carbon dioxide is the very beginning of **pneumatic chemistry's** fruitful practice of releasing different "airs" or gases. After all, Black's "fixed air" was the first (and only) gas discovered for many years.

Here was the first indisputable, scientifically collected evidence that ordinary air is neither an **element** nor even a simple substance. This diverged with theories dating from the ancient Greeks right up to the present moment of phlogiston. Accepting Black's findings (and his interpretations of them) meant accepting that air was made up of no less than two substances—"ordinary air" and "fixed air."

Next, Black turned his attention to the unique properties of his special gas. He knew all too well that it stubbornly extinguished flames. Black also found that a bird made to breathe the "fixed air" died suddenly. He returned to chalk ($CaCO_3$)—only this time with the intent of studying not the substance itself, but his gas. He was fascinated to find that the chalk released carbon dioxide on being dissolved in a rather wide variety of gases. But when Black diffused the gas into limewater (CaO), he was downright amazed. The mixture turned cloudy and formed an insoluble chalk.

When he exhaled into fresh limewater, Black noted similar (though less intense) effects. Years before Lavoisier demonstrated the role of carbon dioxide in animal respiration, Black published his reasonable belief that his gas was somehow present in the process. Black tested his hypothesis in an ingenious way. In a Glasgow church, he hung a vessel containing limewater. After 1,500 sets of lungs respired during a continuous 10-hour service, Black found that the limewater had formed the insoluble chalk. Just to be certain, Black hung the limewater vessel for 10 hours on a day when the church did not hold services. As he expected, the limewater was largely unchanged. Therefore, Black concluded that the respired air caused the change in the limewater. He realized that the proportion of "fixed air" in the respired air of animals must be in greater proportion than it is in ordinary air.

Almost immediately, the industrial applications of Black's fixed air were apparent. To begin with, Black's geologist friend James Hutton (1726–1797) articulated the tangential discovery that pressure affects chemical reactions in which gases are involved. After reading Black's work, Hutton understood why subterranean limestone was often less caustic than formerly expected. Hutton had long been frustrated that phlogiston could not explain this reaction. Once he understood Black's finding, Hutton explained that fixed air could not escape the tremendous pressure far below the earth's surface.

Today, carbon dioxide serves many uses. In baking, it is released by yeast or baking powder to make batter rise. It also adds fizz to a variety of drinks. Many fire extinguishers exploit one of the first properties Black noted about carbon dioxide—that it quells combustion. Also, at $-78°C$, carbon dioxide becomes a solid, which is colloquially called "dry ice." Instead of melting to a liquid, dry ice sublimes, changing directly from a solid to a gas. Dry ice is widely used as a refrigerant for certain medicines (and other chemicals) that would be ruined by water. A fitting tribute to Black, dry ice also is used in most laboratories (often with acetone or isopropyl alcohol), whereby it cools a wide variety of chemical reactions.

SEE ALSO Law of Conservation of Mass; Phlogiston; Substance: Elements and Compounds.

Selected Bibliography

Asimov, Isaac. *A Short History of Chemistry*. New York: Doubleday, 1965.

Bensaude, Vincent, and Isabelle Stengers. *A History of Chemistry*. Translated by Deborah van Dam. Cambridge, MA: Harvard University Press, 1996.

Berry, A. J. *Henry Cavendish: His Life and Scientific Work*. New York: Hutchinson of London, 1960.

Crowther, J. G. *Scientists of the Industrial Revolution: Joseph Black, James Watt, Joseph Priestly, Henry Cavendish*. London: Cresset Press, 1962.

Moore, F. J. *A History of Chemistry*. New York: McGraw-Hill, 1931.

Partington, J. R. *A Short History of Chemistry*. New York: St. Martin's Press, 1965.

Riedman, Sarah R. *Antoine Lavoisier: Scientist and Citizen*. New York: Abelard-Schuman, 1967.

Stillman, John Maxson. *The Story of Early Chemistry*. New York: D. Appleton and Company, 1924.

Charles's Law of Gases
Jacques Alexandre Cesar Charles, 1787

In the early part of the eighteenth century, German chemist George Ernst Stahl (1660–1734) developed the **phlogiston** theory of matter. This theory held that a hypothetical substance, phlogiston, was released into the air during **combustion**. Different chemicals were thought to contain different amounts of phlogiston, as evidenced by how well they burned. Though it was eventually rejected, the phlogiston theory explained many well-documented phenomena and scientific experiments. For much of the century, the phlogiston theory was widely accepted.

By the middle of the century, English phlogiston chemists were making the basic discoveries leading toward **pneumatic chemistry**. In 1756, Joseph Black (1728–1799) discovered **carbon dioxide**, the first distinct gas to be identified. Black named his gas **fixed air** and noted that it would not burn or support combustion. He also identified it in the exhaled air of his own respiratory system. Joseph Priestley (1773–1804) investigated fixed air and became convinced he was releasing different kinds of "airs" (or gases), instead of manipulating a single air with elastic properties. Priestley invented the **pneumatic trough**, which greatly improved the methods for freeing and collecting gases. Priestley was first to isolate the gases "nitrous air" (nitric oxide or nitrogen monoxide), **nitrous oxide** (dinitrogen monoxide), and **ammonia**.

Priestley also shares the credit for discovering **oxygen** with Swedish apothecary Carl Scheele (1742–1786). Each scientist discovered oxygen independently during the 1770s. Each was also a staunch supporter of the phlogiston theory. Scheele called the gas **fire air**, and Priestley called it **dephlogisticated air**, believing it had lost its phlogiston. He thought a candle burned brightly in this gas because the gas hungrily grabbed up the phlogiston released during the candle's combustion.

The discovery of oxygen would eventually help dismantle the phlogiston theory during the oxygen revolution led by French chemist Antoine Lavoisier (1743–1794) in the 1780s. Where pneumatic chemists made the discoveries, Lavoisier made the laws. He delineated the difference between an **element** and a **compound**. He set forth conventions to govern the naming of chemicals based on their scientific characteristics. During many reactions (such as oxidation of tin), Lavoisier found that reactants actually gained mass. He correctly surmised that the phlogistonists had the whole process of combustion wrong. Instead of "separation from" (as with phlogiston), combustion was a process of "joining with" (as with oxygen). Lavoisier's **law of conservation of mass** was a direct conclusion of these trials. It was also a final blow to phlogiston. This law correctly stated that matter can neither be created nor destroyed in a chemical reaction.

Even though phlogiston was replaced by a better theory, it led scientists to many important discoveries. Oxygen was one such discovery. Another was **hydrogen**, which was discovered by the staunch English phlogistonist Henry Cavendish (1731–1810) in 1766. Cavendish called his gas **inflammable air** because it burned so fantastically. For a while, Cavendish thought he had isolated the very substance of combustion, the elusive phlogiston itself. What he had isolated was a gas with some very important properties which were soon to help delineate the basis of a new chemical law.

A year or so after Cavendish discovered hydrogen, another English chemist, Joseph Black (1728–1799), provided a small but surprising demonstration. In front of an audience, he filled a small sack with hydrogen gas, sealed it, and released it. To the astonishment of his guests, the bag rose quickly to the ceiling. At least several guests accused Black of an elaborate illusion involving thin black thread. He had shown that the properties of this gas were vastly different from those of ordinary air. Though this would not be understood for several years, he had dramatically demonstrated that the explosive gas was much less dense than ordinary air. (For more information, please see the entry on Hydrogen.)

During this time, scientists like Black were just beginning to experiment with the densities of different gases. Though they did not understand that heat (rather than smoke or phlogiston) was responsible, two French brothers, Joseph (1740–1810) and Étienne **Montgolfier** (1745–1799), devised the world's first **hot-air balloon** ascents. On June 5, 1783, they lifted a linen balloon 6,000 feet in the air. Later that year, the brothers successfully launched animals. Then on November 20, they launched two humans approximately 3,000 feet over the city of Paris. (For more information, please see the entry on Balloon Flight.)

Also during this time, one of the most respected scientists in all of Paris, Jacques Alexandre Cesar Charles (1746–1823), was also exploring lighter-than-air flight. Charles investigated hot air, which relies on the physical property of lowered density. He also focused on the chemical property of low density, as demonstrated by Joseph Black's hydrogen sack trick. Charles pioneered a method to produce a previously unimaginable quantity of the gas needed for flight. He also invented a balloon that could contain the gas. (His basic design is still followed for traditional **gas balloons** used primarily for sport.) Just 10 days after the Montgolfiers' first human flight, Charles and one other passenger soared high above the city of Paris in a long, tranquil drift.

Charles studied why both kinds of balloons took flight. The Montgolfier trials had shown that the balloon began to rise only when its quantity of air became very hot. From the start, they also found that the most effective form of descent was to simply let the air inside the balloon cool. In experiments over the next several years, Charles found that what made the air apparently lighter was that heat caused it to expand, decrease its density, and increase its volume. On cooling, the volume of the air decreased and the balloon became heavier. However, this was not the case with hydrogen. Charles's method of producing large quantities of hydrogen also produced a lot of heat. Yet Charles found time and again—and most strikingly through his own lengthy gas balloon flight—that hydrogen gas did not lose its own special lighter-than-air quality after it cooled.

These observations led Charles to formulate the law that bears his name. This law states that a gas expands by the same fraction of its original volume as its temperature rises. If the pressure remains constant, its volume

(V) and the temperature (T) do not change. The law is, therefore, written as V/T = constant. Assuming no change in pressure, **Charles's law** states that the volume of a gas is in direct proportion to the temperature of the gas, which Charles had discovered by working first with ordinary air and only later with specific gases. In this enterprise, Charles had some historical predecessors. In 1699, Guillaume Amontons (1663–1705) had actually discovered the relationship between the volume of a gas and its temperature. He demonstrated that the volume of a gas decreased when it cooled and increased when it was heated. For some reason, scientists largely neglected Amonton's finding.

Also, for some reason, Charles did not publish his law. In 1802, Joseph Louis Gay-Lussac (1778–1850) published his own experiments that agreed with Charles's work. Like Charles, Gay-Lussac also ascended in a balloon (he reached an altitude of 23,000 feet over Paris). Because Gay-Lussac published the findings first, Charles's law is sometimes known as Gay-Lussac's law or the Charles–Gay-Lussac law.

Although the law is generally accepted, it was originally proved on ordinary air and extrapolated to all gases. Many years later, it was proven valid only for the "ideal" gases. Along with **Boyle's law**, discovered in 1662 by Irish chemist Robert Boyle (1627–1691), Charles's law forms the basis of the ideal gas law. The ideal gas law applies only to a volume of gas under "ideal" conditions, which do not occur in the natural world. It is therefore generalized from the other two laws, Charles's and Boyle's. However, the establishment of the ideal gas law was an incalculably important step in the formation of modern chemistry because it helped scientists approximate standard pressure and temperature.

Three laws form the approximate basis of how pressure, volume, temperature, and particles of an amount of gas are related: Boyle's law, Charles's law and **Avogadro's law**, after Italian scientist Amedeo Avogadro (1776–1856). Boyle's law states that the volume (V) of a gas multiplied by the product of its pressure (P) remains constant only if there is no change in both the temperature and the number of gas particles inside a container. It is written as VP = constant. Avogadro first proposed his law in 1811, and it states that if equal volumes of different gases all have the same pressure and temperature, then they all contain the same number of particles. This is true regardless of the properties of the gases themselves, but is once again only true for a so-called ideal gas.

Finally, the so-called **universal gas law** combines these three laws into one single statement: PV = nRT (where P is pressure, V is volume, n is the number of moles of a gas, R is a constant known as the universal gas constant, and T is temperature). This law outlines the three ways the pressure of a gas can be doubled: by halving its original volume (and squeezing the same amount of gas into it), by doubling the amount of gas (in the original volume), or by doubling the absolute temperature.

These laws continue to be important in nearly all areas of chemistry and, on a larger scale, nearly all areas of science.

SEE ALSO Balloon Flight; Hydrogen; Law of Conservation of Mass; Oxygen; Priestley's Experiments on Gases; Proust's Law.

Selected Bibliography

Asimov, Isaac. *A Short History of Chemistry*. Garden City, NY: Anchor Books, 1965.
Asimov, Isaac. *Asimov's Chronology of Science and Discovery*. New York: Harper & Row, 1989.
Brock, William H. *The Norton History of Chemistry*. New York: W. W. Norton, 1992.
Greenberg, Arthur. *A Chemical History Tour: Picturing Chemistry from Alchemy to Modern Molecular Science*. New York: John Wiley & Sons, 2000.
Krebs, Robert E. *Scientific Laws, Principles, and Theories: A Reference Guide*. Westport, CT: Greenwood Press, 2001.
Parkinson, Claire L. *Breakthroughs: A Chronology of Great Achievements in Science and Mathematics, 1200–1930*. Boston: G. K. Hall, 1985.
Partington, J. R. *A Short History of Chemistry*. New York: St. Martin's Press, 1960.

Chemical Cells Can Store Usable Electricity
Ewald Jurgen von Kleist and Peter van Musschenbroek (independently),
Leyden jars, 1745–1746
Alessandro Volta, chemical pile, 1796

The ancient Greeks first observed **static electricity**. They discovered that after they rubbed **amber** with a cloth, this fossilized resin attracted small bits of material, such as straw or feathers. The Greeks did not know it, but this attraction was a form of static electricity. When a large number of atoms in an object gain or lose electrons, the object may take on an electric charge. When Greek scientists rubbed amber with a cloth, friction caused electrons to transfer from the cloth to the amber. The amber took on a negative charge because it had extra electrons. This is why Greek experimenters observed the attraction of other objects to the amber. The Greeks also did not know why amber was particularly suited to this kind of electrification; scientists discovered the reason for this many centuries later.

The ancient Chinese also knew of a black rock called **lodestone**. Chinese scientists observed that lodestone attracts objects made of iron. For many years, people made few discoveries about magnetism and static electricity. Most of the knowledge from antiquity was not widely available. Then, in the thirteenth century, magnets came into widespread use in Europe. During this century, Roger Bacon (ca. 1214–1292) made several important records about the properties of amber and lodestone. A few years later, Peter Peregrinus (also called Pierre de Maricourt, ca. 1269) wrote a letter on a battlefield. In his letter, Peregrinus described a piece of lodestone that he nicknamed "terrella" for "little earth." Peregrinus mapped opposite areas of concentrated attraction on the poles of his terrella, just like those of the Earth. In 1551, these inquiries came to a fruitful culmination when Girolamo Cardano (also called Jerome Cardan, 1501–1576) realized that the attracting effects of charged amber and lodestone are different processes. Cardano had articulated the first distinction between electricity and magnetism.

Later, around 1600, Queen Elizabeth's private physician, William Gilbert (1544–1603), found that he could make glass, wax, and sulfur act like amber: When he rubbed them with a cloth, these substances attracted materials such as hair, straw, and feathers. In his treatise *De Magnete*, Gilbert proposed that these materials were "electrics" (or insulators) and that their curious action was due to some kind of electrical fluid.

In the early part of the eighteenth century, Francis Hauksbee (sometimes Hawksbee or Hawkesbee, ca. 1666–1713) invented a vastly improved electric friction machine. This machine created electrostatic effects by means of a 9-inch glass globe and a series of metal rakes that rested atop rubbing pads fastened to the globe. When a person cranked a handle, the globe spun and created powerful bursts of static electricity that could be used for research. Hauksbee's **electric machine** was a great improvement over other methods of producing electricity. However, electricity still had few established facts or laws, and no really practical applications. But this was about to change. (For more on this topic, please see the entry on Electricity Can Produce Light.)

One of the eighteenth century's first major electrical discoveries was made by Stephen Gray (1695–1736). Gray studied **conduction**, the actual flow of electricity. In his inquiries, Gray used Hauksbee-type electric machines to electrify many objects that were previously thought naturally resistant to electrification. He electrified a glass bottle and its cork stopper, a large pine splinter, and a large metal ball.

Following Gilbert, Gray proposed that electricity was a "fluid." He became convinced of this position's validity when he electrified his large metal ball by sending a powerful electric charge to it through a linen cord. This important demonstration was one of the earliest uses of electrical current, which is the flow of an electric charge through a conductor. (Strictly speaking, a linen cord was not a very good conductor, but Gray's charge was so powerful that it satisfactorily overcame this limitation.) Later, Gray was able to send his so-called electrical fluid over 1,200 feet, along a wet hempen cord suspended from silk suspension threads. However, when Gray tried to demonstrate the experiment for some fellow scientists, he was puzzled to find that his charge fell far short of traveling this distance. All of the factors were identical, except for one change: Instead of silk suspension threads, Gray had used brass support wires. When he went back to silk, the experiment worked as it had before.

Gray observed that the electricity seemed to "disappear" into the brass, as if it "absorbed" his charge. He surmised that "electrics" (glass, resin, and silk) hold a charge and that "non-electrics" (such as metals and water) conduct a charge. Gray's 1729 distinction was the first understanding of insulation versus conduction. Gray drew up a thorough list of conductors and insulators. (For more on this topic, please see the entry on Electrical Conductors and Insulators.)

After studying Gray's work in the 1730s, French scientist Charles Dufay (1689–1739) found that objects charged from the same electrified glass tube repelled one another, but they attracted those charged from an electrified resin rod. This set Dufay to wondering about the supposedly singular, unified electrical "fluid." Extending his inquiry, he found that rubbed glass repelled a piece of thin metal leaf, but that rubbed amber, wax, or gum tended to attract it. Dufay concluded that there must be two kinds of electrical "fluids." He called these vitreous (from glass) and resinous (from resin or amber). In two given bodies, Dufay had found negative and positive electrical charges, though he articulated them as two separate "electrical fluids."

In the 1740s, Benjamin Franklin (1706–1790) took up the question of Dufay's two electrical fluids. From the start, Franklin was skeptical that two fluids existed. In a large public demonstration, Franklin had two volunteers absorb opposite charges—one from glass and one from resin. When the volunteers touched fingers, a powerful spark passed between them. Immediately, Franklin demonstrated that both people were completely discharged of electricity—their charges had "neutralized" one another. If there were two separate electrical "fluids," then why should this be? Franklin explained that an object with an excess of fluid "shared" electricity with an object with a "deficiency" of fluid. In other words, Franklin said that there was only one electrical "fluid." (For more on this topic, please see the entries on Electricity Is of Two Types and Franklin's Electrical Researches.)

Many scientists believed that Dufay and Franklin had demonstrated beyond doubt that electricity was a fluid. They did not employ the term *fluid* metaphorically to explain the action of a fluidlike substance. Rather, to most eighteenth century scientists, electricity was a fluid with a specific gravitational mass. In 1745, a Prussian official and scientist, Ewald Jurgen von Kleist (1700–1748), reasoned that he could isolate and collect electricity like any other fluid. In a series of trials, von Kleist connected an electric machine to an metal rod. He submerged one end of the rod in a jar about half full of water. When he held the jar in his hand and his assistant cranked the electric machine, von Kleist found that the bottle received an extremely powerful charge.

Von Kleist also found that he could receive a severe shock even after he disconnected the bottle from the electric machine: When von Kleist touched the metal rod with the finger of his free hand, he felt a powerful jolt of electricity. Von Kleist surmised that he had captured electrical "fluid" in his jar and that, when he touched the rod, some of it flowed to him. Some scientists believe that the shock from Kleist's jar was the most powerful non-naturally occurring shock ever delivered to this point. It was also one of the earliest shocks delivered through means of stored electricity.

At the same time as von Kleist, Peter van Musschenbroek (1692–1761) independently made a similar discovery. Van Musschenbroek connected an electric machine to a heavy piece of metal. This was probably the iron barrel of a pistol that he suspended by means of silk cords into a half-filled jar of water. Van Musschenbroek sent a charge from his electric machine, through the barrel of his gun, and into his jar of water. However, in doing so, he received a violent shock and vowed never to repeat the experiment. (Other scientists did not heed his warning and continued this work.)

Today, von Kleist and van Musschenbroek share credit for the discovery. However, its name, the **Leyden jar**, favors van Musschenbroek's research institution, the University of Leyden. In modern terms, a Leyden jar acts as a capacitor. Simple Leyden jars consist of a glass jar sealed with a cork. Generally, sheets of thin metal foil (rather than water) cover roughly half of the inside and outside of the jar. The foil acts as an electrical conductor, in contrast with the insulating glass. Like the original Leyden jars, a brass rod is inserted through the cork and into the jar, so that the rod touches the foil *inside* the jar. When the brass rod is electrified, electrical current charges the inner foil. Because current cannot pass through the glass, the foil on the outside becomes charged by means of induction (assuming it is properly grounded). This charge is opposite to the charge held by the foil inside the jar. If the inner and outer layers of foil are connected by a conductor, their opposing charges discharge the jar of its electricity.

The Leyden jar was one of the first devices invented to successfully store an electric charge. It paved the way for scientists to empirically test the methods of producing electricity through chemical means. In 1780, Luigi Galvani conducted a famous experiment on electrical "fluid" and animal tissue. Galvani clamped the spinal cord of a recently dissected frog to an iron railing with a set of brass hooks. During a powerful electrical storm, Galvani observed that the frog's muscles twitched. Galvani thought that this meant electricity had somehow been produced in the frog's tissues. During subsequent inquires, Galvani touched the frog's spinal cord with two different types of metal at the same time to make the frog's muscles twitch. This experiment attracted international attention among scientists. Galvani arrived at the famously erroneous conclusion that electricity produced in the moist tissue of the frog had caused the frog's muscles to contract. Galvani thought his discovery was evidence of "animal electricity."

In the mid-1790s, Count Alessandro Volta (1745–1827) began investigating Galvani's claim. He believed that Galvani had misinterpreted his results. Volta was familiar with a previous (dangerous) experiment done by Johann Sulzer (1720–1779). Sulzer reported that when he held "unlike" metals such as lead and silver together on his tongue, a taste like iron sulfate was produced. Sulzer also reported a curious "tingling" sensation on his tongue, which was unique to anything he had ever experienced. Sulzer found that he could only recognize this taste and this sensation when the different metals were actually touching.

Volta believed that the electrical charge (that caused the frog's muscles to contract) was produced by a similar combination of different metals, not the moist tissue of the frog. To investigate his hypothesis, Volta layered pairs of different metal disks. After several trials, he found that the best combination was a wide sheet of silver following a wide sheet of zinc. To increase conductivity, Volta carefully placed pieces of cloth saturated with a briny saltwater solution between each metal layer.

Volta found that chemical action occurred when his moist cloth was in contact with the two different metals. In modern terms, this was the first successful electrolyte. The chemical action resulted in a tiny electric current. Volta quickly surmised that there was a relationship between the "pile," or number of metal layers, and the strength of electrical charge rendered. In 1796, Volta discovered that when he increased the number of plates to 60, he could effectively create a powerful charge. This demonstrated two important scientific discoveries. First, Volta had produced electricity without need of animal tissue. He had disproved Galvani's hypothesis. Second, and more importantly, by piling up his stack, Volta had constructed the first chemical battery, known as a **voltaic pile**. He had shown that electricity is produced through the combination of different moist metals. This had the practical effect of creating a revolution in the use of small amounts of electricity for both scientific research and technological application.

The voltaic pile contained the basic elements of modern batteries—several cells, connected together (by means of a positive and a negative terminal) with conducting materials (or wires) that carry electrical current. During the nineteenth century, small amounts of electricity, such as those found in voltaic piles, were put to use in mechanical clappers on electric bells, telegraph machines, and modern dry and wet cells. Forty years after Volta's first pile,

Michael Faraday (1791–1867) demonstrated that the source of electricity was really neither the moist tissue of animals nor the contact between different metals. Faraday found that the source was chemical action, as when zinc is eaten away and continuous energy is liberated in the form of electricity. Faraday's theory did not violate the **law of conservation of mass**, which explains that energy must be released in a specific kind of reaction.

Today, electrical energy produced through the chemical action of batteries powers many products of everyday life. Most of these batteries and other sources of electromotive force are labeled according to their voltage. In the modern **metric system**, these labels are expressed in the unit named after Alessandro Volta, the **volt** (V). A volt measures the potential difference between two points (where 1 joule of work is done in moving a charge of 1 coulomb between the points). In other words, the volt measures the force or pressure of a current: volts = amps × ohms.

SEE ALSO Electricity Can Produce Light; Electricity Is of Two Types; Franklin's Electrical Researches.

Selected Bibliography

Bordeau, Sanford P. *Volts to Hertz: The Rise of Electricity*. Minneapolis: Burgess, 1982.

Canby, Edward Tatnall. *A History of Electricity*. New York: Hawthorn Books, 1968.

Heilbron, J. L. *Electricity in the 17th and 18th Centuries: A Study of Early Modern Physics*. Los Angeles: University of California Press, 1979.

Krebs, Robert E. *Scientific Laws, Principles, and Theories: A Reference Guide*. Westport, CT: Greenwood Press, 2001.

Meyer, Herbert W. *A History of Electricity and Magnetism*. Cambridge, MA: MIT Press: 1971.

Rowland, K. T. *Eighteenth Century Inventions*. New York: David & Charles Books, 1974.

Chlorine Is a Distinct Gas That Can Be Easily Produced in a Laboratory
Carl Scheele, 1774

In scientific research, discoveries are often made by several scientists working together, augmenting each other's skills. An excellent example is the **Montgolfier** brothers' invention of the **hot-air balloon**. Another is the French "task force" on chemical nomenclature. Yet another is the 28-person commission that developed the **metric system** in France. (For more information, please see the entry for each of these topics.)

Sometimes, however, scientific discoveries follow a different process—that of a genius researching alone in his or her laboratory. Swedish apothecary Carl Scheele (1742–1786) is an example of this kind of researcher. Scheele's modest disposition and self-taught background made him an unlikely genius.

Nonetheless, a review of his contributions to science suggests the strange feeling that this solitary apothecary was somehow born for a life in chemistry. Scholars have found that Scheele made many unpublished discoveries before other chemists who are generally credited for discovering them. For this reason, many chemical historians have recently designated Scheele as arguably the most prolific chemist of the eighteenth century.

By the time Scheele was a teenager, he had taught himself the **phlogiston** theory of substances. This theory states that **combustion** releases the hypothetical substance phlogiston from a burning object. Combustion ceased only when an object had released all of its phlogiston. As the first rational theory of chemistry (and the last permutation of **alchemy**), phlogiston explained both everyday phenomena and the findings of scientists inside their laboratories. With little formal education, young Scheele went to work as an apprentice pharmacist and pursued what would become his life's work. In these early years, he also established a curious routine that he would follow throughout his professional life: By day, Scheele ran the pharmacy, closely attending the needs of his customers. By night, he conducted chemical research in the solace of the pharmacy's laboratory.

By the time Scheele took a position as pharmacy manager in the small town of Köping, his chemistry talents had become internationally respected. Though many chemists of Scheele's skill and reputation might have thought running a pharmacy beneath them, Scheele quietly embraced his responsibilities and contentedly looked after his customers. Scheele was so devoted to his curious routine that he turned down prestigious offers from many great academies of Europe. Frederick II of Prussia personally offered him a Berlin position, and the English government offered him an outlandish salary for his services. Still, Scheele doggedly clung to his routine of running a pharmacy by day and discovering some of the cornerstones of modern chemistry by night.

Throughout his professional life, Scheele left his pharmacy only once—to take his long-postponed pharmacy examination in 1777. To no one's surprise, he passed. Just as his customers benefited from his loyalty, so did the world of chemistry. In fact, as best chemical historians can piece together, Scheele was the first to discover such chemicals as hydrogen sulfide, arsenic acid, lactic acid, arsine, malic acid, barium oxide (baryta), manganese and magnates, benzoic acid, molybdic acid, calcium tungstate (scheelite), oxalic acid, copper arsenite, permanganates, gallic acid, silicon tetrafluoride, glycerol, tartaric acid, tungstic acid, uric acid, and hydrogen fluoride. Scheele also shares credit for one of the most important elemental discoveries ever made. He and Joseph Priestley (1733–1804) independently discovered **oxygen** in the early 1770s. Actually, chemical historians have shown that Scheele isolated the gas several years before Priestley. (For more on this topic, please see the entry on Oxygen.)

Only one of Scheele's other discoveries rivals that of oxygen in lasting scientific importance. In 1774, Scheele turned his nighttime attention to a piece of pyrolusite (or manganese dioxide) brought to him by his friend and fellow chemist Johann Gottlieb Gahn (1745–1818). After a series of tests, Scheele reported that Gahn's specimen contained lime, silica, iron, and another substance. This other substance was more difficult to separate from the piece of pyrolusite. Many years later, Gahn demonstrated that it was the element manganese. But Scheele did not yet know this.

Scheele treated the pyrolusite with a number or reagents in an attempt to identify the mysterious substance. After many unsuccessful tries, Scheele treated the pyrolusite with hydrochloric acid over the steady heat of his sand bath. After time, Scheele was fascinated to find that the acid-treated pyrolusite yielded a yellow-green gas. When he produced a larger quantity, Scheele noticed that the gas had a pungent odor. Because of its smell, Scheele originally thought that he had unremarkably produced **aqua regia**, a mixture of nitric acid and hydrochloric acid capable of dissolving metals such as gold and platinum. However, when he tested the gas further, Scheele confirmed that it was markedly distinct from anything he had ever seen, including aqua regia. The pungent odor was unique.

Scheele began regularly producing the gas in this simple manner. He also continued to explore its intriguing properties. Scheele found that the gas readily dissolved cinnabar and did not precipitate silver out of its solutions. When the gas sank to the bottom of an open bottle, Scheele noted that it was significantly more dense than ordinary air. When he filled the bottle with distilled water, he noted that the gas was not water soluble. These were important discoveries.

In his notes, Scheele was even more impressed with another of the gas's properties. When he filled the empty bottle, he noted that the gas turned his cork a distinct yellow. In subsequent tests, he found that the gas dissolved nearly all the color from a piece of moistened blue litmus paper. It similarly diluted the color from several dried flowers. Scheele was completely unable to restore color to either the paper or flowers—they were permanently bleached. With this observation, Scheele had described one of the gas's most distinct and useful properties—its ability to permanently bleach a wide variety of substances.

Scheele named his gas "dephlogisticated acid of salt." But before long, most of his contemporaries referred to it as "marine acid air" because it was made with "marine" (or hydrochloric) acid. The French founder of modern chemistry and modern chemical nomenclature, Antoine Lavoisier (1743–1794), renamed the gas "oxy-muriatic acid." This was due to his erroneous belief that the gas contained oxygen. Finally, in 1810, Sir Humphry Davy (1778–1829) proved that the gas contained no oxygen. Davy gave the gas its

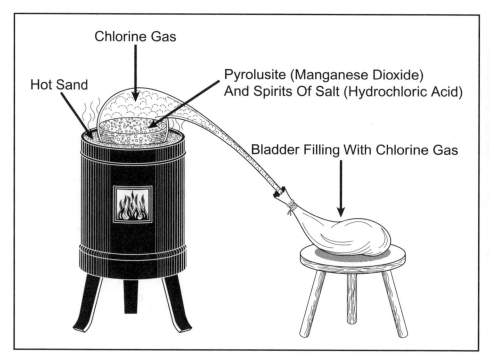

Figure 5. Scheele's discovery of chlorine. By treating manganese dioxide with hydrochloric acid, Carl Scheele first isolated chlorine in 1774. He named it "dephlogisticated acid of salt." In 1810, Sir Humphry Davy called the gas chlorine, from the Greek for "greenish yellow."

present name, **chlorine** (see Figure 5), which he formed from the Greek word for "greenish yellow."

For the first time, Scheele had added manganese dioxide to hydrochloric acid in order to produce chlorine gas in his laboratory. Countless chemistry students have since repeated his method and have similarly experimented with the gas's properties and smelled its pungent odor.

In fact, breathing a modest amount of chlorine causes irritation to the nose, throat, and lungs, as chlorine is a highly toxic gas. The World War I battle of Ypres, France, on April 22, 1915, marked the first use of chemical weapons in active combat. German soldiers released large quantities of chlorine into the wind, which carried the gas across enemy lines. Because it was much more dense than air, the chlorine poured into the soldier-filled trenches. Many French soldiers died of inhaling the poisonous gas. Those that did not found themselves choked and blinded. As a result, they were easily overrun by German soldiers wearing gas masks. After this battle, soldiers and their animals on both sides were routinely issued gas masks to defend against this deadly application of chemistry.

Today, most chlorine is used in industries that often exploit one or more of the gas's properties first identified by Scheele. Most manufacturers produce chlorine gas by passing an electric current through a solution of sodium chloride in water. This process forms the basis of the chlorine–caustic chemical industry—one of the largest industries of its kind. Chlorine is used to produce paper, certain plastics, insecticides, cleaning fluids, antifreeze, medicines, paints, and several important petroleum products. Many municipal water treatment plants and nearly all swimming pools exploit chlorine's toxic properties to purify water. In small quantities, chlorine is an effective, safe, and inexpensive antimicrobial agent. Finally, just as Scheele observed, chlorine is an effective bleaching agent. When chlorine is dissolved in sodium hydroxide, it becomes a mixture of sodium chloride and sodium hypochlorite. This mixture is widely used as a disinfectant, as well as a bleaching agent.

SEE ALSO Chemical Nomenclature; Elements and Compounds; Oxygen.

Selected Bibliography

Gillispie, Charles Coulston, ed. *Dictionary of Scientific Biography*, Volume XI. New York: Charles Scribner's Sons, 1972.

Lockemann, Georg. *The Story of Chemistry*. New York: Philosophical Library, 1959.

Porter, Roy, ed. *The Biographical Dictionary of Scientists*. New York: Oxford University Press, 1994.

Sootin, Harry. *12 Pioneers of Science*. New York: Vanguard, 1960.

Urdang, George. *The Apothecary Chemist: Carl Wilhelm Scheele*. Madison, WI: American Institute of the History of Pharmacy, 1958.

Citrus Prevents Scurvy
James Lind, 1747

The idea that diet plays a key role in human health is a very old one. Ancient Egyptian practitioners believed that excessive food caused disease and prescribed vomiting after meals for a variety of illnesses. The Egyptians also used liver to cure night blindness. Modern scientists know that liver supplies vitamin A, which can often alleviate this disease. The ancient Greek philosopher Socrates (ca. 470–399 B.C.) warned against eating and drinking except when one feels hungry or thirsty. He also urged moderation at mealtime. The most famous ancient doctor, Hippocrates (ca. 460–380 B.C.), prescribed specific diets for individual ailments. Hippocrates claimed that certain illnesses could be cured or prevented by finding the correct combination of simple foods, such as linseed, millet, eggs and wheat flour.

During the European Renaissance, scientists such as Francis Bacon (1561–1626) studied many of the ancient medical texts. These texts spawned new theories of the relationship between food and health. One such theory was outlined in the book by Luigi Cornaro (1464–1566) in 1558, *The Sure and Certain Method of Attaining a Long and Healthful Life*. Cornaro himself had become so obese that he nearly died. However, he greatly reduced his food intake until his health returned. Cornaro reportedly lived to be 102 years old.

About this time, the French explorer Jacques Cartier (ca. 1491–1557) explored the area that is today Quebec. There, North American Indians showed him how to treat the disease **scurvy** with an infusion probably made from pine bark and needles. However, poor documentation and a general lack of scientific understanding largely prevented this information from coming into practice in Europe, where scurvy would remain a scourge for several more centuries.

Modern medicine has shown that scurvy results from the inability of humans to synthesize their own ascorbic acid or vitamin C. Humans and a few other animals must therefore obtain all of this vitamin from the foods they eat. After 30 or so weeks of dietary ascorbic acid deprivation, the classic signs of scurvy begin to appear. These signs may include bleeding and spongy gums, loose teeth, poor healing of wounds, excessive bruising, a lack of appetite, soreness in the joints, anemia and eventually death. Most of these symptoms result from a weakening and rupture of the capillary walls.

In terms of documentation, scurvy is a very old disease. The oldest medical references are found on clay tablet fragments from ancient Babylonia. Some may go back to 1700 B.C. The first known attempt to categorize and diagnose scurvy is found in the ancient Roman writings of Pliny the Elder (23–79). In Europe, scurvy was described by Jacques de Vitry (ca. 1160–1240) in the thirteenth century. However, much greater social attention was paid to communicable diseases, which were more prevalent.

Because of its relatively long period of required deprivation, scurvy was not particularly devastating in Europe until the fifteenth century. During this century, the exploding areas of trade, empire and exploration required European powers to send military and merchant ships to sea for long periods of time. Nearly all sailors survived on rations of salt beef and dry biscuits, with no fresh fruit or other sources of ascorbic acid. The horrors of scurvy were described by the Portuguese explorer Vasco da Gama (ca. 1469–1524) in 1498. Da Gama lost nearly 100 of 170 men from an outbreak of scurvy while at sea. Almost 250 years later, the British sailors of Lord George Anson's (1697–1762) fleet suffered affliction rates in excess of 75%.

An important attempt to treat scurvy was made by the British physician John Huxham (1692–1768). In 1747, Huxham put 1,200 scurvy-ridden sailors on a vegetable diet. This treatment met limited success, largely because of the ascorbic acid in vegetables such as uncooked cabbage. In this same year, the Scottish physician James Lind (1716–1794) was founding the practice of military hygiene. On land, Lind had found great success in preventing communicable outbreaks through measures such as burning recruits' clothing, regular washing of uniforms, chemical delousing, and locating latrines away from groundwater and camp kitchens. However, during this time, the Royal navy had suffered several embarrassing abortive blockades due to widespread outbreaks of scurvy.

Lind was charged with treating a ship of sick sailors. He knew that many people believed carbonated water alone could cure scurvy. (Of course, it cannot.) Several physicians claimed to have cured cases of scurvy using a variety of other diets and dietary changes. Lind also knew of reputable historical accounts detailing how the powerful East India Company had stumbled onto a treatment. In 1600, it had provided several vessels with rations of lemon juice. The accounts claimed that, to their surprise, these ships had far fewer cases of scurvy-like symptoms than their sister vessels. The problem was that this historical data had been ignored for a century and a half. Social conditions were also different, as general nutrition had been gradually improving since the second half of the seventeenth century.

In 1747, Lind undertook a scientific inquiry into the problem. On May 20, he began treating a dozen sailors aboard the HMS *Salisbury*. All showed signs of advanced scurvy. Lind broke the men into six groups of two and attempted to control all external factors. He made sure they all ate the same meals and slept in the same part of the ship. Lind gave two patients each a quart of cider every day. He gave two patients 25 drops of elixir vitriol (a mixture containing sulfuric acid) three times a day. Another two received two spoonfuls of vinegar three times a day. Lind had two others drink sea water. Two others took a complicated prescription of chemicals that Lind mixed each day. Finally, Lind fed the remaining two men a lemon and two oranges each day.

To Lind's surprise, only the men who ate citrus fruit showed any signs of improvement. As he tracked their progress, Lind was amazed to find that the men made a complete recovery after little more than a week of treatment. Their gums stopped bleeding and the swelling receded significantly. Many festering wounds began showing signs of healing, and the men had normal appetites.

After his initial tests, Lind only partially recognized the value of citrus fruit in curing scurvy, demonstrating the high degree of confusion over this disease. Part of Lind still clung to the stubborn, old belief that moist sea air significantly contributed to the outbreaks of scurvy. Nonetheless, Lind conducted further tests during a 10-week cruise in 1746. He later published his findings in the 1753 book, *A Treatise on the Scurvy*. In this book, Lind recommended that the Royal navy include fresh citrus fruit, fruit juice, sauerkraut or salted cabbage in its rations.

In his famous voyage to the South Pacific between 1772 and 1775, Captain James Cook (1728–1779) followed Lind's rules, enforcing strict hygiene discipline and providing rations of sauerkraut and lemon juice. Though this voyage lasted three and a half years, Cook lost only 1 of 118 men—and that was to consumption, soon after leaving England.

In 1795, over 40 years after Lind's discovery (and a year after his death), Sir Gilbert Blaine (1749–1834) made lime juice a mandatory ration in the British navy. On these ships, scurvy disappeared immediately. Suddenly, the mighty British navy could circumvent the globe with relatively few men lost to illness. During the Napoleonic wars that followed, the British navy gained the enormous advantage of having far less scurvy than its French enemies.

During the next century, the British government made lime juice a mandatory ration for all seafaring ships, even nonmilitary ones. This earned British sailors the familiar nickname "limey." However, in 1860, the Royal navy switched from expensive lime juice from Malta to cheaper lime juice from West India. As a result, at least two British ships in the Arctic were crippled by widespread outbreaks of scurvy. This fueled the stubborn, incorrect belief that fresh fruit and vegetables do not prevent scurvy. It was not until 1919 that a group of women scientists at the Lister Institute in London figured out that the West Indian lime species contains nearly no antiscorbutic factor after preservation.

During this same time, biochemical research had begun to uncover the mysteries of how citrus fruits and juices prevent scurvy. The Lister Institute's scientists found that different fruit juices contained significantly varying amounts of the chemical responsible for preventing scurvy. This chemical was isolated in 1928 by the Hungarian chemist Albert von Szent-Györgyi (1893–1986) and was called vitamin C. Von Szent-Györgyi later won a Nobel prize for his discovery.

Today, scurvy is extremely rare and exists mostly in babies and elderly people with inadequate diets. In infants, scurvy is called **Barlow's disease**. This condition sometimes occurs when a baby is being weaned from breast milk. It is prevented (similar to Lind's method) by administering orange or tomato juice after the first month.

SEE ALSO Animal Respiration; Immunization Prevents Smallpox.

Selected Bibliography

Bettmann, Otto L. *A Pictorial History of Medicine*. Springfield, IL: Charles C. Thomas, 1962.

Garrison, Fielding H. *An Introduction to the History of Medicine*. Philadelphia: W. B. Saunders, 1917.

Hudson, Robert P. *Disease and Its Control: The Shaping of Modern Thought*. Westport, CT: Greenwood Press, 1983.

Loudon, Irvine (ed.). *Western Medicine: An Illustrated History*. New York: Oxford University Press, 1997.

McGrew, Roderick E. *Encyclopedia of Medical History*. St. Louis: McGraw-Hill, 1985.

Porter, Roy (ed.). *The Cambridge Illustrated History of Medicine*. New York: Cambridge University Press, 1996.

Rhodes, Philip. *An Outline History of Medicine*. Boston: Butterworths, 1985.

Singer, Charles. *A Short History of Medicine*. New York: Oxford University Press, 1928.

Comets Follow Predictable Orbits, Cycles, and Returns
Edmond Halley, 1705

Since antiquity, the sky has been a predictable place. Ancient Egyptian, Chinese and Greek astronomers all developed a considerable understanding of the regular, heavenly changes that occurred. Constellations seemed anchored in place, rotating in relation to one another. The sun rose and set with predictable certainty. The moon went through an easily charted cycle of shape-shifting changes. Every object had its place, and every place had its cycles. However, on rare occasions, the steady, reassuring regularity of the heavens was shattered by the sudden, untold appearance of a **comet**. Most civilizations throughout the ages interpreted a comet as a foreboding sign of Earth-bound catastrophe.

Throughout antiquity and the Middle Ages in Europe, many people associated comets with the deaths of rulers, such as the assassination of Julius Caesar, the death of Agrippa in 12 B.C., and the death of Titus Flavius Vespasian in the year 79. Comets also signaled the collapse of empires, such as the destruction of Jerusalem in 66, Attila's defeat in 451 and the Norman invasion of England in 1066. Comets brought calamities like Noah's biblical flood; any number of bad growing seasons, medieval plagues and cold winters; and especially the London fire of 1666. But more than any single event, comets were supposed to signal the end of the world. In the European Middle Ages, the Church's position was that a comet was a warning shot fired across the bow of Earth by an angry God. This position often led to widespread panics at the mere sight of a comet. So pervasive was this position that, on sighting a comet in 1456, Pope Calixtus II ordered a terrified, xenophobic prayer to be spoken daily, "Lord, save us from the Devil, the Turk, and the Comet!"

By the fifteenth century, the ancient science of astronomy had unlocked many mysteries of the heavens. Beginning in this century, astronomers began to believe that comets might just follow some strange pattern, after all. In 1472, Johann Muller (1436–1476) marked a comet's path across the night sky. After several such studies, Muller became the first scientist to believe that comets follow a distinctly straight line—once through the solar system, never to be seen again. Almost 70 years later, Girolamo Fracastoro (1483–1553) published findings that a comet's tail always points away from the sun. And in 1577, Tycho Brache (1546–1601) used **parallax**—the observed change in an object's position in relation to earth—to find that comets were a very great distance away from Earth.

Astronomers then understood several important facts about comets. Rather than near-miss, Earth-centered warnings, comets were far away in space. They seem to always "point" in the same direction. Comets also seem to travel in a straight line. From these rudimentary (and only partly true) findings, astronomers began to form a body of knowledge showing that comets follow a distinct set of rules. They even began the ardent task of delineating what some of them were.

By the time the young astronomer Edmond Halley (1656–1742) observed the bright comet of 1682, scientific thought about comets was really taking hold of astronomers and other scientists. In fact, many were beginning to move far beyond both Church doctrine and Muller's straight-line theory of comets. Sophisticated studies suggested that one of three mathematical models correctly explained a comet's journey—an ellipse, a parabola or a hyperbola. But which model was correct? This question was misleadingly complex, for only a small sample of a comet's path was visible from Earth. After a very short window of visibility, the comet was gone.

The young Halley had many friends who were influential scientists. One was the great English physicist and astronomer Isaac Newton (1642–1727), with whom Halley often consulted. Halley brought Newton his question about the journey of comets. Much to the young astronomer's surprise, Newton had already figured a complete theory of the paths of astronomical bodies, in relation to the gravitational effect of other bodies. Newton called his theory the law of universal gravitation. With Halley's help, Newton published his recent findings in a book known as *Principia*—one of the most important books ever published.

Following Giovanni Borelli (1608–1679), many scientists were inclined toward the parabolic model of comet journeys. This model stated that a comet passed through the solar system one time and never returned. Halley disagreed with this theory. Using Newton's formulas and findings, Halley arrived at an hypothesis: The paths of comets are nearly (though *not completely*) parabolic. Halley's proposal had several striking implications. If the paths of comets were extremely elongated ellipses, then comets must be in orbit around the sun—just like planets and other highly predictable heavenly bodies.

Halley began the difficult task of calculating the orbit for the comet he had observed in 1682. This task eventually led him to study the orbits of many other, previously observed comets. One of the accounts Halley studied was written by Johann Kepler (1571–1630), who studied a comet in 1607. Halley was curious to find that Kepler's comet traveled in the same part of the sky as the comet Halley had studied in 1682. Halley delved into the matter further.

Halley was amazed to read reputable accounts of comets from 1531 and 1456 that put these comets in the same part of the sky. Looking back through the archive, Halley even had some encouraging findings that observers in 1380 and 1305 saw comets in the same region of the sky, as well. It did not take Halley long to calculate that these comets appeared at fairly regular intervals, with one sighting roughly every 75 years. This was excellent support for his theory of comet orbits as extremely elongated ellipses. But even Halley was reluctant to publish this finding, for even the most novice astronomer could draw only one revolutionary conclusion. Each of these comets was, in fact, the same object. *They* were actually *it*—a single, individual comet returning at extremely regular intervals.

Before publishing, Halley wanted to make certain he was correct. After all, the realization that people were seeing the same comet again and again flew directly in the face of scientific theories that had comets on a straight line, a parabola, or a hyperbola. Further, how would people react to being told their fear of comets (even now) was wildly—even absurdly—unfounded, that they were panicking at the same, regular visitors that people might have seen hundreds or even thousands of times before? Halley's idea that comets orbit the sun (as the planets do) meant that they had been fantastically misinterpreted in various creative ways since antiquity and right up to his own day.

Halley spent several years doing calculations of heavenly bodies in relation to comets. These were based mainly on Newton's law of universal gravitation. Finally, in 1705, Halley published his findings in a book, *A Synopsis of the Astronomy of Comets*. This book had an immediate effect on the scientific world. However, it was right away apparent that ultimate judgment would have to be reserved for one single event not scheduled to take place for another 53 years. In Halley's book, he shared his ultimate frustration with many of his readers. Halley predicted that the comet of 1682 (and 1607, 1531, 1456, 1380, and 1305) would follow its predictable cycle of roughly 75-year returns in 1758. Halley knew he and many of his fellow scientists had little hope of being around to see if his magnificent calculation was correct.

Indeed, Halley died in 1742. When the year 1758 finally arrived, many scientists and commoners alike publicly doubted Halley's famous prediction. Several respected scientific journals even cited a May 1719 comet prediction from renowned scientist Jacques Bernoulli (1654–1705), which never materialized. Many people publicly stated that Halley's comet prediction was a hoax—the stuff of superstition and hocus-pocus, rather than real science.

Nonetheless, astronomers began an intensive search for the long-awaited comet throughout 1758. So great was his fear of missing the ballyhooed comet that French astronomer Charles Messier (1730–1817) did not sleep nights during the entire year, but in the end, the sighting was not to be his. Since the earliest days of stargazing up to today, many of astronomy's best contributions have been made by dedicated amateurs. The first sighting of the comet took place on Christmas, when a determined German farmer, Johann Georg Palitzsch (1723–1788) spotted it with his telescope. It was only when the clouds parted over his observatory that Messier became the first professional astronomer to sight it.

Almost immediately, Halley was universally recognized as a genius. More importantly, scientists accepted his theory that comets follow a predictable cycle of returns. Sixteen years after his death, Halley offered irrefutable proof of Newton's law of universal gravity, which had itself fostered the prediction so profoundly. In the end, Halley got his comet and Newton got his universe.

Today, scientists have shown that comets comprise mostly dust and gases. Generally, comets are visible from Earth for only a small window period, ranging from a few days to several months. Some comets appear to actually have parabolic orbits. However, most return to the inner solar system in highly elongated elliptical orbits ranging from tens to thousands of years in length (see Figure 6). Some comets are visible from Earth in intervals of less than 10 years. After they reach **perihelion**—their position closest to the sun—these comets take less than 10 years to reach **aphelion**, their position farthest from it. These comets reach aphelion near Jupiter because they have been "captured" into smaller orbits by the large planet's powerful gravitational attraction.

Scientists are unsure about the origins of comets. Current theories suggest that comets were created during the formation of the solar system and, as such, are permanent members of that system. In 1950, Dutch astronomer Jan Oort (1900–1992) suggested that a great "cloud" of more than a trillion comets hovers far beyond the edge of the solar system. Comets in this cloud move very slowly. However, they can be disturbed and sent into the inner part of the solar system by a "passing" star's gravity. A year later, Gerard Peter Kuiper (1905–1973) proposed a region outside the orbit of Pluto as a source of short-period comets. Kuiper suggested that this region acts like a reservoir for short-period comets the same way the **Oort cloud** acts as a reservoir for long-period ones. In 1992, astronomers gave the Kuiper theory some validation when they discovered the first of more than 70,000 trans-Neptunian objects.

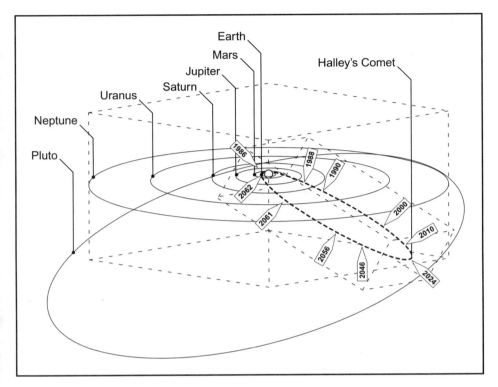

Figure 6. Elliptical orbit of Halley's comet. Edmond Halley first explained that comets follow a highly elongated, elliptical orbit around the sun. Specifically, Halley's comet will reach aphelion during 2024 and perihelion during 2062.

Using modern calculations and telescopes, astronomers now discover a new comet at an average rate of one every two to three weeks. Yet, this "first" comet is still the most famous, its discoverer still remembered. Since 1758, Halley's comet has made three regular, predicted visits—in 1835, 1910 and 1986. Scholars have also uncovered recorded sightings of this comet going all the way back to 240 B.C. in China. As the accompanying illustration shows, Halley's comet is scheduled to make its next visit in the year 2061.

SEE ALSO Proper Motion of the Stars and Sun; Stellar Aberration; Uranus Is a Planet.

Selected Bibliography

Asimov, Isaac. *Asimov's Guide to Halley's Comet*. New York: Walker and Company, 1985.
Calder, Nigel. *The Comet is Coming!, The Feverish Legacy of Mr. Halley*. New York: Viking Press, 1980.

Cook, Sir Allen. *Edmond Halley: Charting the Heavens and the Seas.* Oxford: Clarendon Press, 1998.

Etter, Roberta and Stuart Schneider. *Halley's Comet: Memories of 1910.* New York: Abbeville Press, 1985.

Gillispie, Charles Coulston, ed. *Dictionary of Scientific Biography.* Volume XI. New York: Charles Scribner's Sons, 1972.

Lancaster-Brown, Peter. *Halley & His Comet.* New York: Sterline, 1985.

Yeomans, Donald K. *Comets: A Chronological History of Observation, Science, Myth, and Folklore.* New York: John Wiley & Sons, 1991.

Corpuscular Model of Light and Color
Isaac Newton, 1704

Over the centuries, people have had many ideas about the nature of light. For many years, they thought light was a force or substance that traveled from a person's eyes to a focal object and back again. Much more recently, people thought that perhaps light traveled in waves, analogous to those of sound. Still others thought that perhaps light was made up of tiny, impenetrable particles, which would later be called "corpuscles." The corpuscular model of light goes all the way back to the ancient Greeks, who believed that all matter was composed of such unbreakable particles.

Until the mid-seventeenth century in Europe, most scientists believed that light moved instantaneously across even the most massive distances. However, French astronomers studying the moons of Jupiter found a significant delay between actual and perceived times of each moon's emergence from behind the planet. They realized that light does take time to travel, and they made a fairly accurate estimate of its speed. Light was no longer regarded as instantaneous.

Later in the seventeenth century, renewed enlightenment interest in antiquity resulted in a conceptual acceptance of the existence of minute particles. By way of rendering tiny worlds visible, the new compound microscope stimulated further interest. During the same century, the idea that tiny particles could unite to form a tangible solid roused widespread interest. This interest resulted in a rebirth of the old mechanical or corpuscular philosophy among scientists.

However, scientists did not immediately develop a **corpuscular theory** for light. Rather, Robert Hooke (1635–1703), a major contributor to development of the compound microscope, and Christiaan Huygens (1629–1695) developed a sophisticated theory of light as a wave or undulatory motion. They proposed that, just as a water wave moves across a body of water, light moves across the "body" of air. This was a powerful model, one that explained many experiments and observations. Yet, this model would soon be upended by one that seemed, on the surface at least, completely opposite.

In 1666, Sir Isaac Newton (1642–1727), English astronomer, mathematician and one of the greatest scientists in history, discovered the visible spectrum of light. In 1704, he published many of his prolific experiments and urbane theories in the book *Opticks*. This book, which was first published in English (rather than in Latin), eventually became the model of experimental physics in the eighteenth century. In it, Newton rejected the wave model of light. He reasoned that if light were made up of waves, it would travel around corners, as sound does. Instead of waves, Newton said light is composed of tiny, discrete particles moving in straight lines at a finite velocity. He also believed the particles moved in the manner of inertial bodies. Newton's theory became widely known as the corpuscular or particulate theory of light.

Opticks also detailed Newton's experiments on the nature of light, which were crucial to his theories. Guided by the scientific–philosophical writings of Johannes Kepler (1571–1630) and Rene Descartes (1596–1650), Newton concentrated on the actions of prisms on white light. Aristotle (384–322 b.c.) and other scientists of the ancient world thought that white light was "pure" and homogenous. They reasoned that colored light was therefore secondary. They also believed that prisms alter or modify—in a sense "taint"— white light, rather than separate it.

Newton disagreed with this model entirely. He had seen lenses that, when held at just the right angle to white light, appeared to throw colored circles outward from their edges. Newton's most famous experiment became known as the **experimentum cruces** (or the crucial experiment; see Figure 7). In a darkened room, Newton directed a narrow beam of sunlight

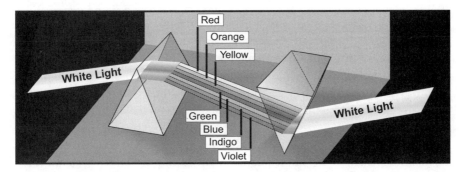

Figure 7. Newton's *Experimentum cruces*. Newton split sunlight through a prism and then "reconstituted" it through a second one. Contrary to ancient belief, colored light comprised white light—not the other way around.

through a prism. Through refraction (or "bending"), this separated the white light into the color spectrum of sunlight. When he passed the spectrum through a second prism, Newton saw the individual colors converge back into the sun's white light. He had successfully "reconstituted" white light from colored.

The ramifications of the crucial experiment were enormous. White light—and not colored light—is heterogeneous and secondary. That meant that colored light was homogenous and primary. The Greeks (as well as many contemporary scientists) had it exactly backwards. Newton confirmed his findings in subsequent experiments in which he was unable to break individual spectral colors into any constituents. He was absolutely convinced that sunlight was made up of multiple colored rays of light. However, these findings did not prove the corpuscular theory of light.

Nonetheless, Newton was able to explain many vital optical phenomena with his corpuscular theory in conjunction with his laws of mechanics. Through application of these theories and careful study of the moons of Jupiter, Newton realized that all light travels at a uniform speed. With this insight, he used refraction angles to determine the relative sizes of light particles for different colors.

For the next 100 years, Newton's corpuscular model of light was widely and greatly favored over the wave model. Yet, at many different points, Newton thought that perhaps there was an "ether," or ethereal substance, in which light traveled. Of course, there was not (as scientists would soon learn). Still, Newton often considered the wave model of light in regard to an ether concept. In effect, many of Newton's theories combined the wave and corpuscular models of light. From the perspective of modern physics, this kind of thinking was far ahead of its time.

The wave theory gained many proponents and even came to be dominant in the nineteenth century. In 1800, astronomer William Herschel (1738–1822) discovered infrared light by using a **thermometer** to measure the heat in different colors of separated sunlight. He found that an area "beyond the red" (hence the term *infrared*) contained the most heat. This led to new, limited interest in the wave model and to the hypothetical ether in which the light waves supposedly traveled. Many learned scientists believed that, because light can travel in a vacuum, a medium other than air (or matter) must indeed exist.

In 1864, British physicist James Maxwell (1831–1879) proposed his electromagnetic theory, also known as the mathematical theory of electromagnetism. He found that magnetic and electric fields occur together. What's more, Maxwell also demonstrated that the speed of electric and magnetic fields is identical to the actual speed of light. Almost at once, scientists

understood visible light to be a particular kind of electromagnetic radiation—
a small part of the electromagnetic spectrum. The hypothetical ether was
completely unnecessary for the propagation of light. The wave model of light
was now much more than just a theory—it had a solid empirical foundation. In a larger sense, Maxwell's electromagnetic model was validated with
the discovery of radio waves (and later x-rays, short radio waves and
microwaves). The pendulum of scientific sway had swung from Newton back
to Huygens. Scientists were more convinced than ever of the validity of the
wave model.

However, in 1900, German physicist Max Planck (1858–1947) found
evidence that electromagnetic radiation, including light, travels as tiny packets
of energy called "quanta." During the next five years, Albert Einstein
(1879–1955) devised many experiments in which light behaved as a particle,
rather than as a wave. Einstein's particles came to be known as "photons."
He used them to explain the photoelectric effect and other questions that
plagued scientists of the early twentieth century. Several years later, the
Danish physicist Niels Bohr (1885–1962) found that the energy of atoms was
also quantized, or emitted and absorbed in tandem with certain energy values.
Bohr's finding explained how atoms became excited (on absorption) and also
de-excited (on radiation or emission). This work led to the field of physics
known as quantum mechanics, which studies how atoms of light are
quantized.

At first, it looked as though quantum mechanics would eventually favor
Newton's corpuscles over Huygen's waves. In the end, however, modern
physics changed not the answer, but the question. It was no longer a question
of either–or. Light was not only a wave, but also a corpuscle; it was not only
a corpuscle, but also a wave. Light became both—or, more correctly, light
became neither. In some experiments, light acts most like a wave. In others,
it behaves most like a particle. Indeed, the question was no longer about which
model was correct, but rather only which model best explained which particular set of experiments. In studying the processes of quantum theory, the
particle model of light is most important. In studying the transmission of light,
the wave model is most useful and appropriate.

For many people, the various theoretical models and historical conceptions
of light can get very confusing, very quickly. By way of simplification, the
following device may help keep track of the modern theories of light over
the last several centuries: wave, particle, wave, compromise. In the seventeenth
century, light became a wave. In the eighteenth century (with Newton), it
became a particle. In the nineteenth century (with radio), it again became
a wave. Finally, in the twentieth century and beyond (with quantum
mechanics), it became a grand compromise. An oft-recited joke in physics

circles offers frustrated students the following advice: On Monday, Wednesday and Friday, consider light a wave; on Tuesday, Thursday and Saturday, consider it a particle; on Sunday, consider it both.

SEE ALSO Electricity Can Produce Light; Principle of Least Action; Stellar Aberration; Three "Primary" Colors.

Selected Bibliography

Brock, William H. *The Norton History of Chemistry*. New York: W. W. Norton, 1992.
Farber, Eduard. *The Evolution of Chemistry*. New York: Ronald Press, 1952.
Krebs, Robert E. *Scientific Laws, Principles, and Theories: A Reference Guide*. Westport, CT: Greenwood Press, 2001.
Parkinson, Claire L. *Breakthroughs: A Chronology of Great Achievements in Science and Mathematics, 1200–1930*. Boston: G. K. Hall, 1985.

Cotton Gin
Eli Whitney, 1793

Today, cotton use exceeds that of any other single plant fiber. Nearly every person uses cotton in one way or another, as it is woven into fibers for clothing, carpets, bedding and many other items. Products made from cotton are often soft, durable, inexpensive and attractive. The cotton plant makes up the *Gossypium* genus, of which there are about 40 species. Four of these are widely cultivated.

The cultivation of cotton predates ancient history. Historians do not know where its useful properties were discovered or even where cotton was first grown. Cotton developed in both hemispheres. Fossilized plants from 2900 B.C. have been found in Mexico. Peruvian Indians used cotton for fishing nets around 2500 B.C. By the year 1000, Indians in both areas had elaborate uses for cotton. It was heavily exported by the Mayan empire. Cotton was also the basis for an important Aztec industry at the time of Spanish colonization in the early sixteenth century.

In the East, archaeologists have also identified fossilized bits of cotton string and fibers at a site in Pakistan. These fibers may be as old as 4,000 years. A site on the Eastern coast of Africa has turned up cotton seeds of an ancient variety native to India. Some historians therefore believe that ancient Hindus may have been the first people in the Eastern hemisphere to domesticate this plant. The Greek historian Herodotus (ca. 485–425 B.C.) described Indian clothing made from cotton in 445 B.C. Alexander the Great (356–323 B.C.) invaded India and brought cotton clothing to Greece. Demand for cotton spurred several of the early land routes to the East. By the tenth century, the cotton trade had spread across most of Europe. By the seventeenth century, England's growing cotton **weaving** industry became heavily dependent on cotton imported from its southern New World colonies.

In a period of about 75 years during the eighteenth century, English demand for cotton soared to levels that surpassed even those of wool. The wool industry had changed little and had little need for innovation. The burgeoning cotton industry demanded nothing but change. It was as though cotton production were perfectly timed to coincide with the industrial revolution.

Previously, seeds from raw cotton were removed by hand. This was no longer feasible for several reasons. First, hand labor was a slow, expensive process, even where large numbers of slaves were forced to work with no remuneration. Second, new technologies, such as the **water frame spinning machine** of Richard Arkwright (1723–1792), allowed cotton manufacturers to produce huge quantities of cotton thread and even cloth at greatly reduced costs. (For more on these topics, please see the entry on Spinning Machines as well as the one on Weaving Machines.) Finally, after the Revolution, Americans became increasingly unhappy having to export raw cotton to England for manufacture of goods that they would have to then import. The federal government was especially interested in achieving not just political, but also economic independence. Soon, textile mills began operating in the United States.

All of this generated a tremendous need for raw cotton, such as the world had never before seen. The southern colonies were easily able to grow the plant, but removing the seeds was still an enormous obstacle. Many people believe that Eli Whitney (1765–1825) invented the first **cotton gin**, but this is not the case. In the 1740s, a version of the roller gin had reached the colonies (*gin* is short for "engine"). The roller gin, or **churka** (a Sanskrit word that describes the machine's jerking motion), has been in use in India since antiquity. In India, the growing of cotton was part of everyday life. The churka allowed women to spin their own yarn and cloth at home. This machine employed two hardwood rollers and worked "like a wringer washer" (Britton 1992, 10). The cotton was fed through the dual rollers, which grabbed the fiber tightly and pinched free the seeds. The seeds were trapped by long grooves in the rollers and deposited onto the floor, while the now-clean fiber exited the rollers.

By the 1770s, several improvements had been made to the churka, and it was in limited use in the United States. However, there were still several problems with this machine. Each churka could produce only about 5 pounds of lint (or cleaned fiber) per day. This was nowhere near the desired level of production. An even bigger problem was that the churka worked well with the variety known as long-staple cotton, but the kind of cotton American plantations wanted to grow was short-staple cotton, which is full of tightly clinging seeds and other "trash" that the churka and improved roller gins could not remove.

Eli Whitney was about to change everything. The story goes that, in 1792, the young Whitney was living in Georgia as an extended house guest of Catherine Littlefield Green (widow of Revolutionary War General Nathanael Green). Whitney often performed odd jobs on her property, and Green became very impressed with his skill. One night, several prominent guests complained to Green that they could not grow cotton because of the prohibitive costs of cleaning it. Green volunteered Whitney to design a machine for cleaning short-staple cotton.

A Yale-educated New Englander, Whitney had never before seen a cotton boll. On examining one, he grasped the problem of cleaning it when he noted that its seed was "covered with a kind of green coat, resembling velvet" (quoted in Green 1956, 46). According to Whitney, he somehow had a model for his machine 10 days later. Beginning in November, 1792, Whitney built his machine in secret. By April of the following year, Whitney had a fully working model of his machine locked away in the plantation house.

The machine consisted of a square wooden frame around a wooden cylinder fitted with metal spikes, which entangled themselves in pieces of raw cotton placed inside the frame's hopper (see Figure 8). A hand crank on the outside of the frame drove the cylinder. As the cylinder rotated, the spikes meshed with a grid of thin wire. Cotton fiber slid through the grid, but seeds were too large to pass through. The seeds fell into a chamber separate from the clean fiber. Whitney's gin thereby offered the added advantage of saving the cotton seeds for future sale.

In 1794, Whitney entered into a partnership with a businessman, Phineas Miller. Almost immediately, the two were overrun with orders for their new gins. When they could not meet demand quickly enough, other businessmen began selling copies. Whitney and Miller spent years involved in court battles to enforce their patents.

On seeing the cotton gin, many people were surprised at its simplicity, in contrast to the catastrophic effect this simple machine had on the world. Whitney's cotton gin was, within only several years' time, employed on a virtually universal level. One person using this machine could do the work of fifty people cleaning by hand in the same amount of time. Hugely profitable plantations sprung up throughout the South.

Before the gin, the whole concept of a powerful South was languishing. After silk failed, landowners had trouble turning a profit from cotton. Now, nearly any person was employable for running a gin—from young boys to old men. The cotton industry of the South exploded to unforeseen, even unimagined levels of production and wealth. The plantation system, which had come under recent fire as outdated and unprofitable, was re-invigorated with unmitigated gusto. Cotton surpassed tobacco as the chief crop. Quickly, the

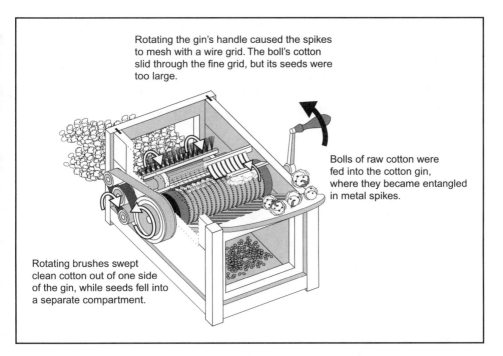

Rotating the gin's handle caused the spikes to mesh with a wire grid. The boll's cotton slid through the fine grid, but its seeds were too large.

Bolls of raw cotton were fed into the cotton gin, where they became entangled in metal spikes.

Rotating brushes swept clean cotton out of one side of the gin, while seeds fell into a separate compartment.

Figure 8. Whitney's cotton gin. Whitney's cotton gin was capable of cleaning the species of cotton grown most widely in the United States. Its effect on the agrarian South (and later on the industrialized North) was catastrophic. It helped give the South its "king" crop, cotton.

United States became rich on trade with England. During the 20 years following Whitney's invention, land that could support cotton tripled in value.

Accordingly, the Atlantic slave trade, so closely tied to the economy of the American South, expanded as well. Plantation owners no longer needed many slaves to clean cotton, but they needed more than ever for planting and bailing. The idea of slavery, which had been a major public controversy as recently as the Continental Congress, became deeply rooted in the American psyche and way of life. With the profit of cotton, relatively few Americans would consider the thought of emancipation. By 1825, the slave population in South Carolina and Georgia was ten times that of the early 1790s.

The rapidly industrializing northern states also benefited tremendously, as cotton provided raw materials for both manufacture and export. By 1820, the United States grew almost a third of the world's cotton—in excess of 180,000,000 pounds annually. Five years later, nearly 171,000,000 pounds were exported to England alone.

Cotton is still a very important crop. In 1960, cotton accounted for three-quarters of all fiber in the United States. In the late 1970s, new synthetic fibers began appearing in great numbers on the market. By 1977, cotton accounted for only one-third of all fiber in the United States. However, during the 1990s, a combination of new cotton blends, competitive pricing, fashion trends, and collective advertising made cotton account for one-half of American fiber.

Cotton is currently ginned in methods that stem, in principle, from Whitney's machine. Modern machines called "saw gins" grab the fiber of short-staple cotton and forcibly pull it away from the seeds. As in antiquity, other forms of cotton are cleaned by simple machines such as roller gins.

SEE ALSO Spinning Machines; Weaving Machines.

Selected Bibliography

Britton, Karen Gerhardt. *Bale o' Cotton: The Mechanical Art of Cotton Ginning.* College Station, TX: Texas A&M Press, 1992.

Green, Constance McLaughlin. *Eli Whitney and the Birth of American Technology.* Boston: Little, Brown, 1956.

Mirsky, Jeannette, and Allan Nevins. *The World of Eli Whitney.* New York: Macmillan, 1952.

Rowland, K. T. *Eighteenth Century Inventions.* New York: David & Charles Books, 1974.

World Book (encyclopedia) CD-ROM. San Diego: World Book, 1999.

Coulomb's Law
Charles Coulomb's Rule of Electrical Forces, 1785

Along with advancement of practical knowledge, the eighteenth century spawned many important new scientific theories. In chemistry, the **oxygen**-based theory of Antoine Lavoisier (1743–1794) was replacing the old **phlogiston** system of Georg Ernst Stahl (1660–1734). In time, a modern understanding of chemistry developed. A similar formation was taking place in electricity, where Charles Dufay (1698–1739) and Benjamin Franklin (1706–1790) theorized that electricity was a fluid. The former proposed a two-fluid theory of positive and negative energy, while the latter refined it into a single-fluid theory. In addition, Luigi Galvani (1737–1798) found that the leg of a dead laboratory frog contracted violently when he ran electrical current through it. Galvani's theory claimed that electricity was somehow produced by moist organic tissue, such as animal muscle. Though incorrect, this theory was important to the work of Alessandro Volta (1745–1827). Eventually, Volta would hypothesize that electricity was produced, not in the tissue of animals, but in the combination of different metals. This theory eventually led Volta to the first practical battery, known as the **voltaic pile**. (For more information, please see the various entries on these topics.)

Another scientist who developed several important theories of electricity was the Frenchman Charles Coulomb (1736–1806). Following Dufay, Coulomb set forth a "two-fluid" magnetic theory—one fluid he called "boreal," and the other "astral." He took these terms from the north and south of the magnetic compass. While this explained polarity, it did not explain how cut or broken magnets retained this property of magnetism. Therefore, Coulomb later proposed that, in noncharged metals, tiny magnetic particles were dispersed in no particular order of direction. On magnetization, he thought that the particles lined up to point in an aggregate direction—somewhat similar to the theory accepted today. Coulomb also believed that his magnetic findings were exactly applicable to contemporary theories of electricity.

In a paper presented to the French Academy of Science around 1784, Coulomb reported that the force of torsion on a thin metal wire depended on the kind of metal. From this finding, Coulomb was able to figure out the definite torsional characteristics of specific metal wires. He had thereby invented a method for measuring the actions produced by very small forces. So long as the tiny forces—such as magnetic attraction and repulsion—twisted the fine wire, Coulomb had a way to measure them.

He put this method into practical use with his invention of the **torsion balance**. This was the first instrument for precisely measuring electricity. In the next decade, Henry Cavendish (1731–1810) would use a torsion balance to estimate the density of the earth. Coulomb's torsion balance consisted of a glass cylinder that he covered with a glass disk. Two uncharged pith balls were suspended so that they "touched" one another through the glass, with one outside and one inside. When both pith balls were touched with a charged pin, they abruptly repelled each other. One of the balls was suspended by a torque arm at the top. When the magnetic repulsion caused it to move, the pith ball twisted the wire to a degree in balance with the force of repulsion. By measuring the degree of twist, Coulomb could therefore measure the force of magnetic repulsion.

Coulomb made several important findings, and the torsion balance was a vital invention in its own right. Yet, the inverse square law was the most important theoretical breakthrough Coulomb made with his balance. He made careful measurements of the force of repulsion in relation to different positions of his pith balls. When he moved a ball outside the glass two times the distance from the one inside the glass, the force applied to the inner ball did not coincide exactly as one might expect. It was not one-half, but one-forth. The electrical charge's effect "decreased as to the square of the distance between the centers of the charged balls" (Krebs 2001, 70). This formed the inverse square law, which can be expressed as the proportion $1/r^2$, where r is the distance between balls.

This was an important finding. But by far the most influential theory proposed by this scientist is appropriately known as **Coulomb's law** (also called the **rule of electrical forces**). Coulomb's law grew out of his work with the torsion balance and from the inverse square law. Coulomb's law also states that any magnitude of a force—be it repulsion or attraction—is directly proportional to the product of the magnitudes of the charge. That is, Coulomb demonstrated that the force between any two magnetic poles is in direct proportion to the pole strength's product. Accordingly, he showed that this same force is also inversely proportional to the square of the distance between the poles. The force between any two electrical charges is therefore proportional to their product and inversely proportional to the square of the distance between them. Fundamentally in tandem with his previous inverse square law, Coulomb's law explained principally that the force between two charges varies inversely as the square of their distance.

Dissemination of this law fostered understanding of electricity as a measurable phenomenon with specific properties. For the first time, enlightenment scientists began to realize that (like gases and chemicals) electricity followed rational, specific, quantifiable laws. In fact, Coulomb's law could be written as an equation:

$$\propto \frac{qq'}{r^2}$$

where q and q′ are the charges applied to the balls (Whitmer 1952, 6). Coulomb's law would eventually foster a significant improvement in the theoretical understanding of electricity as a current (rather than a fluid) carried by an "electric" or conductor.

In honor of this law's founder, the unit of electric charge in the metric system is the coulomb (c). Appropriately, this unit measures the specific amount of electricity that flows through an electric current of 1 ampere in the time of 1 second. The ampere is the unit of electrical rate, 1 coulomb per second. The coulomb also has a relationship with the **volt** (v), which is 1 joule per second. Also, the farad (f, named for Michael Faraday [1791–1867]) is 1 coulomb per volt. Modern scientists have established that the coulomb is equal to the charge of 6.24×10^{18} electrons.

An instrument, the coulometer, also bears this scientist's name. This instrument is also called a voltameter (after Volta). Generally, it consists of a platinum dish (the cathode), an extremely fine wire made of silver (the anode) and a simple silver nitrate solution (the electrolyte). During a process known as "electroplating," the weight gain of the platinum dish is easily converted to a reading in amperes.

Henry Cavendish would eventually offer more convincing and definite proof of Coulomb's law. To do so, he conducted an ice bucket experiment that used insulating rings to demonstrate the inverse square law. The insulated spheres were separated from one another until they were connected by an inner wire and thereby charged. This established that subtler, less direct experiments are more accurate than those that attempt to measure the force between charges as a direct function of the distance between them (as Coulomb's torsion balance had).

In the following century, Faraday expanded Coulomb's theories of electrical action to full-blown, rigorous investigations of electrostatic forces. Building on Coulomb's work, Faraday postulated that electrostatic action is in fact a field of force. This contention helped him discover the principle of electric induction, along with physicist Joseph Henry (1797–1878). Following this principle, both scientists produced electricity by moving a coil near a magnet. Both the electric generator and electric engine operate on this principle.

SEE ALSO Chemical Cells; Electrical Conductors and Insulators; Electricity Is of Two Types; Franklin's Electrical Researches.

Selected Bibliography

Bordeau, Sanford P. *Volts to Hertz: The Rise of Electricity*. Minneapolis: Burgess, 1982.

Canby, Edward Tatnall. *A History of Electricity*. New York: Hawthorn Books, 1968.

Efron, Alexander. *Direct Current Electricity: Franklinian Approach*. New York: Chapman & Hall, 1960.

Krebs, Robert E. *Scientific Laws, Principles, and Theories: A Reference Guide*. Westport, CT: Greenwood Press, 2001.

Meyer, Herbert W. *A History of Electricity and Magnetism*. Cambridge, MA: MIT Press, 1971.

Rowland, K. T. *Eighteenth Century Inventions*. New York: Harper & Row, 1974.

Schwarz, W. M. *Intermediate Electromagnetic Theory*. New York: John Wiley & Sons, 1964.

Whitmer, Robert M. *Electromagnetics*. New York: Prentice-Hall, 1952.

D

Digestion Is Primarily a Chemical Process
René-Antoine Ferchault de Réaumur, 1752
Lazaro Spallanzani, 1782

Many archaeologists believe that, along with metallurgy, pottery and the making of paints and perfumes, cookery became a highly evolved practice of prehistoric peoples. Indeed, several even suggest that gastronomy may qualify as the earliest science (before even astronomy). In this context, it seems almost inevitable that great ancient civilizations made studies of human **digestion**.

The ancient Greeks and Romans began delineating the debate over whether human digestion was a mechanical or a chemical process. Around 400 B.C., the Greek physician Hippocrates (ca. 460–380 B.C.) broadly focused medical attention on the complex relationships between parts of an organism. Centuries later, Galen (ca. 129–210), a Greek physician in Rome, published the book *Anatomical Procedures*, which served as the standard textbook on anatomy for over a thousand years in Europe. Galen thought digestion occurred mainly in the liver, where food was changed into blood for circulation throughout the body. He was unsure whether this was due to a grinding, mincing and pulverizing action, or whether it was part of some mysterious vital force of bodily secretions. Along with others, Galen thus engendered a debate that would last, on and off, until the eighteenth century in Europe.

The debate was heating up just as Belgian scholar Johann Baptista van Helmont (1579–1644) began a study of digestion, which he published in 1648. Van Helmont constructed a general theory of **transmutation** involving fermentation on a broad scale. He thought that specific "ferments" (or acids) in the stomach and liver brought about physiological changes such as digestion. Later scientists would compare his "ferments" to the work of enzymes. Van

Helmont proposed a series of six fermentations that take place in the body. Several types—such as those of the stomach, gall bladder and duodenum— were roughly on the right track. Others—fermentations of the heart and brain—were completely outlandish. Nonetheless, van Helmont had proposed a rudimentary theory of chemical digestion.

By the start of the seventeenth century, the debate was raging in full swing. By mid-century, anatomists had made substantial studies of the digestive system's glands. The pancreatic, submaxillary, and parotid duct had all been identified. A student of van Helmont, Franciscus Sylvius (1614–1672), outlined a more sophisticated theory of chemical digestion. Sylvius argued that fermentation involved a mixture of food, saliva, bile, and juices from the pancreas. His was a theory of warfare between acids (such as pancreatic juice) and alkalines (such as bile). He thought that digestion ended when each chemical was neutralized and gastronomic balance was restored. Sylvius also thought digestive problems came from imbalances in acidity or alkalinity. Many apothecaries exploited this theory for commercial purposes. Sylvius' theories also became important to the doctrine of iatrochemistry, which tried to explain all infirmities and treatments entirely in terms of chemistry.

As powerful as it was, iatrochemistry was a reaction to the dominant seventeenth-century metaphor for bodily function, which stated that the body was a complex machine. Mechanical theorists proposed a theory of iatrophysics. Italian physician Santorio of Padua (1561–1636, sometimes known as Sanctorius) studied the mechanics of his own body and published his records in 1614. William Harvey (1578–1657) published *On the Movement of the Heart and Blood in Animals,* an authoritative text that provided powerful new arguments for the mechanical model of the body, including digestion. Finally, though he did not provide anatomical experiments, philosopher–mathematician René Descartes (1596–1650) compared the body to a perfect clockwork mechanism that functioned according only to mechanical laws.

By the eighteenth century, both chemical and mechanical models had been well articulated. However, even the most convincing proponents lacked definitive experimental evidence for their claims. In 1752, René Réaumur (1683–1757) was not fully convinced by either side and undertook a series of experiments on gastric digestion in birds. Réaumur knew that hawks ordinarily swallow prey in large pieces, digest what they can use, and regurgitate the remainder. He induced a pet hawk to swallow small metal cylinders with mesh wire over each end. Inside each cylinder was a piece of meat. Because it could not digest the metal cylinders, the hawk regurgitated each one (see Figure 9).

Réaumur found that the meat inside was partially digested. Because the metal cylinders were intact, the hawk could not have digested the meat

Figure 9. Réaumur's digestion experiment with his pet hawk. Réaumur placed a piece of meat inside a small metal cylinder. He somehow enticed his hawk to swallow the cylinder. Since it could not digest the cylinder, the hawk regurgitated it. Réaumur found the meat partially digested. He concluded that gastric juice had entered the cylinder and begun the primarily chemical process of digestion.

mechanically. He correctly concluded that the meat must have been digested chemically when the hawk's stomach juices filled the cylinders. He began to believe that digestion might be entirely chemical. In a series of definitive follow-up experiments, Réaumur somehow persuaded his hawk to swallow a small part of a dry sponge. When the hawk regurgitated the sponge, Réaumur squeezed the liquid out and saved it. When he submerged a piece of meat in the regurgitated liquid, he found that the meat slowly dissolved. This reinforced his conclusion that mechanical actions (while possibly beneficial) were far from necessary. He had also shown that digestion could occur outside the confines of the stomach.

Réaumur repeated his experiments with other animals, including dogs, and drew the same conclusions. In subsequent trials, he also found that stomach liquid more fully digested meat than it did starchy foods. Skeletal workings might resemble machines, but digestion was indisputably primarily a chemical process.

Later in the century, a versatile Italian scientist, Lazaro Spallanzani (1729–1799), took up the question where Réaumur had left off. So far, no one had conclusively studied human digestion. In order to do so, Spallanzani constructed a number of small, hollow wooden blocks. Then he inserted various foods into the blocks and (always one to put science first) swallowed them. After a set number of minutes, which he varied for each experiment, Spallanzani induced himself to vomit so that he could study the changes in the food. (Spallanzani ceased the experiments only on developing horrible nausea.) Spallanzani coined the term **gastric juice** and found that Réaumur had been correct about digestion being a primarily chemical process. However, he also found that the churning action of the stomach (which he synthesized by agitating tubes of gastric juice) was an aid to the chemical process of digestion. But it was far from necessary.

Spallanzani also investigated the theory of fermentation, first raised by van Helmont and Sylvius. His experiments showed that gastric juice caused milk to curdle but not to ferment. Many others thought this showed digestion to be a form of putrefaction. But Spallanzani showed that digestion was no more a matter of putrefaction than it was fermentation. Through a series of anatomical studies, Spallanzani demonstrated that gastric juice was secreted by the stomach and not introduced by another of the body's organs. After Spallanzani and Réaumur, most scientists agreed that digestion was primarily chemical.

All along, Spallanzani had suspected that gastric juice contained a strong acid. However, it was not until 1823 that English physician William Prout (1785–1850) identified free hydrochloric acid through the procedure of chemical distillation. He showed that this acid was necessary for chemical digestion, but that the actual process was the work of some other agent.

In June of the previous year, a U.S. Army surgeon in Michigan, William Beaumont (1785–1853), had become the first person to actually observe human digestion. A young Canadian fur trapper named Alexis St. Martin had accidentally been wounded with a shotgun blast. Beaumont did not expect him to live past 36 hours. Surprisingly, the strong young man did live—but the deep wound in his stomach never healed shut. Eventually, Beaumont placed a bandage over the wound, and St. Martin was able to eat.

Beaumont quickly realized the rare opportunity and convinced St. Martin to let him perform some experiments. These lasted for over 7 years. In Beaumont's time (as today), his experiments on a human subject were deeply controversial. Nonetheless, Beaumont took samples of partially digested food from St. Martin's stomach. He also tied silk threads around various pieces of food and inserted them into the wound. After varied amounts of time, Beaumont withdrew them and studied the effects of gastric juice on them. He made temperature studies of St. Martin's stomach when it was empty and full. He also sent partially digested food samples to laboratories in Europe for chemical analysis. Beaumont published his findings in a book, *Experiments and Observations*. He showed, beyond doubt, that Réaumur and Spallanzani were correct about digestion being a chemical process. Most of Beaumont's conclusions are still valid.

Later in the century, the French physiologist Claude Bernard (1813–1878) showed that digestion in the stomach is really only a preparatory act. The bulk of digestion, he correctly explained, takes place in the small intestine. Today, scientists know that gastric juice is only partly made of hydrochloric acid. It also contains an important enzyme called *pepsin*. Gastric juice begins digesting protein—foods like eggs, milk and meat. Just as Réaumur found over two centuries ago, the stomach does not digest starchy foods. It also does not digest sugars or fats. From the stomach, partly digested food called **chyme** passes into the small intestine, where pancreatic juice, intestinal juice and bile complete the process of chemical digestion.

SEE ALSO Animal Physiology; Animal Respiration; Microscopic Organisms Reproduce.

Selected Bibliography

Asimov, Isaac. *Asimov's Chronology of Science and Discovery*. New York: Harper & Row, 1989.
Bettman, Otto L. *A Pictorial History of Medicine*. Springfield, IL: Charles C. Thomas, 1956.
Brock, William H. *The Norton History of Chemistry*. New York: W. W. Norton, 1993.
McGrew, Roderick E. *Encyclopedia of Medical History*. St. Louis: McGraw-Hill, 1985.
Partington, J. R. *A Short History of Chemistry*. New York: St. Martin's Press, 1957.
Porter, Ray (ed.). *The Cambridge Illustrated History of Medicine*. New York: Cambridge University Press, 1996.
Rhodes, Philip. *An Outline History of Medicine*. Boston: Butterworths, 1985.
Singer, Charles. *A Short History of Medicine*. New York: Oxford University Press, 1928.

Discovery of Nitrogen
Daniel Rutherford, 1772

Like water, **nitrogen** is everywhere. It is in the tiny cells of a person's body, because it is found in all amino acids, the building blocks of protein. Nitrogen, a gas, is also the principal part of the air on Earth. In fact, nitrogen makes up about 78% of atmospheric dry air (air with no water vapor)—more than three times the percentage of **oxygen**. Nitrogen consists of two atoms of nitrogen, which are bonded together to form a molecule. Nitrogen does not combine easily with other elements. However, it is frequently condensed to useful liquid form, which boils at −195.8° **Celsius** (C) and freezes at −209.9°C.

Nitrogen was discovered in the late eighteenth century. Immediately, scientists recognized it as an important contribution to the chemical revolution. It became one of the most significant gas discoveries, along with **oxygen** and **carbon dioxide**. In fact, scientists who discovered the latter gases spurred the discovery of nitrogen. One of the discoverers of oxygen, Joseph Priestley (1733–1804), actually isolated nitrogen while chasing the elusive **phlogiston**—the hypothetical substance supposedly released during **combustion**. Because of its propensity for great burning, Priestley named oxygen **dephlogisticated air**, supposing that it hungrily filled its phlogiston debt through absorption during combustion. On the other hand, Priestley called nitrogen "phlogisticated air," which suggests he may have thought of it as the union of oxygen and phlogiston. In other words, Priestley may have thought nitrogen was a product of combustion because it would not burn. Regardless, he certainly did not realize nitrogen was an independent gas.

Nonetheless, Priestley was one of the first to recognize the importance of nitrogen. He wrote that the nitrogen work of fellow chemist Henry Cavendish (1731–1810) was the most important work ever done on air. One of Cavendish's discoveries was that the atmosphere is almost entirely composed of one substance. Cavendish called this majority the "phlogisticated part of our atmosphere" (quoted in Jungnickel and McCormmack 1999, 368). Then Cavendish demonstrated that nitrogen was more than simply the air left after combustion had taken place. Curiously, Cavendish also found that roughly 1/120 of the atmosphere was a substance he could not identify, but which he knew was not nitrogen.

Today, Swedish apothecary Carl Scheele (1742–1786) is almost universally credited with first discovering oxygen. He recognized that ordinary air was a mixture of at least two gases. Scheele discovered a lot about one of these gases, oxygen, but little about the others. Like Priestley and Cavendish, Scheele was not able to make nitrogen combust. For this reason, Scheele called it "spent air."

To this point in history, the majority of knowledge about gases came from studies of two specific ones—oxygen and carbon dioxide, which was discovered by Joseph Black (1728–1799). In his experiments, Black put a candle in an enclosed container until it had used up all the oxygen. In place of oxygen, Black found that the candle had formed **fixed air** (or carbon dioxide) that would no longer support a flame. Black understood this, but became puzzled when he absorbed the carbon dioxide with other chemicals. His measurements showed that a specific amount of gas had been left behind. He found that this peculiar gas would not support combustion, but did not behave exactly like carbon dioxide. At this point, many scientists had independently recognized that some "other" gas was present, but none had undertaken a comprehensive investigation. And now Black had the most puzzling results to date. What substance made up this quantity of gas left behind? Had all of the original gas become carbon dioxide (after all), or had a new gas somehow been formed? Or, was this residue a different substance altogether?

Black turned the question over to his doctoral student, Daniel Rutherford (1749–1819). Rutherford wanted to isolate this mysterious air that was left over after oxygen and carbon dioxide were purged from ordinary air. He began by confining a live mouse in a container until the mouse died. Next, he burned a candle inside the container until it would no longer burn in the air. Finally, he burned phosphorus in the container until even it extinguished. Rutherford was absolutely convinced that the air inside the container would no longer support combustion.

He then turned his attention to his teacher's special gas, carbon dioxide. He forced the container's remaining air through a solution of chemicals he knew would absorb all the carbon dioxide. Just as he expected, Rutherford found that there was still a large quantity of air remaining in the container, even though he had effectively removed its oxygen and carbon dioxide.

This air would not support combustion or a mouse's **respiration**. It was clearly not oxygen. Rutherford also confirmed that it was not carbon dioxide. The phlogiston theory stated that this air was saturated with phlogiston and would no longer accept any more (from combustion or respiration). Rutherford had therefore demonstrated that this gas was wholly separate from both carbon dioxide and oxygen. Rutherford named his gas **mephitic air**—a name that suggests an unpleasant odor, though nitrogen actually has no odor. The gas's modern name, *nitrogen*, comes from the Greek for "niter producer." This name came to pass because niter—or potassium nitrate—contains nitrogen.

Previously, the French founder of modern chemistry, Antoine Lavoisier (1743–1794), called nitrogen *azote*, meaning "without life." He coined this name after discovering the presence of this gas in animal respiration. In this process, oxygen is consumed, carbon dioxide is produced, and nitrogen passes through the lungs unchanged.

Modern scientists have shown that *all* organisms must have nitrogen to live, because it is part of amino acid structure. Plants make all the amino acids they need. In fact, nitrogen is so basic to plant life that nitrogen fertilizer, air, sunlight, and water are enough, by themselves, to support many modern crops. Nitrogen fertilizers— sometimes called *mineral fertilizers*—are produced mainly from **ammonia** gas and supply the soil with large amounts of nitrogen. Many farmers inject nitrogen fertilizers in the form of ammonia gas directly into the soil.

When farmers utilize nitrogen's crop-growing capabilities, they are relying on a process called the *nitrogen cycle*. In this cycle, nitrogen gas is converted by bacteria and certain yeasts to nitrogen compounds that can be used by plants. Plants use the nitrogen compounds in the soil in order to make proteins they need. When they (or animals that eat them) die or excrete wastes, nitrogen compounds are returned to the soil. From there, nitrogen gas is returned to the atmosphere and the cycle begins anew.

However, human involvement has altered the nitrogen cycle and thereby caused pollution. When rainwater carries unused nitrogen fertilizer into streams and lakes, water plants and algae multiply. When they die and decay, these organisms use up oxygen, which endangers aquatic animal life. Additionally, the burning of fossil fuels releases nitrogen oxide pollutants into the air, causing smog and acid rain.

A different risk from the use of nitrogen fertilizer is the easy propensity with which nitrogen can be used as a main ingredient of powerful "homemade" explosives. In April 1995, the Murrah Federal Building in Oklahoma City, Oklahoma, was, tragically, destroyed by a bomb consisting of little more than 1,300 kilograms (kg) of nitrogen fertilizer mixed with diesel fuel. This crude weapon was the product of nitrogen's incredible power unleashed through a relatively simple chemistry of explosives. Therefore, nitrogen fertilizers represent both a risk of terrorist attack and a source of pollution. But they also represent a staple of modern agricultural practices, which feed billions of people.

Finally, the discovery of nitrogen led to research and the eventual discovery of many other gases that exist as minuscule parts of the earth's atmosphere. When Henry Cavendish first found the mysterious gas in atmospheric proportion of 1/120, he identified it as a "bubble" in his apparatus. Over a century later, William Ramsay (1852–1916) and Lord John Rayleigh (1842–1919) found that atmospheric nitrogen was too dense to be pure nitrogen. They found that Cavendish's bubble accounted for the added density, as it was actually the chemically inert gas **argon** mixed with atmospheric nitrogen. In recognizing this, Ramsay and Rayleigh inaugurated a new chapter in the chemical study of the Earth's atmosphere. In the end, the Earth's atmosphere turned out to be much more than just oxygen and nitrogen, though these gases make up the vast majority.

SEE ALSO Atmospheric Composition; Carbon Dioxide; Hydrogen; Oxygen.

Selected Bibliography

Asimov, Isaac. *A Short History of Chemistry*. New York: Anchor Books, 1965.

Berry, A.J. *Henry Cavendish: His Life and Scientific Work*. London: Hutchinson, 1960.

Brock, William H. *The Norton History of Chemistry*. New York: W. W. Norton, 1993.

Jungnickel, Christa, and Russell McCormmach. *Cavendish: The Experimental* Life. Lewiston, PA: Bucknell University Press, 1999.

Moore, F.J. *A History of Chemistry*. New York: McGraw-Hill, 1931.

Riedman, Sarah R. *Antoine Lavoisier: Scientist and Citizen*. New York: Abelard-Schuman, 1967.

E

Electrical Conductors and Insulators
Stephen Gray, 1729

The eighteenth century was a time of terrific advancement in the scientific study and understanding of electricity. It was also the first century to glimpse practical applications of this study.

Earlier practical application was limited to magnetism, as in the form of the electric compass. This invention was in use in third-century China. The Chinese also knew how to magnetize certain metals from a **lodestone.** It was during the thirteenth century that the magnet came into widespread use in Europe. In this same century, Roger Bacon (ca. 1214–1292) made the first brief observations about the natural electrical properties of **amber** and lodestone. He was followed by Peter Peregrinus (also known as Pierre de Maricourt, ca. thirteenth century, as he definitely lived during 1269). Peregrinus was a French soldier, philosopher and engineer, who wrote a letter on the battlefield of Lucera. In this letter, he described a piece of lodestone that he called a "terrella" (for "little earth"). On this lodestone, Peregrinus used a needle to map opposite areas of concentrated attraction on the poles (like the earth itself).

Around 1600, knowledge of electricity as a specific force or phenomenon was just beginning to emerge. Queen Elizabeth's private physician, William Gilbert (1544–1603), found that glass, wax and sulfur acted electrically similar to amber when they were rubbed with a cloth. They attracted materials such as feathers, hair and dry straw. Gilbert called these materials "electrics" (or insulators). In addition, Gilbert proposed that the curious action of these objects must be due to some kind of fluid. Gilbert wrote a treatise on his observations, *De Magnete*. This enormously successful treatise was read by royalty and scientists alike. Following the work of Peregrinus, Gilbert correctly suggested that, just as a piece of lodestone is like the earth, the earth is like a piece of lodestone. He compared it with a gigantic magnet, which, he explained, caused compasses to point north.

Many seventeenth-century scientists would discover rudimentary electrical phenomena and would take to rubbing electrics (like amber) after reading *De Magnete*. Among these scientists were Galileo Galilei (1564–1642) and Isaac Newton (1642–1747). However, despite the enormity of scientific advancement during the seventeenth century, the fundamental laws of electricity stayed just out of reach.

One of the eighteenth century's earliest electrical observations was made by Francis Hauksbee (ca. 1713), a member of the London Royal Society. Hauksbee greatly improved on earlier **electric machines** that created electrostatic effects through friction. With a globe of roughly 9 inches across, Hauksbee was able to generate enough light for reading. Noticing this action, he began to wonder about the relationship between electrification by friction and production of light by friction. The more he observed, the more he thought the two were related. Hauksbee broadly termed this relationship *phosphorescence*. Despite this important observation, however, electricity was still a fledgling science with relatively few established facts and no practical applications. All this was about to change. Before the century was over, electricity would become a popular science with a significant body of knowledge and many practical applications. The eighteenth century's first major contribution was made by the English scientist Stephen Gray (1695–1736), who studied **conduction**, the actual flow of electricity. One of his first experiments entailed charging (by means of a powerful electric machine) a glass bottle and its cork stopper. After this success, he moved on to electrify a large pine splinter. Thinking that electric fluid could flow between two bodies, Gray was able to electrify a metal ball by running electricity to it through a linen cord. (It was a much more powerful charge than today's household current.) This was one of the first uses of "wire" to convey an electric charge. Soon, Gray was using pieces of reeds to transmit charges to various rooms of his house. He did this long before residential wiring became a possibility.

This set Gray to wondering about how far he could make his mysterious fluid travel. After many trials, Gray strung 800 feet of hempen cord through silk suspension threads and sent a powerful charge from one end to the other. However, when Gray attempted to repeat the experiment, he replaced the silk suspension threads with thinly spun brass wire and was puzzled when he no longer could make the electricity pass through the hempen cord. When he replaced the brass wire with the silk suspension threads, his experiment again worked as before. Time and again, Gray found that the electricity seemed to "disappear" into the brass support wires, almost as through they "absorbed" the charge. He correctly surmised that "electrics" such as glass, resin and silk hold a charge. On the contrary, he found that "non-electrics" such as metal and water conduct a charge. Demonstrating this, Gray used the silk support threads to hold a wet hempen cord and successfully sent his charge a distance of 1,200 feet (with the water aiding in conduction).

Gray's distinction between "electrics" and "non-electrics" was the first articulation of **insulation** versus conduction. Taking the work of Gilbert (from the prior century) an important step further, Gray drew up a list of insulators and conductors. His list was significantly useful to electrical scientists during the next century. In discovering conduction especially, Gray had fallen on an important principle. Electric charges do not "fill up" an object. They remain only on its "outer surface." Gray publicly demonstrated this characteristic by electrifying two oak cubes, which were identical except that one was hollow and one was solid. Both were charged exactly the same way. As far as electricity was concerned, the cubes were identical. Later, this important principle was vividly demonstrated through the use of the "Faraday cage" (after Michael Faraday [1791–1867]). Today, this principle is in play when a metal airplane hit by lightning still protects its occupants.

Performed with the simplest of instruments, Gray's important work profoundly shaped the electrical research of the eighteenth century, especially the work of Charles Dufay (1698–1739). After studying Gray's work on conductors and insulators, Dufay became convinced of a "two-fluid" theory of electricity. Benjamin Franklin (1706–1790) would later coin the terms *positive* and *negative* (although he did not subscribe to Dufay's important but incorrect theory). Both Dufay and Franklin correctly claimed that all objects (except metals and those materials too soft or too fluid) could be electrified. After this discovery, the distinction between conductors and insulators became all the more crucial, as electricity was put to its first practical uses in inventions such as the **lightning rod**. (For more information, please see the entry on Franklin's Electrical Researches and also the one on Electricity Is of Two Types.)

Today, practical application of insulators and conductors (often paired, as in wiring) has helped make residential and commercial electricity use nearly universal in developed nations. Modern insulators (also called "dielectrics") conduct nearly no electricity. These insulators include glass, mica, plastics, and rubber. In some applications or inventions, very dry air or specifically formulated oils also act as insulators; sometimes they provide cooling, as well. Insulators conduct electricity very poorly, because their electrons are so *tightly* bound to their nuclei that they cannot move freely between atoms. Therefore, when a source of electricity is connected through an insulator, too few electrons can move through the insulator to produce a current.

Modern conductors, on the other hand, include metals such as aluminum, copper and silver. Their electrons are very *weakly* bound to their nuclei and travel almost freely, resulting in a strong flow of electricity. This was why Gray's charge dissipated so significantly when he used brass (an alloy of copper and zinc and therefore a good conductor) to support his hempen cord.

Some of the most significant technological breakthroughs of the late twentieth century involved the science of conduction and insulation. *Semiconductors* conduct electricity better than insulators but not as well as conductors. One enormous advantage is that, at low temperatures, semiconductors lose resistance to the flow of electric current. Therefore, they produce very little heat. Common semiconductors are germanium, cuprous oxide, gallium arsenide, indium arenide, lead sulfide and the most widely known semiconductor, silicon. Found naturally combined with **oxygen** in sand as silicon dioxide, silicon is the second most plentiful element on earth. Semiconductors are capable of what is known as *rectifying* (changing alternating current to direct current, which is necessary for many devices), as well as *oscillating* (making alternating current or even radio waves) at a wide variety of frequencies. Semiconductors, therefore, make possible such technologies as television camera tubes, transistors, solar cells and computer chips.

SEE ALSO Chemical Cells; Electricity Can Produce Light; Electricity Is of Two Types; Franklin's Electrical Researches.

Selected Bibliography

Bordeau, Sanford P. *Volts to Hertz: The Rise of Electricity*. Minneapolis: Burgess, 1982.

Canby, Edward Tatnall. *A History of Electricity*. New York: Hawthorn Books, 1968.

Crowther, J. G. *Famous American Men of Science*. New York: W. W. Norton, 1937.

Heilbron, J. L. *Electricity in the 17th and 18th Centuries: A Study of Early Modern Physics*. Los Angeles: University of California Press, 1979.

Meyer, Herbert W. *A History of Electricity and Magnetism*. Cambridge, MA: MIT Press, 1971.

Electricity Can Produce Light
Francis Hauksbee, 1705
Jean Antoine Nollet, 1740s

Today, the electric light is considered one of the watershed inventions of the nineteenth century, largely because of the pioneering work of English inventor Sir Joseph Wilson Swan (1828–1914) and American inventor Thomas Edison (1847–1931). The invention's association with both Swan and Edison is correct. Swan invented the first **incandescent light bulb**, and Edison achieved electric light with a carbonized thread. However, neither Swan nor Edison was the first scientist to discover light produced by electricity. Nor were their explorations the first to learn about its properties. During the eighteenth century, a body of electrical knowledge formed that included several important discoveries about the electric production of light.

One of the earliest scientists to note the phenomenon of light from electricity was the French scientist Jean Picard (1620–1682). Picard was an important astronomer who also proposed a scientific unit of measure 100 years before postrevolutionary France adopted the **metric system**. In his laboratory experiments, Picard used the mercury barometer, which was invented in 1643 by Evangelista Torricelli (1608–1647). This device, which measured atmospheric pressure, consisted of a glass tube filled with mercury and capped at one end. When a scientist inverted the tube (with the open end below the level of mercury in a cup), the mercury's level dropped to the point at which atmospheric pressure sustained it. When Picard used this instrument in a darkened room, he noticed an incredible occurrence. On agitation, brief flickerings and flashes of light appeared in the newly emptied space directly above the mercury. Picard made notes of this occurrence and was puzzled as to its cause. He never realized the light was electrical in nature.

Important as it was, Picard's work was more observation than it was rigorous scientific exploration—at least in the eighteenth-century sense. In 1705, it was Francis Hauksbee (sometimes Hawksbee or Hawkesbee, ca. 1666–1713) who would conduct the first methodical inquiries on the phosphorescent light of a Torricellian barometer. Hauksbee was an important figure in early eighteenth century applied science. He was already expert in scientific apparatuses, having invented the first efficient air pump.

Hauksbee undertook a series of investigations that were rigorously empirical and employed many different variables. He created a number of glass vessels containing mercury and other liquids. He shook some of the vessels and carefully maintained calm in others, and he also left air in some of the glass vessels. In others, he entirely exhausted all air from the space directly above the mercury. Hauksbee was trying to isolate the conditions that produced the glow. Ultimately, he wanted to figure out its cause. After many trials, Hauksbee found that the vessels that contained air gave off only faint or intermittent flashes of light. On shaking the vessel, he found that those containing only mercury gave off a gentle, uniform glow. At first, Hauksbee thought that the mercury was somehow giving off the glow, but after more experimentation, he began to speculate that the glow was electric in origin.

Hauksbee successfully showed that this "barometric light" was caused by the electric friction generated between the mercury and the glass. Once he (correctly) realized that the light was electric in nature, he realized that mercury was incidental (and thereby replaceable) to the real cause, electrical friction. He was able to create similar effects on vessels that contained no mercury by simply rubbing their outside surfaces with his hands. Here was indisputable evidence that the light was electric in nature. Equally important, when Hauksbee brought one of his evacuated glass tubes near a tube electrified by rubbing, he found that he could make the empty tube glow. He even learned how to manipulate the spectacular colors and intensities of light given off by his glass tubes.

This work in electrical friction provided the foundation for Hauksbee to greatly improve the electric friction machine (see Figure 10). This machine played an important role in many of the eighteenth century's electrical explorations that were to follow. It was operated by a person turning a handle to quickly rotate a glass sphere. Then a person laid his or her hand on the rotat-

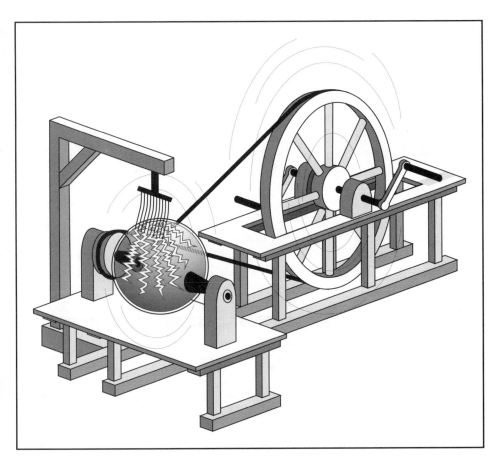

Figure 10. Hauksbee-type Electric Machine. The electric machine was invaluable to early electrical inquiries. A person turned the crank, which spun the glass ball. When the ball rubbed against a metal comb, it effectively created a charge of static electricity.

ing surface of the glass, creating electric friction. On using a 9-inch-diameter blown glass globe in the machine, Hauksbee became the first person to read by generated electric light. Bringing together two glass spheres (as he had done with glass tubes), Hauksbee demonstrated rudimentary examples of electro-magnetic induction. This is the process by which an object with electric properties transfers them to or produces them in an object held in nearby proximity. However, Hauksbee only partially understood this phenomenon.

The scientist who picked up where Hauksbee left off was Jean Antoine Nollet (sometimes known as the Abbé Nollet, 1700–1770). A charismatic figure due to his public demonstrations of electricity, Nollet worked with the electrical pioneer Charles Dufay (1698–1739) in the 1730s. In fact, Nollet helped Dufay prove that the latter was an excellent conductor when he (dangerously) electrified Dufay's body with a Hauksbee-type **electric machine**.

In the 1740s, Nollet began a series of examinations of Hauksbee's discoveries. He first improved Hauksbee's electric machine to run more efficiently, more smoothly, and with fewer breakdowns due to equipment failure. Though his experiments ran the electrical gamut, Nollet's work on electrical light took Hauksbee's work a step further "towards the ultimate practical invention that was to follow in the next century" (Rowland 1974, 113).

Nollet's most important discovery was that he could make Hauksbee's electric machine power a series of flasks connected to it by a conductor. Nollet evacuated the air in each flask through a vacuum as Hauksbee had done. But his improvement was that inside each flask, Nollet placed a metal conductor. He linked each conductor to a metal chain connected to the electric machine's globe. As he expected, when he cranked the machine, its rotating sphere glowed brilliantly, as it had for Hauksbee. However, for the first time, so did the electrical conductors inside the glass flasks spread around the room. Nollet had successfully used a conductor to carry electricity to a series of dispersed sites.

Here was not only production, but conveyance of electricity. Nollet had therefore also successfully applied the principle of **conduction**, the actual flow of electricity, to the technology of electric lighting. Nollet's dispersed electric machine was so efficient that it rendered all objects in the darkened room clearly visible, and it did so many years before such technology was widely understood.

Many scientists built on the work of Hauksbee and Nollet during the nineteenth century. In 1809, Sir Humphry Davy (1778–1829) became the first person to exhibit an **electric arc**, in which an electric current leaps between two electrodes. This dazzling invention relied heavily on the work done with electric machines during the previous century. Today, modified electric arcs are used in particle accelerators.

The electric arc was only one of several early kinds of electric light. Another created a glow from partially exhausted glass globes that were excited by electrical discharge. The most famous electric light is the kind employing a glowing filament. In this kind of light, electrical current encounters resistance in crossing the filament. This heats the filament, causing it to glow and give off light. Of course, this is the kind of light produced in incandescent bulbs.

Today, incandescent bulbs are used in homes, businesses and many other buildings. Car headlights and flashlights employ incandescent bulbs. These types of bulbs have three main parts: the filament (made from tungsten), the bulb (made of glass) and the base (which holds the filament). Today's bulbs also contain gas (usually a mixture of **nitrogen** and **argon**), which lengthens the filament's life.

The other large variety of electric light is gaseous discharge lamps, which produce light by passing current through a gas rather than a filament. Fluorescent, neon, mercury vapor, and low-pressure sodium lamps all fall into this category.

Another kind of electric light, the light-emitting diode (or LED), was discovered in the latter half of the twentieth century. LEDs use a semiconductor material to give off colored light. These lamps use very little power and can last for extremely long periods of time. LEDs are most widely used in calculators, watches, clocks and other digital displays, where (in aggregate) they form numbers or words.

The production and manufacture of electric lights is one of the world's largest and most important industries. Beginning in the nineteenth century, this industry also helped make the United States an economic, industrial and technological power.

SEE ALSO: Chemical Cells; Electrical Conductors and Insulators; Electricity Is of Two Types; Franklin's Electrical Researches.

Selected Bibliography

Bordeau, Sanford P. *Volts to Hertz: The Rise of Electricity*. Minneapolis: Burgess, 1982.
Canby, Edward Tatnall. *A History of Electricity*. New York: Hawthorn Books, 1968.
Meyer, Herbert W. *A History of Electricity and Magnetism*. Cambridge, MA: MIT Press, 1971.
Rowland, K. T. *Eighteenth Century Inventions*. New York: Harper & Row, 1974.
World Book (encyclopedia) CD-ROM. San Diego: World Book, 1999.

Electricity Is of Two Types: Positive and Negative
Charles Dufay, 1733
Benjamin Franklin, 1747

Many students are confused when it comes to positive and negative electricity. Often, they believe that the "positive" determines the flow of the current, and they reason that the "negative" is made to follow. However, the opposite is true: Negative electrons mandate the direction of current flow. Most *positively* charged objects actually have a deficiency of electrons, while most *negatively* charged objects actually have an excess of electrons. Therefore, the negative electrons determine the direction of the current's flow because the direction is toward the positive, which lacks electrons. This modern understanding of electricity is contrary to the belief of many of today's students, as it was to the eighteenth-century scientists that founded modern electric theory. Like many of today's students, these scientists were confused by the elusive terminology of positive and negative electricity.

One of the most important scientists to study electricity in the eighteenth century was Stephen Gray (1695–1736), who discovered **conduction**, the actual flow of electricity. In doing so, he proved that electricity could flow some 800 feet through ordinary twine suspended in the air by silk threads. (These were much higher voltage charges than today's lower voltage household currents. Also, when Gray later wet the twine, he successfully sent his charge a distance of about 1,200 feet.) Observing this, Gray became the first person to suggest that electricity must be some kind of "fluid" capable of "flowing" over great distances.

However, when Gray substituted metal wires in place of the silk threads (to support the twine), he found that the charge quickly dissipated. Gray eventually arrived at the conclusion that materials either conduct or insulate electricity. The "non-electrics," as he called them (like metals and water), conducted charges, whereas the "electrics" (such as silk, glass and resin) tended to hold charges. This distinction, made in 1729, was the first clear articulation of insulators versus conductors. Gray compiled an important list of conductors and insulators.

Gray also settled on an elegant and effective principle: Electricity does not fill up a body, but is instead held entirely on its outermost surface. Today, this principle is in play when a metal airplane is struck by lightning and the occupants are still protected.

Gray's work—especially this latter principle—caught the attention of Charles Dufay (1698–1739), a retired military officer and member of the French Academy of Sciences. Dufay quickly became interested in expanding Gray's work, most especially in the area of different types of conductors. One night before the French court, Dufay actually had himself suspended by insulating silk chords (like Gray's twine) and connected to an early type of **electric machine**. When anyone touched him, Dufay emitted long, powerful sparks. This was a very dangerous experiment, though Dufay proved once and for all that he was an effective electric (or insulator).

Dufay's more important work was of a less flamboyant nature. He investigated the claim made by Jean Desaguliers (1683–1744) that all objects could be classified as either "electrics" or "non-electrics" when he demonstrated that all bodies (except metals and those materials too soft or too fluid) could be electrified. In fact, Dufay took pleasure in electrifying many of them to great public awe and admiration, for electricity was quickly becoming all the rage in France. Dufay also conclusively demonstrated that a charged body attracts another body—which (in turn, at the moment of contact) receives a similar charge and is then repelled. He demonstrated this by often electrifying the actual bodies of volunteers.

However, Dufay's most important discovery was to come only once he noticed that all objects (regardless of their nature or makeup) charged from the same glass tube repelled one another, but attracted those charged from an electrified resin rod. Extending his inquiry, Dufay found that rubbed glass would repel a piece of metal leaf, while rubbed amber, wax or gum tended to attract it. Why should this be the case? After all, previous scientists had focused attention toward the material receiving the charge rather than the material doing the charging. Dufay concluded that "there are two kinds of electricity, very different from one another; one of which I call vitreous, the other resinous electricity. The first is that of glass . . . the second is that of amber . . . they repel themselves and attract each other" (Dufay, quoted in Crowther 1937, 65).

Here, for the first time, were two kinds of electrical fluid, each named for its source: vitreous (from glass) and resinous (from resin or **amber**). Dufay quickly fleshed out his two-fluid theory. He claimed that, in two given bodies, two electric fluids exist in equal amounts. The change took place on charging when, according to Dufay, some of one fluid is removed and an excess of the other is thereby left behind.

It all fit together for Dufay, as this two-fluid theory explained everything he had discovered in his experiments. The two fluid electricities act together harmoniously, he explained, like the poles of magnets (a metaphor that would later lead to many incorrect theories). Perhaps, said Dufay and his growing legion of supporters, this was a universal law, perhaps *the* universal law that explained the relationship between electricity and magnetism.

Dufay was first to suggest a theory that explained the different qualities of electricity that scientists were noting in their experiments. However, throughout his researches, Dufay never used the terms *positive* and *negative*. The first recorded use of these electrical terms came in a letter written by American statesman and inventor Benjamin Franklin (1706–1790).

Early on, Franklin became skeptical of Dufay's two-fluid theory, which he believed was overly complex and probably incorrect. Franklin favored a one-fluid explanation after finding that when he rubbed a piece of glass with a cloth, the glass received an electrical charge exactly the same strength as that of the cloth. The charge was of an "opposite" kind (the glass now attracted objects the cloth had repelled, and vice versa). If two fluids were in play, why would one object receive an "opposite" charge of electricity, in *exactly* the strength lost by the other?

In a large public demonstration, Franklin showed that two volunteers who had absorbed opposite charges could produce a powerful spark between them. Immediately thereafter, Franklin demonstrated that both people were totally discharged of electricity, as their charges had neutralized one another. Though he still considered electricity a fluid, Franklin now theorized that only one electricity existed.

Like Dufay, Franklin believed that *all* matter contained electricity in differing degrees. He claimed, therefore, that objects could be measured on a scale from "positive" (with an excess of electricity) to "negative" (with a deficiency of electricity). Objects with excess electricity repelled each other, but attracted (by their excess) objects that were deficient. He said these latter objects wanted to "share" (or "borrow") some of the electricity. This, claimed Franklin, is how opposite charges cancel one another out. For Franklin, this fluid acted almost mathematically: A positive charge attracts a negative one, like charges repel one another, and an uncharged body is essentially neutral (with neither too much nor too little electricity). To Franklin and other scientists, it made neat, enlightenment sense.

By today's standards, both theories were incorrect in the sense that they assume that "positive" is the direction of the electric current's flow. Still, both theories were on the right track in several important ways, which culminated in Franklin's realization that there was but one "fluid" with many properties. Many authors have claimed that Dufay's two-fluid theory plagued thinkers for years to come. Yet others point out that the irreplaceable work of Franklin came about as a direct challenge to this flawed theory.

In either event, one must stress that Franklin's interpretation of a single fluid with "positive" and "negative" charges is not the same as the modern notion of positive and negative. On some level, the modern confusion in learning these theories is due to an historical equivocation—or evolution—of terminology.

SEE ALSO Chemical Cells; Electrical Conductors and Insulators; Franklin's Electrical Researches.

Selected Bibliography

Bordeau, Sanford P. *Volts to Hertz: The Rise of Electricity*. Minneapolis: Burgess, 1982.

Bowen, Catherine Drinker. *The Most Dangerous Man in America: Scenes from the Life of Benjamin Franklin*. Boston: Little, Brown, 1974.

Canby, Edward Tatnall. *A History of Electricity*. New York: Hawthorn Books, 1968.

Crowther, J. G. *Famous American Men of Science.* New York: W. W. Norton, 1937.
Heilbron, J. L. *Electricity in the 17th and 18th Centuries: A Study of Early Modern Physics.* Los Angeles: University of California Press, 1979.
Meyer, Herbert W. *A History of Electricity and Magnetism.* Cambridge, MA: MIT Press, 1971.

F

Farmer's Almanac
Benjamin Banneker, 1792–1796

An **almanac** is a book that lists a great deal of varying information. Generally, an almanac may include a calendar with important dates and events, as well as facts about government and history. It might also include important information about weather, movements of planets, and figures on population, agriculture, and industry. Previously, almanacs were published for farmers and ship's navigators who needed to know about events such as lunar and solar eclipses, the rising and setting of the sun, and astrological predictions of the stars and planets. Almanacs also listed general information about the weather, which sometimes included broad predictions. As almanacs became more sophisticated during the eighteenth century, the almanac maker required superior skills in scientific and mathematic calculations, because more sophisticated readers came to expect much more accurate predictions and forecasts—with science (rather than folklore) to back them up. Because almanacs often included quips, phrases and proverbs, the almanac maker also benefited from a healthy sense of humor.

Historians believe ancient astrologers in the Middle East probably published almanacs of heavenly predictions. The ancient Romans published long almanacs on stone and marble slabs. These almanacs listed and explained the rites of Caesar's festival days—an early use of almanacs as calendars. In 1634, historians pieced together and published an almanac originally written in the year 354. Well-documented sources refer to manuscript almanacs in 1150, 1380 and 1386. English scientist Roger Bacon (ca. 1214–1292) probably published an almanac around the year of his death. In England, the oldest existent calendar almanac dates to 1431 and is today housed in the British Museum. Also, the long history of mechanically printed almanacs dates back to the so-called Astronomical Calendar of 1448. This almanac was printed by the early pioneer of moveable type, Johannes Gutenberg (ca. 1395–1468).

Almanacs played an important, even honored role in colonial America. By the mid-eighteenth century, many small towns boasted educated, literate citizens. Yet, these towns had no newspapers. No American city had a magazine before 1740, and even the city of Baltimore did not have a newspaper until 1773. This created a situation for American almanacs that was markedly different from those in England, where many other texts were widely available. In many of these colonial places, an almanac was the only secular text printed. It was also the only source of community information. Some American almanacs became so popular that they won widespread recognition and readership. The most well-known example is *Poor Richard's Almanack*, first published by Benjamin Franklin (1706–1790) in 1733. This almanac was especially loved for its widely quoted proverbs.

Historians often claim that the most outstanding almanac published during the eighteenth century was the *Farmer's Almanac*, published by Benjamin Banneker (1731–1806) between 1792 and 1796. Banneker was a nationally respected African-American astronomer, mathematician, scientist and land surveyor during a time that generally afforded few opportunities for black Americans. After publishing the almanac, Banneker became the most widely known and respected black American in early U.S. history. The *Farmer's Almanac* was distinguished from other almanacs because Banneker based his predictions on newly developed enlightenment scientific principles and his own impeccable mathematic calculations.

Banneker was born a free man (the son of a free man) near Baltimore. His grandmother, an Englishwoman, taught him to read and write. For a few years, he also attended a school open to both blacks and whites. Through this brief formal education, Banneker got his first taste of math and science. While successfully farming his own land, Banneker would constantly develop his exceptional math and science skills for the rest of his life. He became one of the nation's foremost astronomers, though he was entirely self-taught in this field. Banneker also built a clock by hand that was one of the first clocks in the United States. He made it entirely of wood, even though he had only a picture of a clock and a small, inexact pocket watch for reference. Banneker's wooden clock kept accurate time for close to half a century. Because he owned land and was a free man, Banneker was one of only a few African-Americans able to participate in Maryland's elections before 1802. (After this year, new laws restricted that right to white men only.) As a surveyor, Banneker also helped lay out boundaries for the District of Columbia. One of a handful of his era's remarkable commoners, some historians have dubbed Banneker the premier example of a "village Newton."

For his almanac, Banneker performed many scientific calculations, most especially regarding the positions of heavenly bodies and terrestrial weather. Banneker worked from an *ephemeris*, which is a table showing the daily positions of heavenly bodies. He calculated the positions of the sun, moon and planets each year, using preliminary data from the *Nautical Almanac* and reckonings of the calendar. From these data, Banneker calculated the solar and lunar eclipses, rising and setting times of the sun, daily weather forecasting, and tables of high and low tides.

Banneker also used logarithms to figure the longitude and certain anomaly of the sun and moon for specific days each month. These data gave the distance of heavenly bodies in relation to their specific *apogees,* or furthest points away from earth. Banneker figured these distances by recording the sun's or moon's position in relation to a dozen or more constellations. By comparison against a set of known coordinates, Banneker could figure the precise distances between the earth and the sun and the moon.

Banneker's ephemeris also included information about both solar and lunar eclipses. An enormous, even unprecedented amount of work went into calculating each eclipse. Banneker published each calculation to the nearest second in his almanac. The almanac also included meticulous drawings of each stage of the eclipse. Finally, after each of the year's eclipses had been calculated, predicted, and illustrated, Banneker gave a detailed statement about the year's aggregate eclipses.

The *Farmer's Almanac* was the first almanac to openly compare its own scientific methods. Before making his own calculations, Banneker (as was common practice) mastered the scientific and mathematic method of several authoritative sources. Banneker was very much aware of his almanac's limitations, as he compared his chosen methods with other contemporary ones available to him. In several cases, Banneker pioneered improved methods for calculating ephemerides. He detailed the improved methods in subsequent editions of his yearly almanac. Even though it left the *Farmer's Almanac* open to lazy criticism, Banneker freely showed his almanac's imperfections by detailing how he improved on earlier methods of calculation.

In general, this method raised the bar immeasurably for almanac makers, as readers began to expect almanacs to be much more scientifically accurate. One of Banneker's most significant improvements came in 1795, when he changed his method of presenting information about eclipses. Previously, he had stated the degree to which solar or lunar eclipses would be visible from Baltimore, or even from the United States. Now, for the first time, the *Farmer's Almanac* provided the precise latitude and longitude of the ideal location for observing the eclipse. Banneker showed not only that an eclipse would occur on a specific day, but also the location of the earth—and his readers—in relation to the sun or moon.

Banneker's *Farmer's Almanac* was not only hugely influential in development and distribution of scientific knowledge. It also carried enormous social impact. In the 1793 edition, American doctor and statesman Benjamin Rush (1745–1813) forcefully made the case to appoint a U.S. Secretary of Peace. But even before winning acclaim and powerful support for his almanac, Banneker sent a copy of his very first edition to Thomas Jefferson (1743–1826). With it, he sent a bold letter outlining the case for abolishing slavery. Jefferson was so impressed, he sent a copy of the almanac to the French Royal Academy of Sciences as evidence of the equal talent of African-Americans. Abolitionists in England and later in the United States pointed to the *Farmer's Almanac* as evidence of the equal intellectual and scientific abilities of people of African descent.

In the nineteenth century, most almanacs stopped the practice of predicting the weather. However, during this same time, many governments, trade groups, newspapers and professional organizations began issuing specific kinds of almanacs. Today, almanacs such as *The World Almanac* contain reference information of general interest and on a wide variety of topics. Others, such as *The Astronomical Almanac* published by the U.S. Naval Observatory, contain information for specific groups of people. This almanac contains information about the stars, tides, navigation, weather and eclipses. *The Astronomical Almanac* is therefore found on most American seagoing vessels, where it is considered essential for navigation.

SEE ALSO Modern encyclopedia.

Selected Bibliography

Allen, Will W. *Banneker, The Afro-American Astronomer*. Salem, NH: Ayer, 1971.

Bedini, Silvio A. *The Life of Benjamin Banneker*. New York: Charles Scribner's Sons, 1972.

Graham, Shirley. *Your Most Humble Servant*. New York: Julian Messner, 1952.

Sagendorph, Robb. *America and her Almanacs: Wit, Wisdom & Weather, 1639–1970*. Boston: Little, Brown, 1970.

Stowell, Marion Barber. *Early American Almanacs: The Colonial Weekday Bible*. New York: Burt Franklin, 1977.

Franklin's Electrical Researches, Discoveries and Inventions
Benjamin Franklin, 1746–1752

A question sometimes put to beginning students of electrical inquiry is "Could you live without electricity?" as a way of encouraging reflection on the modern electrical age. Excluding the electrical processes of the body (especially the nervous system and brain), a perceptive student might answer, "Yes, I could technically live without it." On further reflection, he or she would probably add the caveat, "But I wouldn't want to." Virtually every object of any practical use is today made with electricity or is dependent on it at one point or another. In addition to the most identifiable examples, such as computers, telephones and electric lights, the most basic necessities of life are dependent on electricity—from the manufacture of clothing to the purification of water to the preservation of food. As much as the contemporary moment is an age of information, it is also an age of electricity. And as much as the eighteenth-century industrial revolution gave the world steam power, it also gave the world its first glimpse of electrical applications.

Since antiquity, scientists such as Pliny the Elder (23–79 B.C.) have known that if they rubbed **amber** with various cloths, the amber attracted materials such as feathers, hair and dry straw. The early Chinese naturalist Koupho (295–324) compared this ability of amber with the power of a magnet—a metaphor that became important for later scientific research during the European Renaissance. In 1551, the Italian mathematician Girolamo Cardano (1501–1576) first drew the distinction between electricity and magnetism. For the first time, knowledge of electricity as a separate phenomenon was beginning to emerge. Around 1600, the English physician William Gilbert (1544–1603) found that other objects (such as glass, wax and sulfur) acted like amber when they were rubbed with cloth. He called these materials "electrics" (or insulators). Gilbert was also the first to propose that the curious action of these objects must be due to some kind of fluid.

This idea that electricity is a type of fluid led to the first comprehensive theory of electricity, proposed by the retired French military officer-turned-scientist, Charles Dufay (1698–1739). Dufay found that when he charged small pieces of glass, they attracted some amber-like substances and repelled others. He concluded that there are therefore two different kinds of electricities, positive (or *vitreous*) and negative (or *resinous*). This conclusion became known as the "two-fluid" theory of electricity.

This emerging understanding was poised for a breakthrough, which came when Benjamin Franklin (1706–1790)—statesman, philosopher, publicist, printer, inventor and scientist—took up the study of electricity. Franklin began his study after witnessing a demonstration of static electricity in Boston in 1743. He began specifically experimenting with a **Leyden jar**, an early type of chemical battery cell. He found that when a piece of glass was rubbed with a cloth, the glass received an electrical charge of *exactly* the same strength as that which was on the cloth, but of an *opposite* kind. Where the glass had either attracted or repelled an object, the cloth now performed the opposite action. Dufay's theory explained this as the diametrically opposite reaction of a pair of fluids, but Franklin was skeptical of this reasoning. If this were the case, then why would one object receive an "excess" of electricity, while one would receive a "deficiency" of *exactly* the same strength?

In 1748, Franklin also experimented with an electrostatic jack, an early kind of electric motor. He attached a metal knob to each of two opposite sides of an insulating wheel. Franklin created a positive charge on a small stationary piece of metal in front of one knob and a negative charge on a piece in front of the other. This effectively electrified the wheel, causing it to rotate. This rotation occurred because the knobs were attracted to the stationary pieces of electrified metal. However, when the wheel's rotation brought the knobs to the pieces of metal, a spark passed and made the charge of the same sign on both. This resulted in the knobs repelling one another and caused the wheel to rotate, once again.

At this point, Franklin was already formulating his theory of electricity. He also was ready for a public experiment with human volunteers. He had two people stand on separate, insulated platforms. One volunteer rubbed a glass tube with a cloth, and carefully absorbed only the charge from the cloth. The other carefully absorbed only the charge from the glass. When the two people brought their fingers slowly together, a strong spark passed between them. Further contact showed that each volunteer was completely discharged of his or her electricity, as the charges had neutralized one another. Because the exact same action had charged and discharged both volunteers, Franklin became convinced that there was only one kind of electrical "fluid."

Franklin published his comprehensive theory of electricity in a series of letters. Some were read before the English Royal Society, others were published in scientific journals and still others were published in popular magazines. Throughout his life, Franklin shared his scientific information freely. In April 1751, Franklin published a collection of his letters in a pamphlet entitled *Experiments and Observations on Electricity*. More than eighty pages long, this pamphlet made Franklin internationally famous.

Franklin's electrical theory stated that all matter contained electricity to varying degrees and could therefore be measured on an electrical scale from "positive" (with an excess of electricity) to "negative" (with a deficiency of electricity). Objects with too much electricity would repel each other, but they would attract objects with too little electricity. In Franklin's theory, electricity was still a substance that "flowed" like a fluid. When an object with an "excess" of fluid touched an object with a "deficiency," the fluid electricity would be "shared." In one fell swoop, Franklin had explained how opposite charges cancel out one another. He had also explained every phenomenon that Dufay had observed: A positive charge attracts a negative one, like charges repel one another, and an uncharged body is essentially neutral (with neither "excess" nor "deficiency"). Like all useful theories, Franklin's explained the wide variety of scientifically documented phenomena, as well as everyday events. An added appeal was that Franklin's single-fluid theory was much neater than Dufay's messy theory of two distinct fluids.

With his theory clear in his mind, Franklin turned his attention toward the sky and to the question of "natural" electricity, or lightning. As much as any other single scientist, Franklin ushered in a new era of practical application of scientific discoveries. As an enlightenment scientist, Franklin believed that scientific knowledge should benefit humanity. At the time, this was still an emerging idea.

During the Renaissance, scientists had considered many wild theories to account for lightning. Beginning in the mid-eighteenth century, however, scientists, including Franklin, proposed that lightning was a fantastically powerful form of electricity. Franklin's laboratory experiments suggested that lightning would strike pointed objects before flat ones, especially if the objects were at a higher altitude. Thus far, he had never actually tested this hypothesis outside of his laboratory. In 1752, Franklin performed his famous kite experiment. During a storm, Franklin charged his Leyden jar from a cloud's static electricity. He deduced that the fully charged jar gave proof that lightning is a form of electricity. Several times, Franklin felt a light shock during this experiment. He was lucky that electricity did not strike his kite. Several scientists in France and Russia had already been killed while undertaking a similar experiment.

For several years, Franklin had been circulating the idea of lightning rods in his *Poor Richard's Almanack*. In the article, "How to secure Houses, Etc. from Lightning," Franklin outlined his plans for the pointed **lightning rod**, which would provide a grounded path away from a building and into the ground. Franklin built several rods right away. Franklin lightning rods have saved countless houses, large buildings, and other structures from destruction during electrical storms. They have also made buildings much safer to occupy. Today, many basic lightning protection codes recommend that lightning rods be built to virtually the exact specifications first drawn up by Franklin 250 years ago.

In these researches, Franklin laid the groundwork for the modern study of atmospheric electricity. He is therefore credited with founding this field of inquiry. In the years before Franklin, electricity was a faltering, infantile science with no practical applications. Franklin generated widespread interest, laid the foundations for the scientific study of electricity and designed electricity's first relevant uses.

Franklin was correct that only one kind of electricity exists. Today, however, the "fluid" metaphor of electricity is, strictly, incorrect. Moreover, Franklin had his theory of "positive" and "negative" electricity *exactly backward*. This fact causes much confusion for many students.

Modern electrical theory employs the concepts of electrons and protons. Franklin said that a positively charged object has an "excess" of electricity and that a negative charged object has a "deficiency." In reality, *the negative electrons determine the flow of an electric current.* Therefore, positively charged objects actually have a *deficiency of electrons.* Negatively charged objects actually have an *excess of electrons.* This is the opposite of Franklin's theory and of what many students initially believe. Because negative electrons determine the flow of an electric current, the current flows toward the positive, because the positive has the deficiency of electrons.

The modern theory of electricity is based on knowledge of electrons and protons that was totally unavailable during Franklin's day. Though flawed, Franklin's theory made the modern theory possible.

SEE ALSO Chemical Cells; Electrical Conductors and Insulators; Electricity Is of Two Types.

Selected Bibliography

Bordeau, Sanford P. *Volts to Hertz: The Rise of Electricity.* Minneapolis: Burgess, 1982.

Bowen, Catherine Drinker. *The Most Dangerous Man in America: Scenes from the Life of Benjamin Franklin.* Boston: Little, Brown, 1974.

Canby, Edward Tatnall. *A History of Electricity.* New York: Hawthorn Books, 1968.

Crowther, J. G. *Famous American Men of Science.* New York: W. W. Norton, 1937.

Heilbron, J. L. *Electricity in the 17th and 18th Centuries: A Study of Early Modern Physics.* Los Angeles: University of California Press, 1979.

Ketcham, Ralph L. *Benjamin Franklin.* New York: Washington Square Press, 1966.

Krebs, Robert E. *Scientific Laws, Principles, and Theories: A Reference Guide.* Westport, CT: Greenwood Press, 2001.

Krider, E. P. "The Heritage of B. Franklin," in *Benjamin Franklin: Des Lumières a Nos Jours.* Edited by G. Hugues and D. Royot. Université Jean-Moulin-LYON III: Didier Erudition, 1991.

Meyer, Herbert W. *A History of Electricity and Magnetism.* Cambridge, MA: MIT Press, 1971.

Rowland, K. T. *Eighteenth Century Inventions.* New York: David & Charles Books, 1974.

Franklin's Practical Household Inventions
Franklin Stove, 1740
Bifocal Spectacles, 1784

Historians frequently consider Benjamin Franklin an example of a "Renaissance Man" on the level of Leonardo Da Vinci (1452–1519). In support of this, they point to Franklin's important contributions in vastly different fields of study, and even occupations. Indeed, he excelled as a statesman, scientist, inventor, editor, publisher, civic leader, revolutionary, philosopher and printer—to name but a few. But historians also consider Franklin a modern scientist concerned with improving the lives of ordinary citizens—a new possibility in his day. Many even consider him more like James Watt (1738–1819) and Thomas Edison (1847–1931) than former scientists such as Isaac Newton (1642–1727). They claim that scientists of previous centuries occupied a position similar to that of natural philosopher, in which the application of knowledge mostly never entered into the equation. With Franklin, however, experimental science is inextricably linked to practical application. His approach was part of a larger enlightenment belief in the scientist as public figure whose work benefits the whole of humanity.

On seeing the first successful balloon flight in 1783, a bystander (somehow failing to see the invention's importance) scoffed, "What good is it?" Franklin replied, "What good is a newborn baby?" As a scientist and inventor, Franklin quickly saw the value of other inventors' creations. In fact, he refused patent and profit from his inventions, instead choosing to offer them freely and enthusiastically for the betterment of all.

Even in light of Franklin's dual placement as a modern scientist–inventor and Renaissance-style intellectual, the breadth of his inventions might seem staggering. Franklin's electrical inquiries (including the famous kite experiment) resulted in both important scientific knowledge and the invention of the **lighting rod**, which saved many buildings from destruction. Franklin realized the importance of ventilation for disease control. He also discovered the importance of lime for controlling acidity in American soil. Franklin published *Poor Richard's Almanack*, which was full of practical wisdom. He even invented a musical instrument, the glass harmonica. Along with the lightning rod, two of Franklin's inventions rank among the most important of the century. They are the **Franklin stove** and **bifocal spectacles**. (For more information, please see the entry on Franklin's Electrical Researches and also the entry on Electricity Is of Two Types.)

Like many inventions, the Franklin stove (also known as the Pennsylvania fireplace) grew out of great need—readily provided by severe New England winters (see Figure 11). Before invention of the Franklin stove, people mostly used wood-burning fireplaces to heat their homes. Even when they had a cast iron insert or stove in the fireplace, a majority of heat was lost through the chimney. Franklin noted this firsthand when he noticed that the side of his body that faced the stove was the only part that was warm. He studied at length why the stove heated the air directly in front of it, but not the entire room.

From his earlier experiments with **electrical conduction**, Franklin knew iron was a good "electric" and would conduct heat. Therefore, he soon realized that a majority of heat was radiating out of the back and sides of the stove. The first, most obvious improvement Franklin made was to move the sides and front of the stove further out into the room. When heat was lost to the metal, it would no longer travel up the chimney, but into the room.

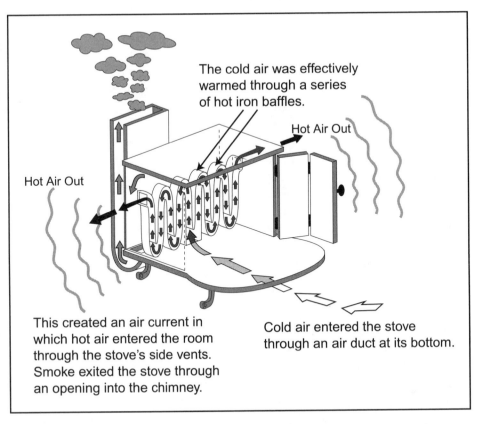

The cold air was effectively warmed through a series of hot iron baffles.

Hot Air Out

Hot Air Out

This created an air current in which hot air entered the room through the stove's side vents. Smoke exited the stove through an opening into the chimney.

Cold air entered the stove through an air duct at its bottom.

Figure 11. Thermodynamics of the Franklin stove. Franklin's was the first type of stove to achieve thermal efficiency through the use of heat exchange.

From his experiments with the radiation of heat, Franklin also knew that warm and cold air exchanged places in the fireplace or stove. His own stove had tiny openings on its side, which prevented the warmed air from entering the room. To combat this problem, Franklin first designed an air box or duct at the bottom of the stove. This box drew cold air out of the room and into the stove, where it was heated by the flue gases. Heating took place in a series of red-hot baffles that Franklin designed on the inside of the stove. By greatly enlarging the stove's side openings to full-blown vents, Franklin forced the warm air created in the baffles to discharge back into the room. This effectively created a current, thereby achieving thermal efficiency. Cold air was drawn in through a duct and was warmed over the baffles. The warmed air was then discharged back into the room. As before, smoke and carbon were vented out the stove's top and into the chimney.

Franklin showed that previous stoves or furnaces were inefficient because they ignored the most basic scientific principles of heat and heating. With his stove, he applied these same principles so successfully that he could easily keep his room twice as warm by burning only a quarter of the fuel.

Even though the governor of Pennsylvania offered him a patent that would have made him very wealthy, Franklin gave away the plans for his stove in a pamphlet he printed in 1744. This pamphlet also explained—in simple terms available to everyone—the scientific principles of effective heating. Though at least several London iron mongers made vast fortunes from his design, Franklin said that "we should be glad of an Opportunity to serve others by any Invention of ours, and this we should do freely and generously" [punctuation in original] (quoted in Ketcham 1966, 79). The Franklin stove was used widely in the United States and England during the next 150 years.

In addition to the lightning rod, another of Franklin's inventions still enjoys widespread use: bifocal spectacles. Though Franklin definitely developed bifocals in 1784, he (and others) probably experimented with them as early as the 1760s in London. Certainly distance and reading glasses existed in separate form prior to this period, though no one knows when or even where they originated. Arab inventors introduced "reading stones," transparent pieces of highly polished beryl or quartz half-spheres, sometime around the year 1000. The Arab scientist Abu-'Ali Al-Hasan Ibn Al-Haytham (ca. 965–1039) made extensive studies of refraction and wrote a comprehensive treatise on lenses. However, the first written record of glasses in the West was not made until the Franciscan friar Roger Bacon (1220–1292) recommended reading lenses for the elderly in 1267. After this date, many written records mention reading glasses. European paintings also show various scholars and religious figures wearing glasses, mostly for reading. The invention of the microscope provides another big clue to the history of glasses. The Dutch-man Hans Jansen (ca. 1580–1638) is generally credited with the invention of the first practical microscope, probably around the year 1600. Books of the period usually list his profession as spectacle maker, providing evidence of their widespread use.

One of the biggest engineering obstacles to the widespread wearing of glasses was the trick of keeping them on the wearer's face, as ear pieces are a modern innovation. Wearers of glasses in prior times often held their glasses up to their eyes when they wanted to use them. Nonetheless, by Franklin's day, polished glass had long been the norm and so had metal wire rims.

As with the Franklin stove, Franklin's own experience led him to invent bifocal spectacles. Franklin himself used both spectacles for reading and spectacles for distance. Reflecting on this, he later wrote that he would become frustrated, especially when traveling. Franklin was a voracious reader, but he also liked to look up from his book or journal to take in his surroundings. This simple change of focus meant he would have to remove his reading spectacles and don his distance spectacles. Then he had to replace his distance spectacles once he was again ready to read.

This problem sparked an idea. Franklin had two pair of lenses ground to the same diameter—one for reading, one for distance. Then he had them cut in half and placed in one pair of spectacles to be held together by the frame. Franklin found that "I have only to move my eyes up or down, as I want to see distinctly far or near, the proper glasses always being ready" (quoted in Crowther 1937, 130). Franklin also noted an unexpected benefit upon traveling to France. At the dinner table, he used the reading half of his glasses for seeing the food on his plate and the distance half of his glasses for seeing his important dinner companions. Seeing their expressions and their lips move helped Franklin better understand what they were saying in his non-native tongue.

In modern terms, the "distance" lens on top corrects nearsightedness, or *myopia*. In this condition, rays of light from a distant object meet in front of the retina (the part of the eye that absorbs light rays and changes them into electrical signals), rather than directly on it. This produces a blurry or out-of-focus image. The lens to correct myopia is concave, or thin in the middle and fatter at the edges. This lens bends rays of light toward the edges or outward, so they may meet directly on the retina.

In many ways, farsightedness, or *hyperopia*, can be thought of as the opposite of nearsightedness. Here, rays of light from nearby objects focus beyond the retina, rather than directly on it. Faraway objects may appear clear, but nearby ones appear blurry. To correct this condition, convex lenses are used. These lenses are fatter in the middle and thinner at the edges, forcing rays of light to meet on the retina.

When a person (like Franklin) has both conditions, he or she can wear bifocal lenses. In fact, many people today wear *tri*focal lenses, which combine a third lens (or modern focal point) in the middle. This lens is for viewing objects at in-between distances.

SEE ALSO Electricity Is of Two Types; Franklin's Electrical Researches; Latent and Specific Heat.

Selected Bibliography

Bowen, Catherine Drinker. *The Most Dangerous Man in America: Scenes from the Life of Benjamin Franklin.* Boston: Little, Brown, 1974.

Crowther, J. G. *Famous American Men of Science.* New York: W. W. Norton, 1937.

Drewry, Richard D. "What Man Devised That He Might See." Web article about the history of eyeglasses. At the time of publication, available at http://www.eye.utmem.edu/history/glass.html.

Ketcham, Ralph L. *Benjamin Franklin*. New York: Washington Square Press, 1966.

Krider, E. P. "The Heritage of B. Franklin," in *Benjamin Franklin: Des Lumières a Nos Jours*. Edited by G. Hugues and D. Royot. Université Jean-Moulin-LYON III: Didier Erudition, 1991.

Rowland, K. T. *Eighteenth Century Inventions*. New York: David & Charles Books, 1974.

H

Heat Is a Form of Energy, Not the Element Caloric
Count Benjamin Thomson Rumford, 1798
Sir Humphry Davy, 1799

Before the time of Aristotle (384–322 B.C.), Greek philosophers had already postulated that four **elements** made up all matter in the universe. These elements were water, earth, air and fire. Building on this theory, Aristotle proposed that **transmutation** was possible. This theoretical process involved changing elements from one to another by adding or subtracting heat or moisture. For example, Aristotle thought that by boiling water under just the right circumstances, a person could end up with solid matter, or "earth."

The forerunners to modern chemists, alchemists spent much of the Middle Ages engaged in attempts at transmuting lead and other ordinary metals into precious ones, such as gold. Though never successful with transmutation, the alchemists did manage to discover a few elements, such as lead, copper and sulfur. However, the ancient four-element theory was mostly accepted with few changes.

In 1700, this chemical order fundamentally changed when German scientist Georg Ernest Stahl (1660–1734) proposed the powerful **phlogiston** theory of heat and **combustion**. This theory held that any object capable of burning contains the hypothetical substance "phlogiston," which combustion liberates into the surrounding air. Burning ceased only once an object had released all its phlogiston. Most European and especially English pneumatic chemists readily embraced this powerful theory, which provided an explanation for most of their experimental results, including calcification (or oxidation). The majority of the century's decisive chemical discoveries—such as **carbon dioxide**, **oxygen**, **chlorine**, and **hydrogen**—were made by scientists working within phlogiston's theoretical framework.

As fruitful as it was, however, phlogiston was not chemistry in the modern sense. It was a sophisticated final gasp of **alchemy**. As such, it was destined to be replaced by a better theory during the century's final quarter. This theory was that proposed by French chemist Antoine Lavoisier (1743–1794). Lavoisier is not important for discovering new chemicals or gases. Rather, he founded modern chemistry by reinterpreting the discoveries of the phlogiston-based pneumatic chemists (many of whom heartily resisted his theory).

In doing so, Lavoisier made unprecedented studies of the minute changes in weight that took place in chemical reactions. In many combustion and oxidation reactions, he found that reactants actually *gained* rather than lost weight. He postulated that this was because phlogiston was not *separating from* the reactant. He said that something else—namely, the newly discovered gas, oxygen—was *joining with it* during the reaction. Based on this finding, he proposed that the weight of the products must equal the weight of the original reactants. This discovery became known as the **law of conservation of mass**—one of the cornerstones of modern chemistry and a major part of phlogiston's destruction. (For more information, please see the separate entry on this discovery.)

Equipped with his new theory, Lavoisier went on to publish a crucial table of chemicals that distinguished between elements and **compounds** for the first time. Lavoisier's table listed many elements such as oxygen, silver, zinc and mercury. However, he also listed the so-called element **caloric**. When a substance was burned or oxidized, Lavoisier said it combined with oxygen and released a kind of imaginary heat fluid he called caloric. Caloric was supposed to flow from an object after the object had been heated or "bruised" by mechanical means. Most scientists who embraced Lavoisier's oxygen chemistry also embraced the concept of caloric—which had largely inherited the hypothetical mantle of phlogiston.

Lavoisier's contributions were cut short when he was beheaded in 1794 during the Reign of Terror in France. Ironically, his widow and collaborator, Madame Marie-Paulze Lavoisier, later married the man who would disprove the existence of heat as caloric fluid, the American-born (but Tory-loyal) scientist Count Benjamin Thomson Rumford (1753–1814). Like other scientists, Rumford knew of certain instances of heat that caloric could explain no better than could phlogiston. Caloric fluid supposedly flowed "down temperature" from hot to cold bodies only, just as water naturally flowed downhill. But what about, for example, when two objects (such as a person's hands) are rubbed together briskly, thereby generating a measurable amount of heat?

The calorists claimed that **latent heat**, as discovered by Joseph Black (1728–1799), was responsible for this frictional heat. As an object was "bruised" or "damaged," as by rubbing, they said that a kind of change of state ensued, in which particles were "ground off" and caloric fluid escaped. However, Rumford supposed caloric could not adequately explain the rubbing together of objects where no heat source was present. Where grindings were present (such as industrial cutting or drilling applications, which created tremendous heat), the grindings had exactly the same characteristics, with no apparent change in state. Rumford knew that other pre-caloric scientists such as Sir Isaac Newton (1642–1727) and Francis Bacon (1561–1626) had conceived of heat as the motion of tiny particles.

While superintending the boring of brass cannon at a Munich factory, Rumford was particularly struck by the phenomenon of frictional heat. In this factory, a team of two horses was used to turn a large drill through a cylindrical piece of brass. Almost immediately, Rumford noticed that the brass became extremely hot during boring—so hot, in fact, that the workers often put pots on it and boiled water for tea. He also found that the heat was produced as long as the horses were working. It seemed to Rumford that a virtually inextinguishable amount of caloric fluid would have to be released in the metal shavings that resulted.

Then Rumford noticed a very peculiar change of circumstances. When the drill's bit became worn and dull, it produced fewer chips as it bored less quickly and efficiently. However, the dull bit seemed to produce *more* rather than *less* heat—the exact opposite of what the calorists claimed (see Figure 12). Rumford correctly reasoned that a *sharp*, not a *dull*, bit should bore through the metal, thus releasing more caloric fluid. But time and again, the opposite happened—the harder the horses worked, the more heat they produced. Rumford began thinking that caloric didn't account for the release of heat at all. But if not caloric, then what?

Rumford was familiar with Lavoisier's experiments to measure heat with an instrument known as a calorimeter. He decided to design what amounted to an enormous calorimeter of his own—one large enough to hold the cannon boring apparatus inside a wooden box. And in a beautifully simple stroke of genius, he submerged the entire box—cannon and all—in cold water at a temperature of 60° **Fahrenheit** (F). Then he set the horses in motion to turn his cylinder at a rate of thirty-two times per minute. After a few minutes, he reported that the water had grown warm to the touch. After an hour of steady boring, he measured the water at 107°F. After an hour and a half, it was 142°F, and after 2 hours it increased to 178°F. At 2 hours and 20 minutes, the water was nearly 200°F. In his 1798 paper, "Heat Excited by Friction," Rumford reported that several bystanders were visibly shocked when the large quantity of water actually broke a boil—entirely from the frictional heat.

This experiment had proved that nothing from the air had joined with the metal and that the friction could produce what seemed like an inex-

Figure 12. Rumford's Discovery that Heat is a Form of Energy. While superintending the boring of brass cannon, Count Rumford observed a lot of heat being produced. Prevailing theory said that boring released heat fluid in the form of elemental caloric. Yet, Rumford recorded *more* rather than *less* heat being produced as the bit dulled. This led him to the formulation that heat is a form of energy—animal energy was being transferred to heat energy in the form of friction.

haustible amount of heat. As long as the horses worked, the water rose in temperature. In today's terms, this experiment had shown that mechanical energy and heat are different forms of the same "thing"—energy. As Rumford put it in 1798,

> It is in hardly necessary to add that anything which any insulated body, or system of bodies, can continue to furnish without limitation cannot possibly be a material substance: and it appears to me to be extremely difficult, if not quite impossible, to form any distinct idea of anything, capable of being excited and communicated, in the manner the heat was excited and communication in these, except it be MOTION.

In other words, Rumford discovered that it was the energy of motion, not the substance caloric, that had created the heat. Caloric fluid was no more an element than was phlogiston. Caloric had no gravitational mass. In the strictest Newtonian sense, therefore, caloric was not (and could never be) an actual substance. Heat is really just another form of energy—a form that should be considered no more inextinguishable than any other form of energy. Once the horses tired and stopped working, heat (or energy in the form of heat) would stop being produced in the friction of the cannon. After all, Rumford knew he couldn't really produce an endless amount of heat, which was why he measured it in the first place.

Moving a step further, Rumford also correctly postulated that the horses had to work harder to turn a dull bit than a sharp one, thereby creating more friction (and heat). In each situation, the amount of heat produced was directly proportional to the amount of work done. By tracking the temperature of the water, Rumford had measured the amount of work being done by the horses, much of which was transferred into heat energy through the friction of the drill bit.

This experiment laid the foundation for the first law of **thermodynamics** and the conservation of energy. The law of conservation of energy states that the amount of energy in the universe is always the same. It can never be increased nor decreased. Heat and work (or "motion") are equivalent and (for these purposes, at least) theoretically interchangeable. The amount of energy "put in" equals the amount of heat "put out" (and vice versa). In Rumford's words, the heat generated was "exactly proportional to the force with which the two surfaces are pressed together, and to the rapidity of the friction."

A year later, another ingeniously simple experiment lent further substantiation to Rumford's findings. His student, Sir Humphry Davy (1778–1829), went outside on a day when the English weather had dropped to precisely 29°F. Amidst a crowd of chilly onlookers, Davy proceeded to vigorously rub two pieces of ice together. After a few minutes of brisk rubbing, he produced liquid water, which he measured at 35°F. He had produced heat in the form of friction where there was no hot object capable of bleeding heat in the form of caloric into his ice. Several skeptics argued that caloric fluid must have somehow leaked from the frigid surrounding air.

In answer to this, Davy created a laboratory vacuum and repeated the ice experiment with the same results inside the vacuum. After this, he performed a fundamental experiment in which he mounted a wheel (driven by a clockwork) and piece of metal on a block of ice in his closed container. At 32°F, Davy proceeded to melt a small piece of wax by rubbing the wheel against the piece of metal for friction. After this demonstration, Davy announced that "heat . . . may be defined as a peculiar motion probably a vibration of the corpuscles of bodies . . . not dependent on a peculiar elastic fluid." He went on to proclaim bluntly that "caloric does not exist" (quoted in Hartley 1971, 95). It had gone the way of phlogiston. Or rather, like phlogiston, it had simply gone away.

SEE ALSO Latent and Specific Heat; Law of Conservation of Mass; Oxygen; Substances: Elements and Compounds.

Selected Bibliography

Darrow, Floyd L. *The New World of Physical Discovery*. Indianapolis: Bobbs-Merrill, 1930.

Greenberg, Arthur. *A Chemical History Tour: Picturing Chemistry from Alchemy to Modern Molecular Science*. New York: John Wiley & Sons, 2000.

Hartley, Sir Harold. *Studies in the History of Chemistry*. New York: Oxford University Press, 1971.

Krebs, Robert E. *Scientific Laws, Principles, and Theories: A Reference Guide*. Westport, CT: Greenwood Press, 2001.

Moore, F. J. *A History of Chemistry*. New York: McGraw-Hill, 1931.

Rumford, Count Benjamin Thomson. "Heat Excited by Friction." Published in *Philosophical Transactions* (vol. 88), 1798. Reprinted editions of this paper are widely available in books or on the web, at sites such as:
http://dbhs.wvusd.k12.ca.us/Chem-History/Rumford-1798.html and
http://dev.nsta.org/ssc/pdf/v4-0937s.pdf.

Hydrodynamics
Daniel Bernoulli, 1738
Jean Le Rond D'Alembert, 1744

Hydraulics is the branch of physics that deals with the behavior of liquids. Scientists generally break hydraulics into two categories. *Hydrostatics* describes the behavior of liquids at rest. **Hydrodynamics** describes their behavior in motion (flowing steadily or unsteadily). Ancient peoples made important basic discoveries in both areas. As a science, hydrostatics began with the Greek mathematician Archimedes (ca. 287–212 B.C.). He discovered that when an object is placed in fluid, it appears to lose an amount of weight equal to the weight of fluid it displaces. Nearly two millennia later, French scientist and mathematician Blaise Pascal (1623–1662) made another key discovery. He found that when a fluid is placed in a container, it transmits pressure equally in every direction. Together, the laws of Archimedes and Pascal formed the basis of modern hydrostatics.

Hydrodynamics had a similarly ancient beginning and later scientific transformation—this time during the eighteenth-century enlightenment. The use of flowing water for the practical benefit of humanity is certainly older than recorded history. In Northern Africa and the Middle East, archaeologists have unearthed sophisticated irrigation canals, very deep wells, and even ceramic conduits that served as pipes. But because the practice of hydrodynamics required extensive experience with water as well as rudimentary practices of science, it only began developing several thousand years later.

By the time of Aristotle (384–322 B.C.), the medium theory of motion had come into prominent favor. This theory explained motion as a "fluid" acting on a body. When an object fell, one of the four elements—fire, air, earth or water—"pushed" it from behind, therefore displacing other objects or elements. The reason for this was because nature abhorred a vacuum. Though erroneous, this was the first theory to explain fluid motion. It would last over a thousand years in Europe.

Aristotle was not a scientist, and he tended heavily toward reason rather than observation. Italian Renaissance genius Leonardo da Vinci (1452–1519) was one of the first scientists to emphasize direct study of nature. He also proposed the first correctly formulated theory of hydrodynamics. Da Vinci studied the flights of birds and the physics of motion. After many observations, he developed the continuity principle, which stated that the velocity of flow varies inversely with a stream's cross-sectional area.

In the early seventeenth century, Galileo Galilei (1564–1642) added scientific experimentation to da Vinci's observation. In his study of gravitational acceleration, Galileo found that a body sliding down an inclined plane attained a specific speed at a specific descent point regardless of incline. Galileo tended toward empirical experimentation so much that he sometimes made findings that he knew were wildly inconsistent with common knowledge. For example, through fundamentally empirical methods, his findings held that an increase in slope should have no effect on a particular liquid's flow over a great distance.

One of Galileo's students, Evangelista Torricelli (1608–1647), made the first strictly scientific studies of hydrodynamics. Torricelli invented the mercury barometer to measure atmospheric pressure. He also applied Galileo's concept of parabolic free-fall trajectories to the geometry of liquid jets. The latter enterprise resulted in Torricelli's law. This law states that the velocity with which a liquid flows through a container's opening is equal to the velocity of a body falling from the liquid's surface to the opening. For example, a stream of water flowing through a hole 15 feet below a dam's water surface has the same velocity as a rock falling the same distance. Torricelli's law primed scientists for the fundamental hydrodynamic discoveries of the eighteenth century.

The first scientist to make such a discovery was Daniel Bernoulli (1700–1782), a Swiss mathematician and physician. Early on, Bernoulli performed medical work on blood pressure and developed a keen interest in the flow of liquids. Later, he extensively studied **Boyle's law**, which demonstrated that the volume and pressure of a gas are inversely related. Bernoulli analytically deduced this law, which had previously had been obtained only through empirical trials.

In 1738, Bernoulli published his book *Hydrodynamica*, which contained **Bernoulli's principle** (sometimes called Bernoulli's theorem). For a steady, nonviscous, incompressible liquid flow, Bernoulli's principle is written: $p + \rho \, v^2$ $\rho \, gy = A$, where p is pressure, ρ is density, v is velocity, g is acceleration of gravity, y is height, and A is constant. This principle explains that energy is conserved in a moving fluid (a liquid or gas). Moving in a horizontal direction, pressure decreases as the fluid's speed increases. Conversely, if the speed decreases, pressure increases. Bernoulli's law explains why water moves more quickly through a narrow part of a horizontal pipe than through its wider part. It also establishes that the pressure will be lowest where the speed is greatest. This was the first correct analysis of water flowing in a container.

Bernoulli's discovery was based on early findings toward a principle of conservation of energy, which Bernoulli had studied, in one form or another, since 1720. As a fluid moves from a pipe's wide part to its narrow part, the work done by the moving volume in both parts of the pipe is expressed by the product of the pressure and volume. According to the law of conservation of energy, the increase in kinetic energy is balanced by a corresponding decrease in the product of pressure and volume (or, where volumes are equal, of pressure only). In this aspect, Bernoulli's law anticipated both the conservation of energy and the kinetic–molecular theory of gases by more than a century.

During the same period, a contemporary of Bernoulli's made a second important discovery. French mathematician and encyclopedist Jean Le Rond D'Alembert (1717–1783) performed a series of complex experiments involving the drag on a spherical object immersed in flowing liquid. He also investigated the basis of potential flow analysis. At the start of his research, D'Alembert assumed that drag would approach a zero mark as he steadily lowered the viscosity of his fluid until it approached zero. However, this was not the case. As he continued his trials, D'Alembert was surprised to learn that, as he approached zero viscosity, the net force of drag seemed to converge on a non-zero value. He had discovered a vanishing point of the net force in potential flow analysis.

In 1743, D'Alembert published his *Traité de Dynamique* (or *Treatise on Dynamics*). His finding soon became known as D'Alembert's paradox (sometimes called D'Alembert's principle) due to the vanishing of the net force. D'Alembert's paradox was a new generalization of Newton's third law of motion, which states that for every action, there is an equal and opposite reaction. This finding demonstrated that Newton's law held not only for fixed bodies, but also for bodies that are free to move, such as flowing fluids. Suddenly, D'Alembert had shown that Newton's law explains a host of additional phenomena.

However, D'Alembert himself was dissatisfied with the paradoxical nature of his results. Though he did not know it, the paradox arose because he had neglected to correctly figure for viscosity in his mathematical equations. When a body moved through a fluid and was slowed by the force of drag, D'Alembert knew that the kinetic energy expended in the process of slowing had to go somewhere. He also knew that, in a viscous fluid, it went to heat. However, in the end, D'Alembert never worked his way out of the fictitious zero force of his own paradox.

In the early 1900s, German engineer Ludwig Prandtl (1875–1953) developed the boundary layer theory of drag. Prandtl found that flow is divided into two regions: the bulk of flow, which has long been studied by mathematicians, and the small (boundary) layer near the body. Prandtl explained that viscous effects are dominant only in the latter region. With one theory, Prandtl had explained how potential flow theory is compatible with exact physics. He had also explained D'Alembert's paradox by demonstrating how an elusively small viscosity and an elusively small viscous region can modify the features of global flow dynamics.

Hydrodynamics—especially as explained in Bernoulli's principle—has many practical applications. Bernoulli's principle is easily demonstrated in a simple exercise. If a person holds a strip of paper below his or her lower lip and gently blows, the paper lifts rather than falls. This action is the result of the increased velocity of the air on top of the paper. Pressure above the paper is reduced, so the paper rises.

In this case, aerodynamics follow Bernoulli's principle in the force known as **lift** (see Figure 13). As in the paper demonstration, this force gives an airplane the ability to climb into the air and to hold altitude during flight. Engineers give an airplane wing's upper surface a curved shape to achieve lift. Because the air travels faster over the curved top of the wing, pressure is reduced. Greater pressure on the wing's bottom (due to the intentionally created differential) results in vertical lift (as well as a horizontal component, known as induced drag). In a sense, the plane is "pushed up" by the greater air pressure on the bottom of the wing.

Bernoulli's principle occurs in many other applications as well. In extremely high winds, such as those produced by hurricanes, pressure on the outside of a window is dramatically reduced. Additionally during a hurricane, there is general low atmospheric pressure. In such a scenario, windows may "pop" or break *outward* because of the high pressure differential. The same mechanism is often responsible for flat roofs being blown off of buildings. In many climates, roofs are designed to withstand a high *downward* pressure, as often created by snow loading. In very high wind, a large *upward* pressure differential literally rips a flat roof off of a building.

SEE ALSO Ballistic Pendulum and Physics of Spinning Projectiles; Charles's Law of Gases.

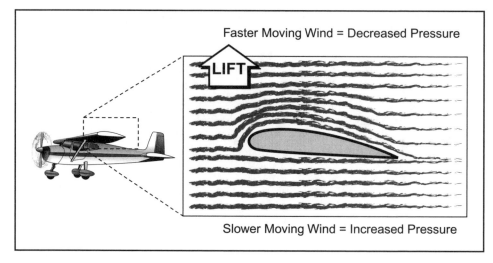

Figure 13. Hydrodynamics of lift. Engineers give an airplane's wing a curved shape. Air travels faster over its curved top than its flat bottom, creating a pressure differential. This achieves vertical lift when the plane is "pushed up" by the greater air pressure below the wing. *See also:* **Magnus effect** illustration, page 19.

Selected Bibliography

Benumof, Reuben. *Concepts in Physics.* Englewood Cliffs, NJ: Prentice-Hall, 1965.

Krebs, Robert E. *Scientific Laws, Principles, and Theories: A Reference Guide.* Westport, CT: Greenwood Press, 2001.

Halliday, David, and Robert Resnick. *Fundamentals of Physics*, third edition. New York: John Wiley & Sons, 1988.

Hydrogen Is an Element
Henry Cavendish, 1766

The **element hydrogen** is present in almost all organic compounds. It is also present in such abundance in the Earth's crust to rank third (after **oxygen** and silicon). Most of the Earth's hydrogen is combined with other chemical elements in the form of chemical **compounds.** Hydrogen is also the most abundant element in the universe, accounting for 90% of the universe's atoms and roughly 75% of its mass. About 75% of the sun's mass consists of hydrogen (the other quarter is helium). This composition is fairly typical of many other stars. However, scientists have only recognized this important element since the second half of the eighteenth century.

Before this time, most scientists thought that fire was an element. This idea goes back to antiquity and was prevalent for many centuries. In the sixteenth century, Swiss alchemist–physician Paracelsus (Theophrastus Bombastus von Hohenheim, ca. 1493–1541) did not know he had produced hydrogen gas by pouring acids over metals. More than a century later, in 1675, Nicolas Lemery (1645–1715) published the important book *Cours de Chymie* (*A Course of Chemistry*), which detailed the chemical knowledge of his time. In this internationally popular book, Lemery described an explosive reaction he caused. Though he did not know it, Lemery produced hydrogen by pouring sulfuric acid over a mixture of water and iron shavings. When he introduced a candle flame, his explosive reaction ensued.

Neither Paracelsus nor Lemery realized hydrogen was an independent gas, let alone an element. In fact, the concepts of gas and elemental chemistry were only still emerging in the second half of the eighteenth century. During this time, the eccentric English chemist Henry Cavendish (1731–1810) began a series of brilliant chemical trials with this substance.

In 1766, hydrogen became the topic of the very first paper Cavendish published. Following the pneumatic methods pioneered by Stephen Hales (1677–1761) and Joseph Black (1728–1799), Cavendish isolated hydrogen gas by pouring acids on copper, zinc and tin. He named his gas **inflammable air** because of its propensity for **combustion**. Cavendish was one of the last scientists to believe in the theory of **phlogiston**. This theory stated that a hypothetical substance, phlogiston, escaped from burning substances. For a while, Cavendish thought that "inflammable air" might just turn out to be the elusive phlogiston itself. (For more information, please see the entry on Phlogiston.)

Soon after beginning his experiments, Cavendish noted that the kind of metal (and not the acid) he used determined the volume of hydrogen he produced. From this observation he concluded that the solution of metals in oxidizing acids was attended with formation of reduction products of those acids. However, Cavendish clung to the belief that the gas was a product released from the metals during combustion. In terms of phlogiston, Cavendish wrote that when the "metallic substances are dissolved in spirit of salt, or the diluted vitriolic acid, their phlogiston flies off, without having its nature changed by the acid, and forms the inflammable air" (quoted in Berry 1960, 48). In other words, Cavendish said that acid meets metal and releases a specific kind of phlogiston—the unique substance "inflammable air."

Many scientists heartily defended the premise that only one type of ordinary air existed. They thought that pneumatic chemists simply manipulated its rather elastic qualities. But in this statement, Cavendish proposed that his gas was not simply an inflammable permutation of ordinary air. Rather, he proposed it was a completely unique substance. In purely modern terms, a completely unique, irreducible substance is an element. Understanding the gas in these terms was still a few years off in 1766. With this realization, Cavendish was (however unwittingly) moving chemistry an important step closer to the end-of-century chemical revolution.

Cavendish set off to test his proposal. He began from the premise that the acid was actually uniting with the phlogiston to produce the hydrogen gas. To test this claim, Cavendish realized he had to figure the amount of "inflammable air" (if any) in ordinary air. For these trials, he greatly improved an instrument known as a **eudiometer**. This instrument allowed him to estimate the loudness of the explosion that resulted when he detonated a sample of ordinary air with "inflammable air." Cavendish proposed a direct relationship between explosion loudness and percentage of inflammable air: The louder the explosion, the higher the percentage.

Using this instrument, Cavendish tested the air in different locations throughout London and Europe. He found that it contained similar amounts of "inflammable air" or hydrogen regardless of location. Previous scientists had reported large, spatial discrepancies in the air's volume of both oxygen and hydrogen. After his eudiometer tests, Cavendish indicted these studies as errant. Eventually, these studies led Cavendish to another investigation and subsequent discovery. He found that the Earth's atmosphere has a constant composition regardless of location. (For more information, please see the entry on this topic.)

Across the English Channel, French scientists were just starting to experiment with lighter-than-air balloons. Early aeronauts Joseph (1740–1810) and Étienne **Montgolfier** (1745–1799) understood little gas chemistry behind the science of flight. Nonetheless, they made the world's first **hot-air balloon** ascent on June 5, 1783.

From his earliest conversations with Cavendish, Joseph Black realized that **gas balloons** were a real possibility. After all, Cavendish had established early on that inflammable air was only about one-fourteenth the density of ordinary air. Black also made several public demonstrations to this effect. Building on the work of Cavendish and Black, the renown French chemist Jacques Alexandre Cesar Charles (1746–1823) began exploring hydrogen's low density. He quickly realized its ready applicability for lighter-than-air flight. Charles pioneered a way to produce a huge quantity of hydrogen, as well as a gas balloon that could contain the highly reactive gas. Just 10 days after the Montgolfier flight, Charles and his assistant soared high above the city of Paris in a long, tranquil hydrogen balloon drift. Because of his foundational work with "inflammable air," most scientists consider Cavendish the scientific benefactor of gas balloon flight. At the same time, Cavendish commandeered the help of aeronauts to investigate atmospheric composition at different altitudes.

Several separate discoveries were a direct result of Cavendish's trials with "inflammable air." In his eudiometer explosions, Cavendish often noted the formation of "dew" on the glass surfaces of his collecting instruments. He found that this "dew" was a lot like water. Through an important series of lengthy trials, Cavendish discovered that water is a compound of hydrogen and oxygen. This shattered the ancient, time-honored belief that water is an element. (For more on this topic, please see the entry on the compound nature of water.)

Cavendish clung to the theory of phlogiston his entire life. Therefore, he never arrived at a modern interpretation of his discovery "inflammable air." This fell to the founder of modern chemistry, French chemist Antoine Lavoisier (1743–1794). Lavoisier reworked the discoveries of phlogiston chemists like Cavendish in the framework of oxygen-based chemistry. He also gave the gas the name *hydrogen* after the Greek for "water former." This name references the gas's scientific origin in its formation of "dew" during Cavendish's eudiometer trials.

One of hydrogen's earliest applications is reminiscent of Cavendish's tests as well. Lavoisier embraced hydrogen's use in place of gunpowder for French fireworks. Today, hydrogen serves a wide variety of purposes in many industries. Unmanned gas balloons still use this gas. However, its use in passenger balloons has been discontinued. On May 6, 1937, the enormous German airship *Hindenburg* exploded into flames while approaching its docking in Lakehurst, New Jersey. Thirty-five people on board and many on the ground were killed. Somehow, the ship's hydrogen gas had ignited. This disaster marked the end of rigid airships. It also marked the end of hydrogen passenger balloons. Today's passenger gas balloons use nonflammable helium gas.

Hydrogen is also an important fuel source. Scientists are currently developing ways to use hydrogen more effectively on a widespread basis. Currently, hydrogen fuel powers the main engine of the U.S. space shuttle orbiter system. Also, a power plant in New York City uses hydrogen fuel to safely and efficiently produce electricity. One day, hydrogen may even be used for fuel cells, in which heat and water are the only by-products of electricity. Hydrogen is also used to produce **ammonia**, primarily by the Haber process. Also, hydrogen serves as a reducing agent, because it can withdraw oxygen and other nonmetallic elements from compounds, leaving only a pure metal behind.

SEE ALSO Atmospheric Composition Is Constant; Balloon Flight; Carbon Dioxide; Substances: Elements and Compounds; Water Is a Compound.

Selected Bibliography

Berry, A. J. *Henry Cavendish: His Life and Scientific Work*. New York: Hutchinson of London, 1960.

Crowther, J. G. *Scientists of the Industrial Revolution: Joseph Black, James Watt, Joseph Priestly, Henry Cavendish.* London: Cresset Press, 1962.

Jungnickel, Christa, and Russel McCormmach. *Cavendish: The Experimental Life.* Lewiston, PA: Bucknell University Press, 1999.

Krebs, Robert E. *The History and Use of Our Earth's Chemical Elements: A Reference Guide.* Westport, CT: Greenwood Press, 1998.

Poirier, Jean-Pierre. *Lavoisier: Chemist, Biologist, Economist.* Translated by Rebecca Balinski. Philadelphia: University of Pennsylvania Press, 1996.

World Book (encyclopedia) CD-Rom. San Diego: World Book, 1999.

1

Immunization Prevents Smallpox
Lady Mary Montagu, Inoculation, 1717
Edward Jenner, Vaccination, 1796

Historians of science are often puzzled by the origin of the disease **smallpox**. Many have even dubbed it a "new" Western disease, because it was not a major killer in Europe until around 1500. Though there is no entirely conclusive evidence, many epidemiologists theorize that the new intercontinental trade—such as the Atlantic trade in slaves—may have introduced more virulent strands of smallpox to Europe. Regardless, solid historical evidence shows that the disease has existed in various parts of the world since antiquity.

Archaeologists have uncovered the oldest evidence, in marks on preserved skin of ancient Egyptian mummies. The Egyptian pharaoh Ramses V (ca. 1150 B.C.) died of smallpox. Ancient Indian texts may directly mention the disease, though significant confusion between smallpox and measles plagued ancient chroniclers of disease. Nonetheless, thousands of years ago, Hindu temples were built for the worship of a smallpox deity. In the West, highly developed medical practitioners in Greece and Rome made no mention of smallpox, and modern historians believe it did not reach either civilization. The first indisputable mention of smallpox is from the Persian physician Rhazes (860–932) in 922. Also, a Japanese medical book written in 982 details well-established isolation hospitals for smallpox patients.

Smallpox probably first came to Europe around the tenth century. It definitely appeared in Germany in 1493 and in Sweden in 1578. Smallpox became pandemic in much of Europe by 1614 and epidemic in England from about 1666 to 1675. The sudden upswing in the disease's infection rates seems to suggest a correlation with the rise of dense populations in large European cities. The nose of Charles IX of France (1550–1574) was brutally scarred by smallpox, and Queen Mary II of England (1662–1694) died of the disease. English and Spanish explorers carried the disease to the New World. Smallpox reached Pennsylvania in 1661 and Charleston, South Carolina, in 1699.

During the eighteenth century in Europe, widespread outbreaks were common and even expected. In addition, though smallpox's mortality rates never equaled that of earlier diseases (such as the bubonic plague), average enlightenment Europeans—and especially the vain aristocrats—were terrified of its disfiguring and crippling effects.

Modern scientists have shown that smallpox (or *variola*, as it was some-times called) is caused by the **poxvirus**. In humans, this virus is highly con-tagious through airborne droplets. Generally 10 to 12 days after exposure, the victim develops severe aches and a high fever. Several days later, he or she breaks out in a rash consisting of thousands of small sores resembling pimples. After 7 to 10 days, the sores fill with pus and then scab over. Within the next month, the scabs fall off, revealing scars that range from mild to disfiguring. In severe cases, permanent blindness may occur. Finally, in about 20% of cases, the victim dies. Virtually all Europeans contracted smallpox during the eighteenth century. Many were calmed only when they had survived a bout with smallpox, because it was well known that even a minor infection conferred immunity.

Most of Europe's knowledge of smallpox came from both the Near and Far East, which had highly developed forms of treatment. Twice during the eighteenth century, Rhazes's treatise on smallpox was translated into Latin for the benefit of European physicians. In fact, for a very long time, an effective technique known as **inoculation** was practiced in many nations, including India, China, Turkey and several in North Africa. Inoculation may have originated many centuries ago in China. This technique involved collecting pus from the sores of smallpox victims and inserting it into a small scratch or similar break in the skin of a healthy person.

Lady Mary Montagu (1689–1762), the wife of the British ambassador to Turkey, witnessed the procedure many times. She made detailed descrip-tions of "inoculation parties" of fifteen or sixteen old women who went house to house collecting smallpox pus from the sores of victims every autumn. Then, the women made a second pass, this time visiting all unaffected indi-viduals of their city or town. One woman used a sewing needle to make a small scratch on a healthy person's arm or hand. Then she inserted into the opening only as much pus as would fit on the point of her needle. After a week, the person contracted a high fever and twenty to thirty pock marks. Generally, he or she was bedridden for 2 or 3 days. When the inoculation worked properly, the fever broke and the sores healed. By contracting a very mild case of smallpox, the person had achieved lifetime immunity from further infection.

Lady Montagu had herself and her three-year-old son inoculated in Turkey. On returning to England in 1717, she became the first person to introduce the concept of inoculation to the West. However, English reaction toward inoculation was not positive. Lady Montagu was aided by the physician Sir Hans Sloane (1660–1753), who urged the scared and skeptical public to embrace inoculation. Doctors undertook controversial tests on both orphaned children and condemned felons at Newgate Prison before finally convincing the Royal family to undergo inoculation.

The New World colonialists also were slow to embrace the procedure. Many had seen nearly 90% of some Native American tribes wiped out by the disease, which also devastated many colonial towns. In New England, Puritan leader Cotton Mather (1663–1728) found that one of his slaves was immune to the disease because of an inoculation received in Africa. Mather became an outspoken proponent of inoculation. He personally assisted in orchestrating trials on slaves, which were similar to those on English convicts and orphans. Later in the century, George Washington (1732–1799) had his family and then his army inoculated. Benjamin Franklin (1706–1790) also advocated the procedure.

Inoculation was the first successful prevention against smallpox. However, problems existed even where the practice was well established. Sometimes the deliberately induced case of smallpox did not turn out to be so mild, and the infected person suffered disfiguring illness, blindness or even death. In several cases, epidemics followed inoculation and probably resulted from it.

It was not until the end of the century that a better procedure, known as **vaccination**, would replace inoculation. Where inoculation involved applying the actual virus, vaccination involved applying a much less harmful one. Edward Jenner (1749-1823), a country doctor in Gloucestershire, heard from a farmer named Benjamin Jesty that milkmaids were naturally immune from smallpox. Jesty claimed to have protected his family from a terrible outbreak of smallpox by having them contract cowpox from his herd. This belief was widely expressed in the saying among milkmaids, "I cannot take the small-pox, since I have had the cowpox." Nearly all milkmaids contracted cowpox, which caused minor sores on their hands (as on the udders of cows), but no other significant symptoms.

Jenner speculated that, if this country wisdom were true, it was due to some kind of natural cross-immunity to smallpox. Jenner relayed his hypothesis to his mentor, the famous surgeon John Hunter (1728–1793). Hunter offered Jenner plain advice, "Don't think, try; be patient, be accurate" (quoted in Haggard 1929, 229). Jenner collected a series of observations, spanning a period of 18 years. He found no cases of smallpox among milkmaids who had first contracted cowpox.

Finally, Jenner was ready to try his finding. On May 14, 1796, Jenner took lymph from the infected hand of Sarah Nelmes, a milkmaid, and injected the fluid into John James Phipps, a young boy. On July 1, Jenner infected the boy with the smallpox virus, as was done during the previous form of inoculation. Absent cowpox, Jenner would have expected the boy to exhibit symptoms of a mild case of smallpox. Phipps did not. In fact, he never contracted small-pox. Jenner conducted several additional trials and found that vaccination was a simple, inexpensive, easily applied preventive against smallpox. Though his paper was rejected by the Royal Society of England, Jenner published his findings in a 1798 book called *An Inquiry into the Causes and Effects of Variolae Vaccinae.*

Once it caught on, Jenner's vaccination fantastically lowered the rate of smallpox infection during the nineteenth century. However, opposition to vaccination—especially compulsory vaccination—remains in many countries to this day. Moreover, the tests on human subjects, be they willing or unwill-ing, remain extremely controversial. Today, many people feel that such tests in the past were unethical. In Jenner's day, many people felt the same way. Due to laws governing informed consent and limiting human subject experimentation, Jenner's tests would be illegal today. Nonetheless, the mere existence of medical tests themselves are evidence of just how far enligh-tenment medicine had progressed by the end of the century. Experimenta-tion was added to observation, with unprecedented success. Also, though scientists did not fully grasp this interpretation until much later, successful immunization provided indisputable proof of the emerging contagion theory of illness.

Today, historians often view smallpox as the first disease conquered by science. Smallpox disappeared entirely from Europe and North America in the 1940s. In 1958, the World Health Organization (WHO) adopted a pro-posal from the Soviet Union that called for the global eradication of small-pox. To this end, Western laboratories developed both a freeze-dried vaccine and a new kind of needle that guaranteed nearly 100 % effective vaccination. Between 1967 and 1977, the WHO made terrific gains in mass vaccination and selective control of smallpox in developing countries. The WHO reported that the last known person to naturally catch smallpox was a 3-year-old Bangladeshi girl named Rahima Banu. She survived the disease, and the WHO declared smallpox eradicated from the planet in May 1980. However, many epidemiologists caution that "pockets" of smallpox exist in rural areas of developing countries. They fear that isolated flare-ups or even larger out-breaks may continue to occur, despite the WHO pronouncement.

In developed nations, the poxvirus exists only in laboratories, where it is strictly controlled. However, in 1978, two British laboratory workers were accidentally infected. Also, many children in developing nations have not been vaccinated. Some epidemiologists even believe that those industrialized nations that have eased vaccinations (such as the United States) may be at risk. Finally, security analysts and epidemiologists warn that rogue nations or terrorist groups may attempt to use the poxvirus as a biological weapon.

SEE ALSO Scurvy.

Selected Bibliography

Bettmann, Otto L. *A Pictorial History of Medicine*. Springfield, IL: Charles C. Thomas, 1962.

Garrison, Fielding H. *An Introduction to the History of Medicine*. Philadelphia: W. B. Saunders, 1917.

Haggard, Howard W. *Devils, Drugs and Doctors*. Garden City, NY: Blue Ribbon Books, 1929.

Hudson, Robert P. *Disease and Its Control: The Shaping of Modern Thought*. Westport, CT: Greenwood Press, 1983.

Krebs, Robert E. *Scientific Laws, Principles, and Theories: A Reference Guide*. Westport, CT: Greenwood Press, 2001.

Loudon, Irvine (ed.). *Western Medicine: An Illustrated History*. New York: Oxford University Press, 1997.

McGrew, Roderick E. *Encyclopedia of Medical History*. St. Louis: McGraw-Hill, 1985.

Porter, Roy (ed.). *The Cambridge Illustrated History of Medicine*. New York: Cambridge University Press, 1996.

Rhodes, Philip. *An Outline History of Medicine*. Boston: Butterworths, 1985.

Singer, Charles. *A Short History of Medicine*. New York: Oxford University Press, 1928.

L

Latent and Specific Heat
Joseph Black, 1761

When Antoine Lavoisier (1743–1794) published his momentous book *Traité Élémentaire de Chimie* in 1789, it listed both heat and light as **elements**. This fact showed both how far the fledgling science of chemistry had come in a short time and how far it still had to go in becoming the science of today. It also showed the special importance placed on the concept of heat by the early founders of chemistry. In fact, according to most chemical historians, modern chemistry was born out of a key debate. On one side were the phlogistonists, who believed that burning material freed a hypothetical **phlogiston** into the air. On the other were the modern chemists, who believed that burning substances combined with **oxygen**. At the center of this debate was the concept of heat.

Study of heat was not a new enterprise. Since the 1500s, scientists had been using various kinds of **thermometers** to measure temperature. In the mid-eighteenth century, renowned scientist Joseph Black (1728–1799) first began applying scientific reasoning to the study of heat. He began to think about heat chemically. At first, Black even suspected that heat itself might be a chemical substance akin to phlogiston.

Like the founders of thermometry before him, Black knew that freezing and melting points in water are standard temperatures. But Black began to investigate a lesser known phenomenon. When ice is heated, its temperature rises to the freezing point of water and remains steady until all the ice has melted. Only then does the temperature begin to rise again. Black also found a similar phenomenon with water's boiling point. In his tests, he found that turning water into steam required what amounted to "additional" heat. This phenomenon was well known in Scottish industry. Liquor distilleries required much more heat than one might initially expect. After realizing this was due to the same phenomenon as the one in his laboratory, Black was surprised to learn that no other scientists had yet studied it. After all, it would have tremendous economic, social and scientific impact.

Black began to believe that heat first "mixed" with the substance being heated. Then, at a certain point, he believed that heat began to literally "combine" with it. Black called this quantity **latent heat**, from the Latin word for "hidden." After this initial discovery, Black began a series of important experiments on latent heats of fusion and evaporation. To his surprise, he found that different bodies have different capacities for heat. Black called this capacity a body's **specific heat**.

Black further proposed that latent heat is the heat absorbed or released by a body during a change of state. He said that this heat brings about no change in **temperature**. Therefore, Black's latent heat of fusion is the heat capacity required to change a substance from a solid to a liquid without a change in temperature. This is the process of melting. With this proposal, Black had made a key distinction. For the first time in history, he clearly distinguished between temperature and heat. Black's distinction laid the groundwork for the principles that temperature is the degree of "hot" or "cold" transferred between bodies, as measured in degrees on various scales. Therefore, Black's perception was that water was liquid because it contained a specific amount of heat. Should this heat flow out of the water, the water would become ice. In other words, Black realized that there was latent heat in water.

Because Black thought of heat as a chemical substance (with Newtonian mass), he thought of latent heat as a kind of equation: ice + heat = water: the substance heat combined with the substance ice, creating the substance water. Black explained that were it not for latent heat, masses of winter snow and ice would suddenly be converted into water as soon as the outside temperature reached the melting point.

Once Black understood the latent heat of melting, he began study of a similar phenomenon, which he called the latent heat of vaporization. This is the heat required to change a substance from a liquid to a gas. Almost instantly, Black surmised that steam contains an enormous amount of heat that does not register on a **thermometer**. This latent heat is responsible for the *state* of water, as steam.

Just as he realized winter snow and ice did not immediately melt, Black reasoned that heat did not immediately turn nearly boiling water to steam. If not for latent heat, boiling water would "explode" into steam all at once. On this question of vaporization, Black studied the **steam engines** of his friend, inventor James Watt (1736–1819), and found that steam was produced slowly in the machine's boiler. He was surprised by the enormous amount of fuel this process required. Through a series of trials, Black determined that converting boiling water to steam required more heat than bringing water all the way to the boiling point. Black found that his original theory was absolutely correct. Steam contains an enormous amount of latent heat. Black hypothesized that latent heat was actually what was powering the steam engines of the industrial revolution. (For more on this point, please see the entry on Steam Engine.)

After his conception of latent heat had become clear, Black began work on **specific heat**. Specific heat is the amount of heat needed to raise the temperature of 1 gram of a specific substance 1° **Celsius** (C). This theory is based on the fact that equal size masses of different substances require different amounts of heat to reach the same temperature. Conversely, if these same two substances receive the same amount of heat, the substance with a *low* specific heat will reach a higher temperature than the substance with a *high* specific heat.

In founding modern chemistry, Lavoisier built on Black's work with heat. In the process, Lavoisier laid much of the groundwork for the field of **thermodynamics**, the study of heat. Lavoisier made extensive use of an instrument known as an **ice calorimeter**. This instrument measured a quantity of melted ice in order to measure the heat produced in specific chemical reactions and in the respiration of different animals. (For more on this topic, please see the entry on Animal Respiration.) Building on Black's work, Lavoisier showed that the heat given off in a reaction is equal to the heat absorbed in an exactly reverse reaction. He also investigated the specific heat of many different substances.

Black's discovery of latent and specific heat contributed greatly to the industrial revolution. Many scientists believe they were key concepts for the widespread employment of steam power. For example, the forerunner to **Watt's steam engine** was the **Newcomen steam engine**. In this crude engine, one chamber heated *and* condensed the water for each and every power cycle. The Newcomen engine was inefficient because it had to overcome the latent heat of steam in order to cool the chamber on every stroke. After Black explained latent heat, Watt realized that a separate condensing chamber could overcome the tremendous expenditure of energy being lost to latent heat with each stroke. In modern terms, this saved tremendous amounts of energy. (For more on this topic, please see the entry on Heat Is a Form of Energy.)

Today, scientists measure specific heat on a scale based on water's specific heat, which is marked as 1. On this scale, 10 calories of heat raise a gram of water by 10°C. However, the same 10 calories of heat raise the temperature of 1 gram of copper by 111°C. Therefore, copper has a low specific heat, marked on this scale as 0.09.

SEE ALSO Heat Is a Form of Energy; Steam Engine; Substance: Elements and Compounds; Thermometry.

Selected Bibliography

Bensaude, Vincent, and Isabelle Stengers. *A History of Chemistry*. Translated by Deborah van Dam. Cambridge, MA: Harvard University Press, 1996.

Crowther, J. G. *Scientists of the Industrial Revolution: Joseph Black, James Watt, Joseph Priestly, Henry Cavendish*. London: Cresset Press, 1962.

Knight, David. *Ideas in Chemistry: A History of the Science*. New Brunswick, NJ: Rutgers University Press, 1992.

Krebs, Robert E. *Scientific Laws, Principles, and Theories: A Reference Guide*. Westport, CT: Greenwood Press, 2001.

McKie, Douglas, and Niels H. de V. Heathcote. *The Discovery of Specific and Latent Heats*. New York: Arno Press, 1975.

Partington, J. R. *A Short History of Chemistry*. New York: St. Martin's Press, 1965.

Law of Conservation of Mass
Michael Lomonosov, 1748
Antoine Lavoisier, 1789

In her novel *A Tree Grows in Brooklyn*, Betty Smith described her heroine's first exposure to chemistry as the realization that "there was no death in chemistry." Nothing, she had discovered, rots or burns away, but instead changes into something else—to live again (at least poetically), as if chemically reincarnated. Appropriately named Francie, Smith's famous heroine was so taken with this beautiful idea, she even became "puzzled as to why learned people didn't adopt chemistry as a religion" (Smith 1947, 389).

Though new to Francie, this idea was not new to the world of science. In the 400s B.C., the Greek philosopher Democritus (ca. 460–370 B.C.), claimed broadly that "nothing can be created out of nothing, nor can it be destroyed and returned to nothing" (quoted in Fine 1978, 21). Many centuries later, the chemist Jean Rey (ca. 1582–1645) correctly recorded that the **calx** (or oxide) of tin is roughly 25% heavier than the original metal. This was the first scientific exploration of this very old concept. The great Irish chemist Robert Boyle (1627–1691) took up the puzzling question of weight gain during calcification in a sealed vessel in 1673. He thought that "igneous particles pass'd thro' the glass and . . . manifestly and permanently adhered" to the metal (quoted in Meldrum 1930, 35).

During the same period, Belgian chemist Johann Baptist van Helmont (1577–1644) carried out his famous "tree experiment." Over a period of 5 years, he found that a tree gained 169 pounds as it grew. At the same time, the weight of its soil stayed exactly the same. Van Helmont thought that water and earth were the only elements that existed. From here, he concluded that the weight gain of 169 pounds must have come from the addition of the so-called **element** water to the tree.

Despite this erroneous conclusion, this and other experiments offer evidence that scientists implicitly assumed matter can neither be created nor destroyed. Why else would they worry about increased mass during calcification or tree growth? By the time of these early experiments, the idea of conservation of matter was never really doubted with any seriousness. Rather, it was nearly always implicitly accepted. However, until the eighteenth century's methods of accurately measuring mass, chemists had no way to test this assumption.

On the cusp of developing these methods was the great Russian intellectual Michael Vassilievitch Lomonosov (1711–1765). A man of several world-class talents, Lomonosov was one of Russia's first true lyrical poets. Lomonosov was born into a peasant life but had the chance for education at Germany's Marburg University and the Freiberg Mining Academy. Lomonosov's reputation would eventually carry him to the pinnacle of success when he became academician for chemistry and mineralogy of Saint Petersburg.

In the prevailing theory of **phlogiston**, the process of **combustion** released the hypothetical substance phlogiston into the air. Lomonosov was one of the first chemists to closely study the curious weight gain during some chemical reactions. Lomonosov actually found that, far from separation of phlogiston, part of the surrounding air actually seemed to be consumed during combustion. In fact, as early as the 1750s, he was studying the nature of heat and the constitution of matter, often noting weight gain of metals as they calcified (or oxidized).

Lomonosov only realized his findings as laboratory results, rather than foundational concepts. Nonetheless, findings like these would eventually undermine the entire theory of phlogiston and charter a new chemical age. They would also send scientists on a journey from implicitly accepting the assumption that matter can neither be created nor destroyed to quantitatively testing this theory.

Though they accepted this assumption, phlogistonists accounted for the weight gain during calcification by claiming that the metal combined with the charcoal fuel during the process and that phlogiston was in the charcoal (and therefore added to the "calx," or the oxide coating). Phlogistonists such as theory founder Georg Ernst Stahl (1660–1734) explained weight gain by postulating that, depending on both the metal and the reaction, phlogiston can alternately have actual weight, no weight, or what amounts to a "negative" weight.

At the century's end, this explanation didn't sit well with several French chemists. Most especially, it troubled the founder of modern chemistry,

Antoine Lavoisier (1743–1794). Lavoisier is not known for discovering new gases or new chemical reactions. Rather, he reinterpreted the discoveries of phlogiston chemists in what became the terms of **oxygen**-based chemistry. Most especially, Lavoisier paid strict attention the concept of mass of chemical reactions. He began a series of experiments on the question of weight gain during calcification.

The old theory of **alchemy** (of which phlogiston was the final gasp) followed the ancient theory of **transmutation** proposed by Aristotle (384–322 B.C.). This theory held that each of the four elements (earth, water, fire, and air) could be changed into one another by adding heat or moisture. One of Lavoisier's early trials challenged this theory. Until this time, it seems that no one had thought of simply boiling water for an extended period of time in a controlled environment to see what happened. Lavoisier boiled water in a sealed glass vessel for 100 days. As expected, he found that the entire mass of the vessel had not changed. Nothing was released, and nothing had been gained from the fire. In short, very little had changed. Lavoisier now doubted the phlogiston theory more than ever.

After this experiment, Lavoisier began paying impeccably close attention to the concept of mass in his experiments. This marked a new and important era for the study of matter. To this end, Lavoisier invented many devices to measure mass. One device trapped and weighed all products of distillation. Another collected the gases produced by fermentation and putrefaction. A third device trapped the gases produced by mercury calcification. Perhaps most importantly, Lavoisier built the most sensitive balance ever produced. This instrument was capable of measuring infinitesimal fluctuations in the mass of reactants. Lavoisier could never have undertaken such meticulous studies of mass without this balance.

Armed with these instruments, Lavoisier focused his investigations on the area of mass in the calcification reactions. For many years, Lavoisier had documented the weight gain of many metals during combustion. What he needed was a way to measure the mass of a calcification in a closed environment. Lavoisier especially wanted to make sure that no outside factors skewed his experiment. In a stroke of genius, Lavoisier enclosed a quantity of tin in a glass container, which he sealed. He weighed the container carefully. Then, he heated the glass, which effectively calcified the tin. When he weighed the container again, he found—just as he expected—that no weight gain or loss had taken place. The mass of the product exactly matched the weight of the reactants.

However, Lavoisier was only partly ready for what happened when he made a small opening in the container: Air rushed in. Lavoisier surmised that the tin had united with "something" in the air, and that that something was the newly discovered gas, oxygen. Quickly, Lavoisier once again sealed the container. When he weighed the container once more, he found he had proved his conclusion. The weight gain of the container exactly equaled the weight gain of the tin. Oxygen had indeed joined with the tin.

The new gas not only supported combustion, but also caused the change in mass during oxygenation. Charcoal no more combined with the tin (via phlogiston) than did Lavoisier's glass vial, his laboratory bench, or his half-eaten lunch. In the end, oxygen was a reactant and, as such, had to be taken into account in the mass of the product. Phlogiston was no longer *separating from* the reactant. Oxygen was *joining with* it.

This finding changed everything. Scientists stopped talking about calcification and began studying oxidation. The simple formulation of *chemical + oxygen → chemical oxide* explained a world of processes that phlogiston could not. Scientists followed Lavoisier and began paying close attention to the mass of chemical reactions. As a guiding principle, Lavoisier formulated his **law of conservation of mass**, which states that the weight of the products of a reaction equals the weight of the reacting substances (see Figure 14). This law has been accepted with virtual universality since Lavoisier's day.

Lavoisier's law certainly applies to oxidation reactions, but did other kinds of reactions adhere to the same law? After all, it was still a logical leap to apply the law universally and to all equations. Lavoisier paid the same close attention to weight and measurement in a series of reactions that he published in *Trait Élémentaire de Chimie*, one of the most important books ever published. One section treats the topic of fermentation of fruit juice into carbonic acid and "spirit of wine" (or alcohol). Lavoisier wrote the simple reaction: must of grapes = carbonic acid + alcohol and showed that here, too, the law was in effect. There is "an equal quantity of matter before and after the operation," wrote Lavoisier. "On this axiom is founded the whole art of making experiments in chemistry" (quoted in Guerlac 1975, 119).

Just several years after founding the chemical revolution and proving the law of conservation of mass, Lavoisier was guillotined during the Reign of Terror, because he was once involved with a company that collected royal taxes. The law of conservation of mass is conceptually analogous to Lavoisier's long-ago firm conducting an audit. If the mass of the product doesn't exactly match the mass of reactants, then something is rotten in the state of France. Often, Lavoisier even compared the law of conservation of mass to a heavily metaphoric kind of algebraic equation.

Lavoisier's law of conservation of mass was an important theoretical piece of the eighteenth century's chemical revolution. Its formulation was foundation to the practice of quantitative measurements and even the most basic concepts of chemical equations. Modern chemistry depends on this law.

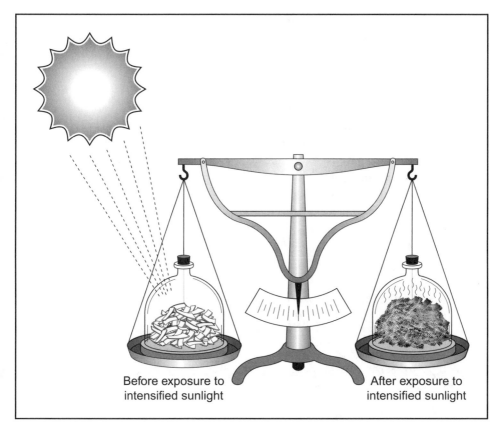

Before exposure to
intensified sunlight

After exposure to
intensified sunlight

Figure 14. Law of Conservation of Mass. This law states that the mass of a reaction's products must equal the mass of its reactants. In the above example, heat energy (in the form of intensified sunlight) has caused metal to oxidize in a simple chemical reaction. The products and the reactants have the same total mass. The law of conservation of mass was foundational to the chemical revolution and the very idea of chemical reactions.

During a 20-year period in the nineteenth century, Hans Heinrich Landolt (1831–1910) performed a number of tests of the law of conservation of mass. Using techniques and equipment far more advanced than Lavoisier's, Landolt determined that any change in mass of his reactants was no more than 1 part in 10,000,000. Today, the law has been established within an astonishing parts per billion ratio.

In Lavoisier's day, scientists unequivocally stated that matter can be neither created nor destroyed in a chemical reaction. However, new discoveries made during the twentieth century have caused scientists to state the law a little differently. Beginning in 1905, Albert Einstein (1879–1955) demonstrated that mass and energy are interchangeable (*E equals m times c-squared*, or energy equals mass times the velocity of light squared).Therefore, if a chemical change gives off energy as heat and/or light, the substances must have lost some mass. In normal chemical equations, the lost mass is far too small to measure.

However, Einstein's theory laid the groundwork for practical application of nuclear energy beginning around 30 years after his original proposal. Measurable quantities of mass are changed into energy in the reactions of nuclear reactors, atomic bombs, and particle accelerators. A million grams of uranium undergoing nuclear fission loses about 750 grams. Therefore, modern scientists now state the law of conservation as follows: mass–energy may not be created or destroyed, but each may be converted into the other. There is some "death," or loss, in chemistry after all, but only in specific cases.

SEE ALSO Oxygen; Phlogiston.

Selected Bibliography

Darrow, Floyd L. *Masters of Science and Invention*. New York: Harcourt, Brace, 1923.

Greenberg, Arthur. *A Chemical History Tour: Picturing Chemistry from Alchemy to Modern Molecular Science*. New York: John Wiley & Sons, 2000.

Guerlac, Henry. *Antoine-Laurent Lavoisier: Chemist and Revolutionary*. New York: Charles Scribner's Sons, 1975.

Holmyard, Eric John. *Makers of Chemistry*. Oxford: Clarendon Press, 1931.

Krebs, Robert E. *Scientific Laws, Principles, and Theories: A Reference Guide*. Westport, CT: Greenwood Press, 2001.

Fine, Leonard W. *Chemistry*, second edition. Baltimore: Williams & Wilkins, 1978.

Lockemann, Georg. *The Story of Chemistry*. New York: Philosophical Library, 1959.

Meldrum, Andrew Norman. *The Eighteenth Century Revolution in Science—The First Phase*. New York: Longmans, Green, 1930.

Poirier, Jean-Pierre. *Lavoisier: Chemist, Biologist, Economist*. Translated by Rebecca Balinski. Philadelphia: University of Pennsylvania Press, 1996.

Riedman, Sarah R. *Antoine Lavoisier, Scientist and Citizen*. New York: Abelard-Schuman, 1967.

Smith, Betty. *A Tree Grows in Brooklyn*. New York: Harper, 1947.

M

Metric System of Measurement
A commission of twenty-eight scientists appointed by the
French Academy, 1790 (unofficially headed by Antoine
Lavoisier)

Until the late eighteenth century, no standard, objective system of measure existed. Internationally, there was no unit that everyone agreed on. Worse yet, there was little uniformity within nations. In most countries, a dizzying array of units was employed in haphazard fashion. Units of measure not only varied widely between regions and countries, but each system even varied its own standards for units. The current system (or, more accurately, nonsystem) was failing in virtually every respect.

In England, the system of measurement grew out of old, local customs dating from the 1200s. The unit of an inch derived from the almost laughable formula "three barleycorns, round and dry." Grains of barleycorn varied greatly in size and shape, and therefore, so did inches. As bad as this system was, at least England had a system available for use.

The sheer confusion in France was quickly becoming unbearable for growing numbers of people. In the late eighteenth century, France employed more units than ever before—and more units than any of its neighbors. People in France had the foot, the inch, the toise (about 6 feet [f]), the line (about 3.175 millimeters [mm]), the ell, the pound, and the hogshead of grain. Getting several communities to agree on the same unit of measure was a feat in itself. But even when agreement did happen, the units themselves were so unstable that they were practically unusable. More than for any other group, the situation was absolutely infuriating for scientists. French science was quickly becoming the most highly developed in the world. For this reason, scientists in every discipline were becoming desperate for an accurate system.

The most famous French scientist, Antoine Lavoisier (1743–1794), had already privately adopted a **decimal** (base 10) **system** for his laboratory work several years earlier.

There was some historical precedent for using a decimal system. Several ancient cultures of the Near East accomplished sophisticated methods of weighing metals precisely. These standardized methods allowed early **alchemy** to develop. Many European scientists also knew about several useful systems of weight and measure used in China for centuries. Most recently, a French clergyman, Gabriel Mouton (1618–1694), had proposed a decimal measurement system with an objective unit of length. This unit was based on 1 minute of the Earth's circumference (1/21,600th). Following Mouton, the French astronomer Jean Picard (1620–1682) designed a unit based on the length of a clock pendulum. For the first time, there were proposals to base standard units on easily demonstrated physical laws. The great advantage of these proposals was that physical laws were applicable to *everyone* regardless of region, political affiliation, allegiance, nationality, or profession. But like the rest of Europe, France was not quite ready for these proposals in their own time.

In fact, it would take the French Revolution to bring about the system of standardized weight and measure so key to the scientific revolution taking place alongside it. When the French provincial assemblies drew up lists of grievances in 1790, the units of weight and measure were so bad that the province of Blois—out of all the grievances it could have filed—filed one recommending the units be completely overhauled. This prompted the National Assembly to request that the Academy of Science create a new system. In turn, the Academy appointed a commission of twenty-eight scientists, which included some of the finest in the world: Joseph-Louis Lagrange (1736–1813), Pierre Simon de Laplace (1749–1827), Gaspard Monge (1746–1818), the Marquis de Condorcet (1743–1794), and, of course, Lavoisier.

The commission quickly outlined three developmental standards to follow in organizing the new system. First, it would agree on a universal physical standard as the basic unit of measure. Second, it would adopt decimal division. And third, it would combine all former units of measure into a standard, straightforward, and simple system.

To begin with, the committee felt it would have been very reasonable to adopt Picard's standard, the single pendulum second, as its basic unit. This unit already had a small but respected scientific following. Also, clocks had long been universally accepted as useful, self-acting machines. However, because the clock was mechanical, its legitimacy could be challenged as arbitrary, potentially unfair, or inaccurate. Even more problematic, the length of pendulum seconds differs according to latitude. This presented a slew of ready-made problems for international (and even regional) use. To gain universal acceptance among nations, Lavoisier proposed that the basis for the system's unit must be easily verified in nature.

As luck would have it, in 1740 the Abbé de La Caille (1713–1762) and Cassini de Thury (1714–1784) realized a perfect unit. They had successfully measured an arc of the Earth's meridian. This unit met the objectives of the committee. Better yet, an arc of the meridian could now be easily measured, thanks to new technology. Jean Charles Borda (1733–1799) had recently invented the graduated circle, which was capable of measuring angles precisely. Adopting this unit of length meant repeating the survey taken by La Caille in 1749, because the new instrument both granted and demanded greater accuracy. Previously, La Caille had surveyed the length of the meridian arc from Dunkirk, France, to Barcelona, Spain. This time, the committee dispatched Pierre Francois André Méchain (1744–1804) to Barcelona and Jean-Baptiste Delambre (1749–1822) to Dunkirk.

The two agreed on an ingenious system of surveying to calculate the meridian arc. They agreed to first carry out linear measurements on the surface. Then, they would make geodetic calculations (for the Earth's curvature) according to the principle of triangulation. The linear measurements consisted of three tall points for triangulation along the meridian's axis. Each point had to be visible to the others (towers, steeples or chateaux) in order to define a succession of triangles. At each point, Méchain or Delambre had only to measure one side of the triangle. Then, using trigonometry, each could calculate the length of the other two sides and mark the results on his map.

This survey resulted in the new unit that would form the basis (and name) of the new system—the meter. This unit of measurement was based on one-fourth the length of a circle passing all the way around the Earth. Stated differently, it was based on a line passing once from pole to equator. The meter was established as exactly 1/10,000,000 of this length. All other units of length would flow from this single measure. They would follow the decimal number system. In one fell swoop, the meter had met all three of the commission's developmental standards.

In this system, each unit would increase or decrease by a base of 10. New units were quickly adopted. In fact, they practically presented themselves to the commission. A centimeter would be 0.01 meter. A millimeter would be 0.01 centimeter, or 0.001 meter. In the other direction, a kilometer would be 1,000 meters. Because all units were decimal, this meant that a meter would be 0.001 kilometer. It was all so simple, so enchantingly beautiful—all according to enlightenment logic, each unit in its place among the others. Lavoisier proposed that the Currency Commission should base French money on the decimal system.

Once the unit of length was established, other units fell into place. For example, the gram was chosen as the unit of weight. The gram was based on the weight of a cubic centimeter of distilled water in a vacuum at 4° **Celsius**. The Academy accepted the commission's units of measure and overall **metric system**. In turn, the National Assembly accepted the plan from the Academy. Lavoisier fashioned official standards for the new system out of platinum.

Just as the Academy's commission had hoped, the official units of measure have been officially preserved. From France, the metric system spread across Europe. Eventually, it spread much farther. Today, most people in every country use the metric system, except those living in the United States and the small, developing nation of Burma. (Even England has long since given up the English system.) Scientists universally use the metric system.

Since its inception in the 1790s, the metric system has gone through several revisions. The present version is the *Systeme International d'Unites* (International System of Units), known commonly as **SI**. In place of the twenty-plus units of the English inch–pound system, SI requires only seven. Four of these are commonly used by ordinary people. The meter is the base unit for length. The kilogram is the base unit for mass. The second is the base unit for time. The **Kelvin** is the base unit for temperature, though Celsius is more commonly used. The units of temperature are equal, but scales begin at different points. Additionally, scientists routinely use the other three units. The ampere is the base unit for electrical measurements. The mole is the base unit for the specific amount of a chemical in a reaction. Finally, the candela measures intensity of light. Also, less frequently used are esoteric units. These include the radian and steradian units for measuring angles.

The U.S. government has made several attempts to switch to SI. Beginning after World War II, it charged educators with teaching students and the public about the benefits of switching to SI. The thinking was that teaching ordinary citizens about the system would make them use it, thereby spurring conversion in government and industry. This approach did not work. Even today, much of the government has not converted.

In the 1950s, however, American pharmacists began filling prescriptions in metric units. In 1957, the U.S. Army and Marine Corps began making measurements in metric units. In the 1960s, Japanese and European automobiles began appearing in large numbers, prompting mechanics to learn metric units for their tools. In the same decade, the National Aeronautics and Space Administration (NASA) began making metric measurements. Also, American companies routinely trade on the international market in SI units only.

In 1975, the U.S. Congress passed the Metric Conversion Act. Conversion officially began at the end of 1992. However, for ordinary Americans, there is little evidence of use. The Office of Metric Programs (in the Department of Commerce) still promotes the use of metric units. However, many scientists feel that the U.S. government should commit to an outright conversion right away—and attempt to make ordinary citizens follow.

SEE ALSO Binary System; Thermometry.

Selected Bibliography

Foster, Mary Louise. *Life of Lavoisier*. Northampton, MA: Smith College Press, 1926.

Guerlac, Henry. *Antoine-Laurent Lavoisier: Chemist and Revolutionary*. New York: Charles Scribner's Sons, 1975.

Holmyard, Eric John. *Makers of Chemistry*. Oxford: Clarendon Press, 1931.

Krebs, Robert E. *Scientific Laws, Principles, and Theories: A Reference Guide*. Westport, CT: Greenwood Press, 2001.

Poirier, Jean-Pierre. *Lavoisier: Chemist, Biologist, Economist*. Translated by Rebecca Balinski. Philadelphia: University of Pennsylvania Press, 1996.

Riedman, Sarah R. *Antoine Lavoisier, Scientist and Citizen*. New York: Abelard-Schuman, 1967.

World Book (encyclopedia) CD-ROM. San Diego: World Book, 1999.

Microscopic Organisms Reproduce, Rather Than Spontaneously Generate
Lazzaro Spallanzani, 1767

Since antiquity, one of the most fundamental questions for scientists has been the origin of life. Aristotle (384–322 B.C.) proposed a theory known as **abiogenesis**, or **spontaneous generation**. This model stated that Aristotle's "lower" forms of life, such as flies, worms, and mice, develop from nonliving sources, like mud and decaying flesh, under specific conditions found in nature. Aristotle never tested this hypothesis empirically. Rather, he thought he perceived this process observationally. Maggots appeared to develop on a piece of rotting meat, just as snakes slithered out from mud and mice appeared on rotting grain.

Versions of spontaneous generation were prevalent for many centuries. Even many of the best thinkers of Europe believed versions of this theory well over a millennium later. They included philosopher Thomas Aquinas (ca. 1225–1274), physiologist William Harvey (1578–1657), and even physicist–astronomer–mathematician Sir Isaac Newton (1642–1727). Many "recipes for life" were well known during the Middle Ages and beyond. A famous recipe for spontaneous generation of mice called for covering stale grain with sweaty clothing in a well-ventilated container and placing the container in a dark corner. After an undisturbed month, the sweat supposedly entered the grain, changing some of it into mice, which spilled forth from the container.

During the seventeenth century, scientists began to seriously doubt the theory of spontaneous generation. In the 1660s, Italian physician Franceso Redi (1626–1697) made the first scientific test of the theory. During this time, naturalists were making significant discoveries about the insect world. A much ballyhooed permutation of spontaneous generation held that maggots arose spontaneously from rotting meat. Redi was skeptical. He had read theories that maggots came from the eggs of adult houseflies. Redi decided to test these different hypotheses.

In his laboratory, Redi put samples of fresh meat in three separate glass containers. He left container 1 open, covered container 2 with fine mesh, and sealed container 3 completely. After several days, Redi returned to the containers. In container 1, maggots had appeared on the surface of the meat. In container 2, maggots had appeared on the surface of the fine mesh. In container 3, maggots had not appeared at all. Redi was not immediately certain of his results and studied the containers further.

Over the next several days, he observed flies laying eggs on the meat in container 1. Redi was surprised to observe them laying eggs on the mesh of container 2. As he expected, Redi identified no flies laying eggs on the sealed top of container 3. Redi correctly concluded that maggots come from eggs, which adult flies routinely lay on or near the surface of rotting meat. Because container 3 was protected from the flies, maggots had not appeared. Redi had already ingeniously preempted arguments about this sealed container by utilizing fine mesh with container 2. Recipes for spontaneous generation almost always called for well-ventilated containers, as ordinary air was a supposedly important ingredient. Redi had therefore provided incontrovertible evidence that maggots arrive not from spontaneous generation, but from the eggs of adult flies.

This was an epoch-making experiment, which introduced a new method of scientific research to the larger body of scientific practices. The controlled experiment has remained a model of experimental biology since Redi's day. Redi had methodologically demonstrated the efficacy of repeating the same experiment in different ways by modifying one parameter at a time. When he published his results in a 1668 book, *Esperienze intorno alla generazione degl'insetti* (*Experiments on the Generation of Insects*), Redi effectively convinced virtually all scientists that complex forms of life (such as flies, worms, and mice) do not come from spontaneous generation. They come from adult animals. However, these same scientists—including Redi—continued to believe that spontaneous generation occurs in some circumstances.

During the next 50 years, the first professional microscopists uncovered an entire world of previously unseen **animalcules** (or **microorganisms**). Because microorganisms hardly seemed capable of laying eggs like Redi's flies, scientists began wondering about the genesis of these tiny creatures. Many scientists thought they were the product of a specific kind of spontaneous generation that, until now, had gone completely unnoticed. Microscopists regularly followed "recipes"—such as soaking hay in warm water or setting warm chicken broth in sunlight—to "produce" their microscopic objects of study.

In 1745, English naturalist John Tuberville Needham (1713–1781) proposed an experiment based on Redi's earlier work. Needham knew that scientists had established that extended boiling kills animalcules. With this in mind, Needham spooned a dollop of boiling mutton gravy into a glass vial, sealed the vial, and waited several days. When he opened the vial and examined the contents under his microscope, Needham found the gravy literally swarming with animalcules. Needham's published findings made him famous throughout Europe. Most scientists interpreted his results as proof that, unlike complex organisms, microscopic ones spontaneously generate. Since antiquity, spontaneous generation had never been a more popular topic of discussion. After all, advocates of the theory had never before been able to claim experimental proof.

During this period, versatile scientist Lazzaro Spallanzani (1729–1799) had been studying the writings of his fellow Italian, Redi. Spallanzani had become deeply influenced by Redi's findings—so influenced, in fact, that Spallanzani began to believe the whole idea of spontaneous generation was preposterous across the board. However, like every other scientist in Europe, Spallanzani knew of Needham's celebrated experiment to the contrary.

Spallanzani had conducted many experiments of his own design. In his inquiries on frog reproduction and on human digestion, he had identified errors in several of his own experiments, as well as those of other scientists. Spallanzani paid close attention to the specific order of Needham's experiment. He began to suspect that the microorganisms had not spontaneously generated in the gravy. Rather, he thought they had entered the vial from the air—during the short window of time between boiling and sealing the container. Several preliminary experiments suggested that this might be the case.

In 1767, Spallanzani designed an experiment to test Needham's findings. He began by carefully spooning cool mutton gravy into four well-cleaned vials. At this point, Spallanzani began a brilliant process of scientifically isolating controlled variables, one by one. He left vial 1 open and did not boil its gravy. He left vial 2 open and thoroughly boiled its gravy. He melted the glass neck of vial 3 shut and drew off the air to create a vacuum, but did not boil the gravy. He melted the glass neck of vial 4 shut, drew off the air to create a vacuum and thoroughly boiled the gravy. Then Spallanzani waited.

After several days, Spallanzani returned to his vials with his microscope. As he expected, vial 1 (which had neither been sealed nor boiled) was teeming with microorganisms. Vial 2 (which had been boiled, but not sealed) had fewer microorganisms, but will still infested. Vial 3 (which had been vacuum sealed, but not boiled) had a roughly equal concentration of microorganisms as vial 2. When Spallanzani cracked open vial 4 (which had been vacuum sealed and thoroughly boiled), he was thrilled with his findings. Vial 4 contained no microorganisms. Spallanzani had proven his hypothesis, that microorganisms had entered Needham's vial from the air and had grown in his boiled mutton. Only the method used for vial 4 had effectively killed the microorganisms and kept them from reentering the vial. Spallanzani was surprised at how unchanged his mutton gravy appeared.

Spallanzani was convinced that spontaneous generation of all organisms—regardless of size or conditions—was a hoax. He published his findings with the conclusion that all organisms come only from other organisms of the same type. As corroborating scientific evidence, Spallanzani also published results of a follow-up inquiry. He sealed meat extracts in separate glass vials and heated each vial for an hour. Several vials of food were actually edible several weeks later. This experiment was the first instance of modern food preservation. Spallanzani suggested generically that the action of animalcules was responsible for putrefaction.

Needham made a career of rhetorically restating spontaneous generation, which he reworked into a full-blown philosophy that he called the "vegetative force." This narrative, which won him international fame, claimed that life could arise haphazardly in the tiniest places. At base, it was a vague mixture of religion, hocus-pocus, and science that supposedly accounted for all forms of life in the universe. Some claimed Needham (who was also a clergyman) had explained the mind of nature's deity. For many people—perhaps caught up in the ironic *zeitgeist* of the turbulent century—the idea that life could arise haphazardly (with little regard for scientific laws or natural order) proved irresistible. With several variations on his experiment, Spallanzani demonstrated that he had already disproved all permutations of spontaneous generation.

Many scientists were unimpressed with Needham's position, as they pointed out that it was long on rhetoric and short on scientific proof. However, these same scientists interpreted Spallanzani's experiments as having conclusively proved what scientists had thought for a century—that air was a key ingredient for spontaneous generation. Citing Redi's use of the fine mesh screen, they said Spallanzani had only proved that spontaneous generation would not occur without air. They agreed that the "vegetative force" was preposterous, but still endorsed spontaneous generation for microorganisms. Spallanzani could never design an experiment to let in air, but not microorganisms. Today, he still gets credit for his discovery, even though it was largely misinterpreted in his own day.

After 1767, many scientists doggedly held on to the theory of spontaneous generation until the middle of the next century. In 1859, the French Academy of Sciences sponsored a contest for the best experiment on the topic of spontaneous generation. The charismatic, young chemist Louis Pasteur (1822–1895) won the competition with an experiment that picked up exactly where Spallanzani's had left off. Pasteur filled two tall flasks with meat broth and boiled each one thoroughly. He left the neck of flask 1 unsealed and open to the surrounding air. Pasteur heated the thin neck of flask 2 until it became soft and pliable. Then he bent it into an S shape. He was careful not to seal the neck. Air could therefore enter the flask. Then Pasteur waited.

After several days, flask 1 became infested with microorganisms, just as Pasteur expected. However, the broth in flask 2 did not become contaminated. Pasteur found no microorganisms growing in this flask. As the final part of his experiment, Pasteur tilted flask 2 so that some of the broth became trapped in the S part of the neck. After several days, Pasteur demonstrated that this broth had become infested with microorganisms, as well. He had effectively allowed air, but not microorganisms, to enter flask 2. When the airborne microorganisms entered flask 2, they settled by gravity in the S curve of the neck.

Pasteur had also demonstrated that microorganisms are all around, even in the air. Just as Spallanzani had shown almost a hundred years earlier, the microorganisms were entering the flask, not spontaneously generating in the food. Proponents of spontaneous generation had no refutation to Pasteur's experiment. In other words, this experiment finally convinced all scientists that living organisms, no matter how small, do not come from nonliving sources. Scientists currently accept the theory of living things, called **biogenesis**, which says that all living things come only from other living things.

After spontaneous generation was disproved, scientists began to wonder about the origin of life from Earth. If living things do not come from nonliving sources, then where did they come from originally? In the early 1920s, Soviet biochemist Aleksandr Oparin (1894–1980) proposed an intriguing theory. He suggested that, before there was any form of life, the Earth was a far different place. The earth's oceans were much hotter, and its atmosphere comprised different gases, such as **ammonia**, methane, and **hydrogen** (with no **oxygen**). Solar rays and lightning could have caused these gases to combine and fall into the hot ocean. When these gases combined in the oceans, they formed simple amino acids, the building blocks of cells. Eventually, these cells developed into primitive cells and microorganisms. Over the course of several billion years, these organisms evolved into more and more complex forms of life. In 1952, American biologist Stanley Miller (1930–) conducted an enclosed experiment that synthesized conditions on primitive

Earth. His findings suggested that amino acids could have indeed formed from nonliving environmental materials.

Following Oparin's theory, some scientists believe that organic compounds first formed from inorganic compounds. In other words, these scientists believe that spontaneous generation actually occurred at least one time— when the first simple living creatures formed. They claim that spontaneous generation is therefore scientifically possible. However, it is no longer occurring for several reasons. The Earth's conditions do not favor such chemical combinations. Also, these primitive creatures would almost surely fail to compete with the highly evolved organisms that now exist.

Not all scientists agree with Oparin's theory for the beginning of life. Some scientists have conducted tests suggesting that amino acids may already have been present on the Earth during its formation. Others scientists believe that energy-producing elements of certain plant and animal cells combined into chemical cells through a complex process of self-organization. Finally, a few scientists believe the theory of panspermia, which states that spores from another part of the universe arrived on Earth, where the conditions were right for their development into living organisms. Panspermia has failed to garner widespread support among contemporary scientists for several reasons. Most scientists strongly doubt that even extraterrestrial spores could have survived a harsh journey through space. They also point out that panspermia begs a fundamental question. If life on Earth arrived from somewhere else, how did that life develop in another part of the universe?

SEE ALSO Immunization Prevents Smallpox.

Selected Bibliography

DeKruif, Paul. *Microbe Hunters*. New York: Harcourt, Brace, 1926.

Hall, Tomas S. *Ideas of Life and Matter: Studies in the History of General Physiology, 600 BC–1900 AD*, volume 2. Chicago: University of Chicago Press, 1969.

Hankins, Thomas L. *Science and the Enlightenment*. New York: Cambridge University Press, 1985.

Millar, David, Ian Millar, John Millar, and Margaret Millar. *The Cambridge Dictionary of Scientists*. New York: Cambridge University Press, 1996.

Olson, Richard, ed. *Biographical Encyclopedia of Scientists*, volume 5. New York: Marshall Cavendish Corporation, 1998.

Travers, Bridget (ed.). *World of Invention*. Detroit: Gale Research International, 1994.

Milky Way Is a Lens-Shaped Distribution of Stars
Thomas Wright, 1750

The **Milky Way** is visible from anywhere on Earth. This is because it is the galaxy that includes the sun and nine planets of its solar system. On cloudless nights, the portion of the Milky Way visible from Earth appears as a milky swath of starlight expanding across the sky. Scientists believe dark blots in the Milky Way are caused by large amounts of light-blocking dust. The Milky Way is much larger than just the sun and its solar system. In fact, the Milky Way has a diameter of about 100,000 light-years. One **light year** is the speed light travels in a year—about 9.46 trillion kilometers (km, or about 5.88 trillion miles [m]).

The Milky Way is one of only two galaxies visible from the earth's northern hemisphere. On clear nights, a nearby galaxy in the constellation Andromeda is faintly visible. In the southern hemisphere, two other galaxies are visible. They are known as the *Magellanic Clouds.*

The concept of the Milky Way has a long history. In fact, this galaxy has held the collective imagination of many different cultures. The ancient Egyptians held that the Milky Way was the celestial counterpart of their land's very lifeblood, the Nile River. Ancient Greek astronomers spun fantastic mythological tales about this galaxy, as part of broader stories about the sky. Many Greek names of constellations, such as Andromeda, Cassiopeia and Perseus, are still in general use. One Greek myth states that Helios, god of the sun, drove a burning chariot across the sky each day. Helios's son, Phaeton, borrowed the solar chariot to drive across the sky. Being a mortal, Phaeton lost control and a scar was branded onto the night sky. This scar was the Milky Way. Later, possibly in Alexandria, the heavily Greek-influenced astronomer Ptolemy (ca. 100–165) made a sophisticated star catalogue in a text known as *Mathematike Syntaxis.* He described the Milky Way in such detail that his observations were not improved on until the eighteenth century. Ptolemy especially noted the mysterious dark spots among the dazzling stars of the Milky Way.

These often-mythologized accounts were useful and durable to the cultures that gave rise to them, as well as those that followed. Certainly, they piqued the interest of many scientists and amateur astronomers. Centuries later, Galileo Galilei (1564–1642) discovered that the Milky Way's stars optically stand out as distinct individuals, due to their proximity to Earth. However, it was not until the 1740s that a scientific approach toward the model and shape of the Milky Way was applied. Even then, it came from an unlikely source.

Thomas Wright (1711–1786) was a British sailor who spent many nights on deck staring up at the Milky Way. Wright had training as a clockmaker and architect, but taught himself practical astronomy. He knew a bit about Greek mythology and the story of the Milky Way. But Wright's night-after-night, naked-eye gazing of the stars led him to ask a question that many others had taken as implicit: Is there actually a greater density of stars in the Milky Way, as its glow might suggest?

Over the course of many nights, Wright had observed the sky, only to have his eye return to the spot where the dots of starlight seemed much thicker. In some places, Wright thought that the appearance of density was actually an optical illusion. Or perhaps, he thought, there were so many stars that their points of light converged into the galaxy's gentle, milky glow. Then, in what was a revolutionary idea, Wright considered the Earth's own placement inside the Milky Way. After all, Nicolaus Copernicus (1473–1543), the founder of modern astronomy, had long since shown that the Earth is a moving planet, with a specific location regarding other moving planets. In 1740, Wright became the first person to suggest that the most highly visible stars—those of the Milky Way—form a single, coherent system. In Wright's original conception, the Milky Way was the cross section of the universe seen when looking toward the center from Earth. This model had many obvious problems and was generally not well thought out.

In 1750, Wright published his "Theory of the Universe," in which he modified his concept of the Milky Way. During the intervening 10 years since his first formulation, Wright had made many observations and careful studies of the galaxy. Now, he suggested that the Milky Way was shaped like a convex lens, with a thick center and thin edges (see Figure 15). Different scientists describe this model as a pancake, a shell, a disc, or a slab, but the key to Wright's model is that it has a thick center with thin edges. Wright put the sun and its solar system in the center of the Milky Way.

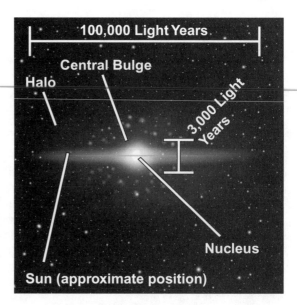

Figure 15. The Milky Way is a lens-shaped distribution of starts. Thomas Wright proposed the Milky Way's shape in 1750. He explained that it is a lens-shaped distribution of stars—thicker in the middle than at its edges.

Wright's new model explained the general placement and appearances of stars from Earth. Looking outward at the lens in a broadside view, one sees relatively few, mostly isolated stars. These are the stars visible in the night sky as individuals. However, when one looks parallel to the tangent plane of the lens—in Wright's terminology, toward its "rim"—many stars become visible. In fact, so many stars become visible that an observer cannot discern the farthest of them separately.

The most striking inaccuracy of Wright's convex model is the placement of the sun and its solar system at the center of the Milky Way. In reality, the sun and planets are far to one side of the Milky Way and about 25,000 light-years from its center. But that is modern knowledge. Wright's conception of the Milky Way changed the perception and inquires of many scientists.

The German transcendental philosopher Immanuael Kant (1724–1804) was an early advocate of Wright's concept of the Milky Way. In 1759, Kant published *General Natural History and the Theory of the Heavens*. In this important book, Kant argued that the Milky Way is a nebula that spins around a mysterious center, deep within the distribution of stars. Kant claimed that the Milky Way is but one of a possibly infinite number of such nebulae spread throughout the universe—a kind of principal galaxy among many lesser ones. Scientists generally agree that Kant's ideas of the Milky Way formed the beginning of the field of cosmology, the study of the structure and development of the universe.

English astronomer Sir William Herschel (1738–1822) expanded the work of Kant and Wright through his practice of "star gauging." In this method, he actually counted the points of light that were visible to him in all directions. Almost immediately, this simple survey confirmed Wright's theory of the asymmetry of stars in the Milky Way. When one counts in directions across, or at an angle to, the Milky Way's plane, the visible stars thin out considerably. However, when one counts in the direction of the line of the Milky Way, they increase quite dramatically. Herschel's important work also showed that, contrary to prior belief, the distribution of stars in space is not spherical, bur rather unidirectional. Therefore, Herschel arrived at a conception of the Milky Way's shape that was nearly identical to Wright's.

An important scientist who suggested that the stars of the Milky Way are in constant motion around a common center was the astronomer Johann Heinrich Lambert (1728–1777). Lambert explained his idea using the metaphor of the solar system. Just as the planets move around the sun, so does the entire solar system move around a common center of the universe. Lambert also suggested that many other nebula far beyond the Milky Way follow a similar pattern of movement.

Figure 16. The Milky Way is a spiral galaxy. The classification of the Milky Way as a spiral galaxy was one of twentieth century astronomy's most important achievements. This classification came from the discovery that dust, gases and even stars unfold from the Milky Way's bulge in a "pinwheel" pattern.

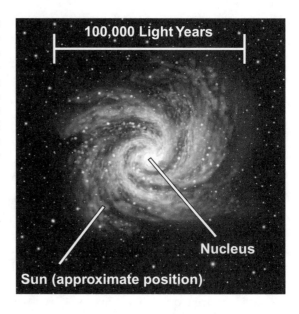

Today, scientists accept Wright's description of the Milky Way's shape as a thin lens with a bulge in its center. Scientists have also discovered that dust, gases and even stars unfold from the Milky Way's bulge in crooked arms that form a pattern of spirals, like an enormous pinwheel. This feature has earned the Milky Way the classification of a **spiral galaxy** (see Figure 16). This classification of the galaxy's shape is generally considered one of twentieth-century astronomy's cornerstones of achievement. Additionally in this modern understanding, the stars of the Milky Way orbit the center of the galaxy. Many modern scientists actually employ Lambert's metaphor of planets moving around the sun to explain this phenomenon.

Because of large clouds of dust and gas, scientists cannot see very far into the Milky Way's center. To study this central region, astronomers penetrate the clouds using radio and infrared rays. Using these techniques, they have discovered tremendous amounts of energy in the central region. Astronomers have also discovered an extraordinarily powerful gravitation force at the galaxy's exact center. Given these discoveries, many scientists theorize that the Milky Way's center is a gargantuan **black hole**. If this is the case, then the energy at the galaxy's center is likely produced when the black hole consumes gas and other matter from the surrounding Milky Way. (For more on this topic, please see the entry on Newtonian Black Holes.)

On a larger scale, the Milky Way is part of a collection of galaxies known as the Local Group. There are about thirty known galaxies in this group, including two other spiral galaxies. Astronomers place this group in part of a larger cluster of galaxies known as the Virgo Cluster.

SEE ALSO Comets Follow Particular Orbits, Cycles, and Returns; Proper Motion of the Stars and Sun; Uranus Is a Planet.

Selected Bibliography

Abetti, Giorgio. *The History of Astronomy*. Translated by Betty Burr Abetti. New York: Henry Schuman, 1952.

Doig, Peter. *A Concise History of Astronomy*. London: Chapman & Hall, 1950.

Hoyle, Fred. *Astronomy*. Garden City, NY: Doubleday, 1962.

Motz, Lloyd, and Jefferson Hane Weaver. *The Story of Astronomy*. New York: Plenum Press, 1995.

North, John. *The Fontana History of Astronomy and Cosmology*. London: HarperCollins, 1994.

Pannekoek, A. *A History of Astronomy*. New York: Interscience, 1961.

Thiel, Rudolf. *The Discovery of the Universe*. Translated by Richard Winston and Clara Winston. New York: Alfred A. Knopf, 1967.

Modern Dentistry
Pierre Fauchard, *Le Chirurgien Dentiste*, 1728

A story often repeated in countries of the Middle East has a man coming upon a crying child. He asks the little boy why he is crying. "A snake bit me," is the reply. "Oh that is nothing," counters the man, "I thought you had a toothache." This brief anecdote points to the history of agony associated with the practice of dentistry. Indeed, until the eighteenth century, dentistry was largely the domain of barbers, tooth-drawers, and full-blown carnival quacks. For the vast majority of human history, the so-called cures for tooth decay were comparable to—and often worse than—the disease. Even today, the whir and buzz of a dentist's drill, somehow amplified in the waiting room, routinely instills patients with varying levels of fear, anxiety, and sometimes even panic. However, even a cursory inventory of previous dental techniques often convinces even the most squeamish patient that the predictable whir and buzz are comparatively reassuring, comforting sounds.

The earliest diagnoses for toothaches fell into three broadly defined categories: tooth demons (products of spells or gods), **humors** (bodily fluids that were out of whack), or toothworms (tiny maggots or gnawing creatures). The toothworm diagnosis would enjoy widespread acceptance for thousands of years before the eighteenth century. Most practitioners believed that toothworms either spontaneously generated inside a tooth or bored their way in. To this day, historians are unsure of how separate cultures in different parts of the world independently arrived at this theory without having contact with one another. Some historians believe that people observationally mistook the limp tissue of inner tooth pulp for a worm, because of its appearance.

The earliest attempts at curing dental ailments involved the relatively benign wearing of magic amulets and charms. Ancient peoples also routinely breathed over an herbal fire in attempt to "smoke" out a toothworm. One remedy older than ancient Egypt was gargling with a concoction made of spiders, eggshells and oil. Another was pouring acid around a decayed tooth. This treatment may have actually given some relief by killing the nerves of the diseased tooth. The ancient Greeks rubbed their gums and teeth with their fingers, but they also relied heavily on prayers to various gods of medicine. Roman physicians saw the efficacy of rinsing out one's mouth every morning with fresh water. But they also cauterized ulcerated gums with red-hot irons. The ancient Romans undertook a plethora of rituals described by Pliny the Elder (ca. 23–79), such as spitting into the mouth of a frog in order to transfer a toothache to the unfortunate animal. In his writings, Pliny is very careful to say that he makes no guarantees that his treatments will actually work.

By the European Middle Ages, all manner of bizarre treatments involving animals and organic products had taken hold. Rub honey on a toothache and grab the toothworm when it appears. Kiss a donkey to transfer the toothache. Run around a church three times without thinking of a fox. Apply a frog to the head. Apply part of a freshly killed hog stomach to the gums. Bite the head off a mouse and wear its body around the neck. Rub a sore tooth with human fat, a skeleton's bone, or the hand of a dead man. Apply a grain of salt to the tooth. Gargle with one's own first-of-the-morning urine. This final "remedy" lasted pretty much unchanged from ancient times until the eighteenth century.

If any of these treatments should (by chance) fail, medieval Europeans had to consult a barber–surgeon, or a tooth-drawer. Though it may seem hard to believe, these practitioners were the forerunners of modern dentists. Until the twelfth century, priests would hear a confession for a toothache, since the Church's official position was that any illness (including toothaches) was a punishment for wrongdoings. After the confession, the priest often took pity on the parishioner and performed an extraction or other agonizing procedure. Commonly, trained physicians considered surgery and body cutting as well below their level of expertise. So the priests enlisted the help of the next closest professional—the town barber. At this point, the problem of tooth decay was so widespread that the Church became convinced that its priests were neglecting their ecclesiastical duties. It banned surgical practice altogether. Surgery then fell to the assistants—the barbers who were already charged with cutting hair, lancing abscesses, letting blood, and giving enemas. Now, they began pulling teeth.

In addition to medieval town barbers, a person could hire a tooth-drawer. Generally, these were itinerant carnival performers (read: quacks). Professional tooth-drawers almost always enlisted deceptive tricks and displays to "prove" their painless extraction expertise. History books are full of well-documented accounts of tooth-drawers tearing out large pieces of bone and jaw while attempting simple extractions. Routinely, tooth-drawers caused permanent facial deformities. Of course, this all took place without anesthesia of any kind for the wretched patient.

At first, it is difficult to see how modern dentistry emerged from this symphony of horrors. However, it somehow did so—in a span of only several hundred years. The early eighteenth century was an extremely fruitful time for the new science of dentistry. By the year 1700, dentists were emerging as a distinct group in Paris—fueled in large part by the rampant vanity of the well-to-do. As luxurious living and narcissistic beauty became the dominant paradigms of nobility (if not the masses), demand for skilled dentists soared to unforeseen levels. Well-funded dentists became the norm. Dentistry became an honorable, even renown profession, and knowledge of human teeth was exploding. However, research was far-flung, disorganized and not disseminated at all between the many cities and capitals of Europe.

The situation was ripe for a big breakthrough, which came in 1728. In this year, Pierre Fauchard (1678–1761) published his outstanding book *Le Chirurgien Dentiste, ou Traite des Dents* (*The Surgeon Dentist or Treatise on the Teeth*). Because of this publication, historians generally consider Fauchard the founder of modern dentistry. Originally, Fauchard studied surgery in Louis XIV's navy under several renowned surgeon-majors. Diseases of the mouth quickly became his special area of interest, and Fauchard spent many years surveying the practices of his dental precursors.

To this day, no single event did more to hasten the arrival of modern dentistry than the publication of *Le Chirurgien Dentiste*. In this two-volume text, Fauchard carefully inventoried virtually all existing dental knowledge. Moreover, *Le Chirurgien Dentiste* made qualitative distinctions between the sciences of dentistry: orthodontia, surgery, implantation, periodontics, reflex pains of the teeth, dental anatomy, fillings, pathology, *material medica*, and prosthetic procedures. For the first time, Fauchard outlined specific distinctions between these practices and the barbarous quackeries of former eras. Fauchard first used the term *dental caries* to correctly describe tooth decay as a breakdown in tissue. This description finally laid the idea of toothworms to rest, after so many centuries of leading people astray. In describing dental caries, Fauchard even stated that all forms of sugar were corrosive and therefore detrimental to teeth and gums. In a few short years, dentistry had arrived as a distinct science after centuries of curses, worms, and spells.

In addition to cataloging and evaluating the practices of dentistry, *Le Chirurgien Dentiste* outlined the clear and concise steps dentists should follow in order to make dentistry a full-blown branch of medicine. Since ancient Greece, doctors have agreed on a code of ethics. In this code, all knowledge is shared rather than withheld, a sacred trust between patient and practitioner is respected, and patient care is privileged above a doctor's remuneration. By enlightenment example, Fauchard showed in *Le Chirurgien Dentiste* that the sharing of dental information for the good of both patient and practitioner should outweigh all considerations of personal gain.

This was a divergence, even in Fauchard's day, from everyday practice. Many of Fauchard's colleagues even vigorously objected to making personal benefit secondary to patient treatment. In response, Fauchard went so far as to state that subterfuge of information—whether by carnival quack or trained dentist—puts the health of all patients in danger. Before *Le Chirurgien Dentiste*, dentistry was a trade in which the profit motive was dominant and in which technical knowledge was the carefully guarded secret of an esoteric group. After the book was published, other dentists began following Fauchard's example. Dental lectures, essays and letters began appearing. Many highly skilled dentists began accepting young interns who they trained—not for personal profit, but for professional courtesy.

Le Chirurgien Dentiste also provided elaborate details about the position of both patient and dentist, especially for performing extractions. Previously, no one had ever thought to pay attention to this important point. A patient was generally made to sit or lie on the floor, while a dentist crouched and kneeled over him or her. Fauchard decided to study the effects of this position in several large dental practices. He found that, after several hours, it hastened fatigue in dentists. Right away, it created fear in patients. In several cases, Fauchard found evidence that the position for extraction had been injurious to a pregnant patient. He even went so far as to argue that the lying-on-the-floor position should be immediately discarded. Soon after, it was. An ancestor of the familiar dentist's chair replaced the floor position for most procedures. Fauchard also narrowed the creative arsenal of hundreds of dental instruments to a basic five (the gum lancet, the punch, the forceps, the lever, and the pelican).

In practices they inherited, many of Fauchard's dental colleagues thought studying techniques of other professions was far beneath their talents. Fauchard did not share their disdain and began an intensive study of the tools of watch and jewelry making. By studying these tools, Fauchard contributed to a revolution in dental instruments, which he used in his dental practices in Angers, Nantes, Tours and Paris (the latter of which he continued until at least the age of 70).

Fauchard had long since realized that extraction had become far too routine a procedure. He preferred drill-and-fill procedures, which had the enormous advantage of saving the diseased tooth. The problem was that he lacked the instruments necessary for perfecting drill-and-fill techniques. The **dental drill** dated back at least to ancient Greece, since the oldest description was from Hippocrates (ca. 460–380 B.C.). But existing drills were too slow and too clumsy to use effectively. Fauchard also needed a permanent, pliable material to use for filling. He began experimenting with tin, lead, and gold fillings. During the same time, Fauchard also adapted several jeweler's drills to his professional use. One of these models—the bow drill—became the first really usable modern dental drill (see Figure 17).

Later in the century, American dentist John Greenwood (1760–1819) built on Fauchard's pioneering work, especially the dental drill. After many years of trials, he invented the foot-powered dental drill in 1790. Taking his cue from other professions, Greenwood ingeniously adapted his mother's

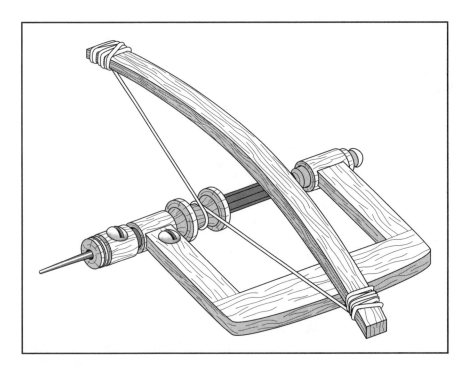

Figure 17. Fauchard's bow-type dental drill. Since antiquity, dental drills were too slow and clumsy for practical use. This made extraction the most common treatment. After Fauchard invented the first usable drill, most dentists began preferring drill-and-fill technique, which effectively saved a diseased tooth.

foot-treadle spinning wheel to rotate his drill. He used this instrument extensively in his own practice with very good results. Strangely, after Greenwood's son used it in his practice, the foot-powered drill was used no further.

Greenwood also revolutionized the quality of false teeth and dentures. He became the favorite dentist of George Washington (1732–1799) who, despite durable cultural myths, never wore a set of wooden teeth. Rather, Greenwood made no fewer than four pair of dentures for Washington. Greenwood fashioned them from hippopotamus and elephant ivory riveted to a gold palate, which was held together with a steel spring. Despite all these enlightenment advances, Washington eventually lost all but one of his teeth to extraction. He suffered from oral pain his entire adult life—a stark reminder of just how far modern dentistry had yet to develop.

Selected Bibliography

Bremner, M. D. K. *The Story of Dentistry, From the Dawn of Civilization to the Present.* Brooklyn, NY: Dental Items of Interest, 1939.

Campbell, J. Menzies. *Dentistry Then and Now.* Glasgow, UK: privately printed (no publisher listed), 1963.

Glenner, Richard A. *The Dental Office, a Pictorial History.* Missoula, MT: Pictorial Histories, 1984.

Hoffmann-Axthelm, Walter. *History of Dentistry.* Chicago: Quintessence, 1981.

Ichord, Loretta Frances. *Toothworms & Spider Juice: An Illustrated History of Dentistry.* Brookfield, CT: Millbrook Press, 2000.

Lufkin, Arthur Ward. *A History of Dentistry.* Philadelphia: Lea & Febiger, 1948.

Prinz, Hermann. *Dental Chronology: A Record of the More Important Historic Events in the Evolution of Dentistry.* Philadelphia: Lea & Febiger, 1945.

Weinberger, Bernhard W. *An Introduction to the History of Dentistry,* volume 1. St. Louis: C. V. Mosby, 1948.

Weinberger, Bernhard W. *Pierre Fauchard, Surgeon-Dentist.* Minneapolis: Lancet Press, 1941.

Wynbrandt, James. *The Excruciating History of Dentistry: Toothsome Tales & Oral Oddities from Babylon to Braces.* New York: St. Martin's Press, 1998.

Modern Encyclopedia
Denis Diderot and Jean Le Rond D'Alembert, 1751–1772

An encyclopedia is a volume or volumes of information about places, events, things, ideas and people. Some general encyclopedias cover information belonging to a wide variety of fields. Other, specialized encyclopedias include detailed information about a specific field or area of knowledge, such as biology, art or medicine.

In various forms, people have published encyclopedias for a very long time. The oldest ancestors of encyclopedias were prehistoric reference texts of borrowed quotations. A text usually ended up as a hodgepodge of information presented in confusing or arbitrary order. The accuracy of original sources was rarely verifiable. It was also never intended for a general audience.

The ancient Greeks made several early attempts that were encyclopedic in scope, if not organization. The philosopher Aristotle (ca. 384–322 B.C.) wrote a series of books in which he attempted to synthesize all existing knowledge. No actual copies have ever been recovered. Another philosopher, Speusippus (ca. 410–338 B.C.), ran Plato's Academy, one of the earliest university-style institutions. While doing so, he wrote a book that many historians consider the earliest quantifiable encyclopedia.

In Rome, Marcus Terentius Varro (116–27 B.C) published *Disciplinae,* a multivolume text on both art and science. Pliny the Elder (23–79) wrote a thirty-seven-volume reference text called *Historia Naturalis,* which dominated European scientific classification for over a thousand years. In this text, Pliny collected thousands of facts about plants and animals, as well as rocks, minerals and fossils.

The oldest known Chinese encyclopedias date from the third century, though many scholars believe they existed much earlier. An encyclopedic text of knowledge required for civil service jobs was published by the scholar Tu Yu (also called Du You, ca. 800) in the ninth century. Historians consider this one of the most important encyclopedias ever published.

Arab scholars produced valuable encyclopedias during the early Middle Ages. In Baghdad, Ibn Qutaiba (828–889) compiled an influential book of history, the *Kitab Uyun al-Ahkbar.* During the tenth century in Persia, al-Khwarizmi (ca. 780–850) published a key scientific reference book. The *Mafatih al-Ulum* exulted fields such as grammar and poetry and downplayed the importance of alchemy. Later, a group of scholars in Baghdad published an important encyclopedia of reconciliation between Greek and Arabic scholarship.

Many medieval European encyclopedias were heavily influenced by the twin authorities of Church and state power. A Parisian theology teacher, Bartholomew de Glanville (ca. 1250), wrote an encyclopedia that he organized according to topic group. In 1264, a friar, Vincent of Beauvais (ca. 1190–1264), wrote *Speculum maius,* which he organized under the headings of politics, academic practices and natural history. Organization by topic in these encyclopedias became much more important to later scientific writers and publishers. In the fifteenth century, moveable type made way for many editions and vernacular translations of encyclopedias.

One of the earliest reference texts in English, *The Mirror of the World,* was a translation from an earlier French edition. In the 1630s, German theologian Johann Heinrich Alsted (1588–1638) published one of the final Latin reference texts, *Encyclopaedia septem tomis distincta*. Despite the language and author's background, this encyclopedia dealt mainly with the science of geology.

During this time, scientists and other writers of encyclopedias stood on the threshold of modernity in their discipline. The period known as the Age of Reason had dawned in Europe. The Age of Reason was characterized by a philosophical belief in the scientific method as the best way to arrive at an understanding of a phenomenon and to organize knowledge. Learned people began replacing faith, superstition, and categorization with experimentation and observation. They also began attacking the many forms of tyranny— whether in governmental practice or dogmatic belief about the workings of nature. Writers and scientists felt an unprecedented urge to circulate the results of scientific knowledge. They also craved a way to organize it for both their scientific equals and a mass audience. A changing medium itself, the encyclopedia was a perfect vehicle for widespread distribution of new ideas during the eighteenth century.

An early example of this new type of encyclopedia was published in 1704 by John Harris (1666–1719), an English theologian. The *Lexicon Technicum* was especially important because all of its articles were arranged alphabetically and because Harris recruited specialists to write articles in their fields of expertise. An even more heavily influential encyclopedia was published by English cartographer Ephraim Chambers (1680–1740) in 1728. The *Cyclopeaedia, or the Universal Dictionary of Arts and Sciences* was first to employ an elaborate system of cross-referencing, which made searching for topic information very efficient. Chambers's title also demonstrates the growing use of the term *encyclopedia* during the eighteenth century. This term comes from the Greek words *enkyklios paideia*, meaning "general education."

In the Age of Reason, knowledge was no longer only the province of a learned few. Rather, it was quickly becoming available to anyone who could read. With articles both approachable and accurate, the encyclopedia was quickly recognized as an excellent format for the Age of Reason. In fact, historians sometimes refer to the Age of Reason as the "the age of the encyclopedia."

Following the success of Chambers's encyclopedia, several French publishers arranged a translation. They put writer, philosopher and scientist Denis Diderot (1713–1784) in charge of the project. However, Diderot quickly "caught fire" and began working on an entirely original encyclopedia instead. This became the twenty-eight-volume *Encyclopédie, ou, dictionnaire raisonné des sciences, des arts, et des métiers* (*Encyclopedia, or critical dictionary of the sciences, arts, and trades*). Almost immediately, Diderot was writing most of the articles on the arts, trades and contemporary culture himself. For the articles on mathematics and physical sciences, Diderot consulted an expert—already his friend and soon to become his main collaborator, Jean D'Alembert (1717–1783).

From the start, Diderot considered the *Encyclopédie*'s central purpose to be the spread of scientific knowledge for the specific purpose of improving the pursuits of ordinary people. He intended it to provide a thorough and complete treatment of rational knowledge. No similarly vast collection of general information had ever been assembled. Both Diderot and D'Alembert therefore saw their project as a record of their civilization's accomplishments to that point—not just distribution, but also preservation of knowledge gained through scientific rationalism. They also understood that this was an acutely rhetorical exercise, meant to cause a revolution in the minds of its readers. Most historians credit the *Encyclopédie* as a major factor of the French Revolution.

When the first volume appeared in 1751, it quickly became clear that the *Encyclopédie* was special indeed. It was the first modern encyclopedia, as it attempted a completely rational view of the world. Most of Diderot and D'Alembert's contemporary intellectuals publicly praised the *Encyclopédie* as the greatest product and most logical extension of the Age of Reason. Never before had such a diversity of subjects—and revolutionary scientific opinions—appeared in a single compilation.

Over the course of many years, D'Alembert wrote the complex entries related especially to physics and mathematics. He also wrote the opening *Discours préliminaire,* a kind of preface that laid out the project's aims and empirical approach, based on the philosophy of science. Here, D'Alembert explained that the *Encyclopédie* would be organized around the specific categories of human knowledge. These divisions had already helped D'Alembert and Diderot divide the chapters. They would also greatly foster the division between human and natural sciences as well as that between natural and mechanical sciences so distinct in the modern world.

By 1772, over 150 experts had contributed to the thoroughly collective work of the *Encyclopédie*. Indeed, recruiting a bona fide expert for each topic quickly became standard practice in the writing of modern encyclopedias. For example, Diderot and D'Alembert recruited pioneer field volcanologist Nicholas Desmarest (1725–1815) for a chapter on columnar basalt's volcanic origin.

In the end, however, Diderot and D'Alembert wrote most of the articles comprising the *Encyclopédie*'s seventeen volumes. A full eleven volumes of plates were in print by 1772. Though D'Alembert succumbed to political pressure and left the project prior to completion, Diderot worked on it for over 25 years. Affectionately nicknamed "Diderot's Encyclopedia," the *Encyclopédie* would be recognized as the age of rationalism's most ambitious publishing enterprise for centuries to follow.

By the 1760s, copies of the *Encyclopédie* had spread across Europe. Not to be outdone, a group of British scholars began publishing an encyclopedia of their own, known as the *Encyclopaedia Britannica*. This encyclopedia caught on fast. By the end of the century, illegal copies were being printed in the United States. It eventually became so popular in the United States that, in the 1920s, an American firm bought the *Encyclopedia Britannica*. (Some debate exists over when the spelling of the name was modernized.) Today, of course, it is one of the most widely read encyclopedias in the English-speaking world. Other important encyclopedias include *The World Book Encyclopedia, Compton's Encyclopedia* and *The Grolier Encyclopedia*. Electronic encyclopedias, such as *Microsoft Encarta,* are also popular. These are mainly available on CD-ROM (Compact Disc–Read Only Memory) format. Also, a large number of web sites offer searchable access to databases of both general and specific encyclopedia articles.

SEE ALSO *Farmer's Almanac.*

Selected Bibliography

Asimov, Isaac. *Asimov's Chronology of Science and Discovery*. New York: Harper & Row, 1989.

Lienhard, John H. "Diderot's Encyclopedia." *The Engines of Our Ingenuity* series. Web article available at: http://www.uh.edu/engines/epi122.htm. Copyright 1988–1997.

Lienhard, John H. "Encyclopaedia Britannica." *The Engines of Our Ingenuity* series. Web article available at: http://www.uh.edu/engines/epi203.htm. Copyright 1988–1997.

Parkinson, Claire L. *Breakthroughs: A Chronology of Great Achievements in Science and Mathematics, 1200–1930*. Boston: G.K. Hall, 1985.

World Book (encyclopedia) CD-ROM. San Diego: World Book, 1999.

Modern Entomology and the "History" of Insects
René-Antoine Ferchault de Réaumur, 1734–1742
Jan Swammerdam, first published posthumously, 1737

Insects live virtually everywhere (and in virtually all climates) on earth. Almost two-thirds of all known animals are insects and up to 10,000 new species of insect are discovered each year. Scientists who study insects are called *entomologists.* They believe that there are many more species of insects yet to be discovered than those they already know.

The oldest evidence of entomology in the Western world is found in the epic poems of Homer (about whose life almost nothing is known). Homer took metaphors from insect life when he compared numbers of ships and soldiers to swarms of bees and wasps. The writings of Plato (429–347 B.C.) offer early insight into ancient classification and the origins of biology. Plato initiated the idea of an **entomology** through the conceptual theory of species. Later, classical biology reached its pinnacle with Plato's student, Aristotle (384–322 B.C.). More than anyone else, Aristotle founded ancient entomology as a systematic practice. Aristotle based his classification on crudely comparative anatomical and physiological features. His system of insects included the category "Entoma," which he said were bloodless animals with four or more feet and (often) with wings. This group probably included modern arachnids, myriapods and worms.

Aristotle's ideas about categorization would last several thousand years. Scientists would only finally overturn his idea of **spontaneous generation** (or **abiogenesis**) in the eighteenth century. This theory stated that insects (along with some plants and crustaceans) spontaneously developed when certain specific conditions were met. It held that mosquito larvae were birthed from the wet mud of deep wells, that house flies came from manure heaps, and that drosophilae (or fruit) flies come from yeasty vinegar. Aristotle's theory wasn't specific to insects, but it depended heavily on philosophical studies of their formation. (For more information, please see the entry on Microscopic Organisms Reproduce, Rather Than Spontaneously Generate.)

Several other ancient thinkers made significant contributions to entomology. Pedanius Dioscorides (ca. 40–90) founded the subspecialty of entomological pharmacology. His classifications of insects for medical remedies lasted almost 2,000 years. In the first century of ancient Rome, Pliny the Elder (ca. 23–79) devoted an entire book of his *Historia Naturalis* to insects. He broadly expanded Aristotle's classifications and compiled a synoptic encyclopedia from the work of perhaps 2,000 other scientists. In the thirteenth century, Albertus Magnus (also called Albert von Bollstädt, 1193–1280) published the zoological text *De Animalibus,* which added seven books to the nineteen he inherited from Aristotle. Magnus's knowledge of classification far exceeded Aristotle's. He listed thirty-three different kinds of insects and detailed the anatomy of a bee. Methodologically, Magnus was still deeply indebted to Aristotle, as he detailed the supposed abiogenesis of many additional insects and followed Aristotle's teleological and causative approaches.

Scientists of the seventeenth century made early attempts away from Aristotle and toward objective scientific observation of insects. A Bolognese physician, Ulysse Aldrovandi (1522–1605) was the first scientist to deal specifically with insects in his 1602 entomological text, *De Animalibus Insectis libri VIII.* This text contains the world's first dichotomic key and **morphology** for "higher" insect groups. Aldrovandi also included a chapter entirely devoted to the form and structure of insect bodies. An English doctor, Thomas Mouffet (1550–1599 or 1604) wrote another important book on insects that was published posthumously. *Insectorum sive Minorum Animalium Theatrum* was a text of enriched scientific knowledge and expanded scope of content, but it repeated mistakes from older texts, such as classifying butterflies as categorically unique from caterpillars due to the existence of wings. The classification of insects was still far from scientific.

Insect entomology would take a decidedly scientific turn only with the publication of the six-volume *History of the Insects* by René Réaumur (1683–1757) between 1734 and 1742. They announced a fundamental break with Aristotle's philosophical classification and ushered in an age of objective observation. Where scientists such as Carolus Linnaeus (1707–1778) would reinvent classification as a modern scientific discipline, Réaumur was strictly an observer. Réaumur knew of many earlier studies of insects and was dissatisfied with classifications of them on the basis of color or broad external features (such as wings). Rather, he proposed to describe the life history and everyday practices of each insect he studied. This was a radical break with burgeoning industrial practice, which tried to study only the most economically useful qualities of an insect (e.g., its ability to make honey, to ripen fruit, or to pollinate flowers). Réaumur's method of curious investigation led him to study the insects *first* in their everyday activities in order to learn how they naturally acted. In *History*, he also urged his readers to pay close attention to insects, to never believe published theories at face value, and to always closely scrutinize with their own eyes.

In his volumes, Réaumur gave detailed descriptions of his insects and their hitherto esoteric lives. He also published many drawings (made by an uncredited artist whom he hired). His many descriptions were often lavish in their intricate knowledge of tiny creatures. In one striking study, Réaumur detailed the recently hatched, minute larvae of blister beetles, which he dubbed "bee lice," because he observed them attached to the tiny hairs of bees. Réaumur expanded his studies to honey bees and was one of the first to use observation hives sandwiched between two panes of glass. He made remarkable studies about aggregate bee behavior. For example, when Réaumur removed the queen, he found that the workers produced a new one by expanding a cell and feeding royal jelly to the larvae. Réaumur also made important observations of insect **respiration**, documenting wild variations of breathing mechanisms between species.

During the years Réaumur was working on his six volumes, another important natural science book was published. *Biblia Naturae* was the 1737 posthumous publication of hitherto unknown, important dissections by Dutch microscopist Jan Swammerdam (1637–1680). Where Réaumur initiated a paradigm shift toward science through learned observation, Swammerdam directly challenged classical practice. One of the first great microscopists, Swammerdam made outstanding dissection studies of small animals, such as insects. Like Réaumur, Swammerdam studied the honey-bee and, in the end, cleared up misconceptions dating all the way back to Aristotle. He conclusively demonstrated, for example, that drone bees were male, that another class comprised neutered females (Swammerdam named them "workers"), and that the ballyhooed "king" bee was actually a queen.

Swammerdam knew of Aristotle's insect studies and his persistent theory of spontaneous generation. Only at the very end of Swammerdam's life were scientists such as Francisco Redi (1626–1697) beginning to chip away at Aristotle's theory through the use of experimental evidence. By boiling food in sealed tubes, Redi would demonstrate conclusively that complex organisms such as insects did not come from spontaneous generation. In the eighteenth century, this was a new idea in the field of complex organisms and a contro-versial one for microscopic organisms. (For more information, please see the entry on Microscopic Organisms Reproduce, Rather Than Spontaneously Generate.)

When Swammerdam dissected tiny insect cocoons under his microscope, he saw elementary insect body parts. Swammerdam thought that all the parts of each insect were present from the very moment of formation. He thought all the adult parts for caterpillars, frogs, or houseflies existed in caterpillars, tadpoles, or pupas, respectively. In Swammerdam's day, scientists were divided on whether the ovum or, less frequently, the sperm contained all the parts of a complex creature. Either way, Swammerdam's theory came to be known as a theory of preformation and was extended to many complex animals, includ-ing humans. Through the careful observational work of Swammerdam and the empirical tests of scientists such as Redi, preformation became widely accepted until the early nineteenth century, when the modern epigenesis "germ layer" theory began replacing it.

Biblia Naturae's original printing included about 300 pages and over 150 lush, folio-sized illustrations. Swammerdam included excellent anatomical descriptions and the first scientific classification of insects into four groups based on his perceptions of their development. Group one included insects that emerged with all their limbs, changed to nymphs and shed their skins once. Swammerdam had actually described arachnids (not insects at all), such as spiders and scorpions in this group. Group two hatched with six legs, grew wings and changed into nymphs. These were ephemeris like dragonflies and mayflies. Group three dealt with "wormlike" animals that were actually larval-stage insects. These insects were born with either many or no legs. Swammerdam thought these insects likely grew legs under their skins, which they shed to became ants and bees. If these same "wormlike" animals did not shed their skin, Swammerdam thought they became flies, his fourth and final group.

Though this classification system was not wholly complete and lacked the concise, parallel structure of later systems, it led the way for classifications based on modern **physiology** through its emphasis on form. Today, entomologists study the physiology, anatomy, behavior, development, life history, ecology, habitat, structure and classification of insects and related animals, such as other arthropods. To date, over one million species of insects have been identified and named.

SEE ALSO Animal Physiology; Plant Physiology; Plants Use Sunlight; Scientific Morphology.

Selected Bibliography

Evans, Howard Ensign. *The Pleasures of Entomology: Portraits of Insects and the People Who Study Them*. Washington, DC: Smithsonian Institution Press, 1985.

Hays, H. R. *Birds, Beasts, and Men: A Humanist History of Zoology*. New York: G. P. Putnam's Sons, 1972.

Krebs, Robert E. *Scientific Laws, Principles, and Theories: A Reference Guide*. Westport, CT: Greenwood Press, 2001.

Parkinson, Claire L. *Breakthroughs: A Chronology of Great Achievements in Science and Mathematics, 1200–1930*. Boston: G. K. Hall, 1985.

Smith, Ray F. (ed.). *History of Entomology*. Palo Alto, CA: Annual Reviews, 1973.

N

Navigational Quadrant (Octant and Sextant)
John Hadley, 1731
Thomas Godfrey, 1731

Throughout the history of astronomy, one sees time and again that astronomy is not just for astronomers. This is especially true in the field of navigation—finding the position of a ship at sea—where dire practical application fueled the scientific search for knowledge of astronomy. In fact, until very recently, the practice of navigation not only fueled astronomy, but also completely depended on it.

The ancient Babylonians developed sophisticated systems of map making based on the position of an observer in regard to the sun. Later, they added a much greater degree of precision by adding lines of latitude and longitude. By the time ancient ships were plying the Mediterranean and Aegean seas, the lines of latitude and longitude had been transformed to a series of great circles in the night sky. A far more sophisticated technique, known as **celestial navigation**, took shape. Because it relied not on the singular sun, but on the varied positions of stars in the sky, celestial navigation made astronomers out of these ancient sailors. Even today, celestial navigation is one of the main navigational methods in widespread use.

Celestial navigation is the technique of finding one's position in regard to the moon, planets, and, especially, stars of the night sky. Since antiquity, sailors have documented an observed change in the paths of celestial bodies that occurs as their ships journey a far distance. Because the position of heavenly bodies appears to change as a ship journeys, the vessel's position may be determined through careful, precise tracking of the apparent heavenly movements.

In Europe, the dawn of the exploration age propelled the need for astronomy to previously unimagined levels. Many sailors had no education other than their own intimate experience with astronomy. As they explored new regions of the globe, they also explored new regions of the heavens. By the fifteenth century, astronomy was no more the practice of professional astronomers than it was an abstract doctrine. For sailors on the sea, it was a living practice of evolving, applied science that began in antiquity and ended somewhere far beyond their own historic moment.

By the seventeenth century, the governments of European powers had adopted a standing practice of fostering astronomy to further their goals of exploration and empire. Eager to expand its holdings and maritime influence to the south, England sent Edmond Halley (1656–1742) to map the skies. He succeeded in accurately cataloging the positions of 350 stars of the previously uncharted southern hemisphere. By the early eighteenth century, England had become the leading maritime power with significant economic interests all over the world. The mighty Royal Navy was also in great need of simple, accurate instruments to aid in navigation.

Since antiquity, various astronomical instruments have aided the practice of celestial navigation. Ancient Chinese sailors probably invented the astrolabe, which measured the angles of celestial bodies above the horizon in relation to an observer's position. This instrument was widely used in antiquity and during the Middle Ages. During the fifteenth and sixteenth centuries, the cross staff came into general use among navigators. This instrument consisted of a wooden frame with scaled measurements on each of its sides. Two wooden cross pieces ran perpendicular to the frame, allowing a navigator to line up the horizon on one end of the cross and a celestial object on the other. This acted as a trigonometric computer for measuring the offset altitudes of celestial bodies (especially the sun).

A little later, the first **navigational quadrants** began to appear in the late sixteenth century (probably around 1594). The quadrant was a large advance in the technical proficiency of instruments available to navigators at sea. It was also used and improved by the Danish astronomer Tycho Brahe (1546–1601). This wooden instrument was generally shaped like a quarter-circle and measured the angle from vertical line of sight to the celestial body. Measurements were made by way of a weighted line that hung down across one of the numbers on the quadrant. To use a quadrant, a navigator lined up the celestial body through a small hole in the quadrant, while someone else read the position of the weighted line across the quadrant's numbers. Even though the various quadrants lacked a 90-degree instrumental arc, most were capable of measuring a full 90 degrees through the use of a double reflection. One of the first great specialists at making quadrants was a famous English clock- and watchmaker named George Graham (ca. 1674–1751).

Graham specially fashioned a large quadrant that was used extensively by Edmond Halley.

Each instrument had drawbacks, however. Few were especially precise under the best of conditions because they were difficult to use with any degree of accuracy. In many cases, their application on the seas gave general or approximate readings. In the worst cases, the readings were wholly mistaken. Ever-increasing need made conditions ripe for improvement of the quadrant.

This improvement came in a new invention, a type of navigational instrument that would become the very symbol of navigation, the sextant. The invention of this important instrument is often the source of much confusion and even controversy. Different historians generally claim credit for at least two inventors. Complicating matters further, historians generally refer to these new reflecting instruments of the late eighteenth century as navigational "quadrants," even though they were strictly octants or (just a few years later) sextants. Nonetheless, both the similarly used octant and sextant developed because of the foundational work of inventor John Hadley (1682–1744) in England. Completely independently and only a few months later, Thomas Godfrey (1704–1749) in the United States duplicated Hadley's groundbreaking work. Therefore, both men are generally credited with the invention.

These navigational instruments were greatly improved because they worked according to **the optic rule**—the principles of light reflection and angles of incidence outlined by scientists such as Halley and Isaac Newton (1642–1727). The optic rule states that when an object is seen through the reflection of two mirrors perpendicular to the same plane, its distance from the eye is two times the angle between the reflecting surfaces. Because the octant followed this rule and employed two mirrors, it measured 90 degrees. This was despite the fact that it actually required an arc of only 45 degrees (one-eighth of a circle, hence the name *octant*).

In use, the octant or the sextant (60 degrees, or one-sixth of a circle) measured the distance in angles between two points (such as the horizon and a star). Like older instruments, these consisted of a tri-shaped wooden frame and a swinging arm. In its so-called zero (or starting) position, the *actual* and *reflected* horizons were mutually aligned. To make measurements, the navigator held the instrument vertical to the horizon and moved the arm until the image of a certain celestial body appeared to "touch" the line of horizon. The double-reflecting instrument thereby "brought down" the celestial body to the level of the horizon. This was one of the biggest benefits of these new instruments, as the navigator no longer had to try the difficult, inaccurate feat of looking two places at once. Instead, the navigator could simply read the celestial body's altitude on a graduated arc on the instrument's side, based on where the arm fell. The navigator used this information to compare a

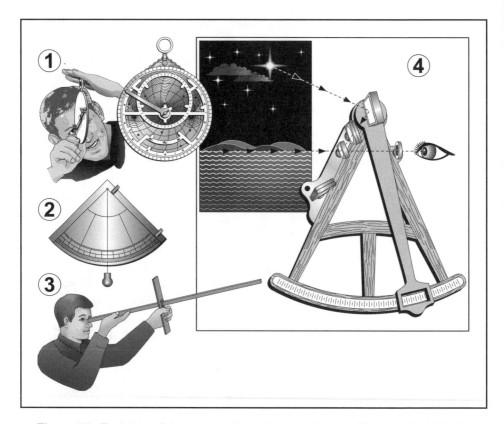

Figure 18. Evolution of instruments for celestial navigation. The astrolabe (1), the mariner's quadrant (2), and the cross staff (3) offered limited precision for celestial navigators prior to the eighteenth century. In 1731, John Hadley and Thomas Godfrey independently invented highly accurate navigational quadrants (4), as well as similar sextants and octants. These instruments ushered in a new age of scientific celestial navigation.

specific body's altitude at varying degrees and then to calculate degrees of the ship's latitude (see Figure 18).

With an octant or sextant, a navigator could calculate a ship's latitude with a degree of accuracy previously unavailable. Equally important, these reflecting instruments were simple to use with a little practice. This allowed a navigator to make quick, frequent measurements. It also allowed several navigators to check one another's measurements.

English navigator James Cook (1728–1779) was the first to use these modern tools of navigation and became a great proponent of them. Their effect on the British Admiralty was profound, as the so-called Hadley quadrants found their way to widespread military production by the middle of the century. Soon they were in widespread use on merchant and scientific ships as well.

In 1734, Hadley made another important refinement to his instruments by fixing a "spirit" or "ghost" level, so that a navigator could make a reading even where the horizon was not entirely distinct. By many accounts, the octant and sextant form was little changed between 1780 and the early twentieth century—perhaps as late as 1920. In fact, the sextant was used until just prior to World War II, when the ball recording sextant replaced it as the choice of navigators. This instrument used no mirrors. Instead, the navigator viewed a celestial object directly, while a dampened steel ball recorded the celestial object's precise altitude. Versions of this instrument became widely used to navigate not only ships, but also night-flying aircraft.

The modern navigational quadrant led to another important innovation, the accurate **chronometer**. A chronometer is an instrument that keeps extremely accurate time, as needed by navigators making precise measurements of their positions at sea. In 1735, English clockmaker John Harrison (1693–1776) invented the first chronometer reliable enough to keep accurate time under the severe conditions of rocky oceans. Suddenly, in the span of less than 5 years, English sailors had gone from a patchwork method of unreliable instruments to the precise combination of chronometer and navigational quadrant. In 1776, French watchmaker Pierre LeRoy (1717–1785) refined the chronometer by making it far less prone to temperature errors, as often encountered at sea. LeRoy's model became the absolute standard for modern chronometers to the exclusion of virtually all others.

Today, celestial navigation is one of the five primary methods of navigation in widespread use. (Dead reckoning, piloting, electronic navigation and inertial guidance are the others.) Celestial navigation is the oldest form of navigation. Because it began utilizing accurate, scientifically developed navigational quadrants in the eighteenth century, it is also the oldest *modern* form of navigation.

During and immediately after World War II, German, British and American scientists made great advances in navigating through radar technology. In the 1960s, the United States pioneered the use of satellites for navigation. The newest form of modern navigation is the Global Positioning System (GPS), which became functional in 1995. GPS utilizes twenty-four satellites in six unique orbits controlled by the U.S. Air Force. A ship or other vessel's receiver (some are even handheld) captures a signal from at least three satellites at one time. The length of transit time for each broadcast tells the electronic receiver its distance from each satellite. With this information, the receiver calculates its precise location in relation to the orbiting satellites.

SEE ALSO Corpuscular Model of Light and Color; Proper Motion of the Stars and Sun; Stellar Aberration.

Selected Bibliography

Abetti, Giorgio. *The History of Astronomy*. Translated by Betty Burr Abetti. New York: Henry Schuman, 1952.

Hoyle, Fred. *Astronomy*. Garden City, NY: Doubleday, 1962.

Motz, Lloyd, and Jefferson Hane Weaver. *The Story of Astronomy*. New York: Plenum Press, 1995.

North, John. *The Fontana History of Astronomy and Cosmology*. London: Fontana Press, 1994.

Pannekoek, A. *A History of Astronomy*. New York: Interscience, 1961.

Rowland, K. T. *Eighteenth Century Inventions*. New York: David & Charles Books, 1974.

Newtonian Black Holes or "Dark Stars" Are Devoid of Light
John Michell, 1783

In 1687, Sir Isaac Newton (1642–1727) published one of the most important books in the history of science, the *Philosophiae Naturalis Principia Mathematica*. Almost single-handedly, the *Principia* formed the basis for classical mechanics when it outlined Newton's three laws of motion and his theory of gravity.

Before Newton, most scientists accepted the ancient theory of Aristotle (384–322 B.C.) that a given object is naturally at rest except when a force is applied to move it. After several studies of planetary motion, Newton realized that this assertion only applies on the surface of a planet where an external force, such as gravity, holds objects down. Newton's first law therefore states that a body at rest remains at rest and a body in motion remains in motion unless a force acts upon it. A body, Newton said, tends to oppose changes in its velocity. This tendency is known as inertia.

Newton's second law explains the concept of acceleration of an object by an external force. Acceleration of a body, this law states, is in proportion to the force acting upon it. *Acceleration* is defined as the rate at which an object's velocity changes either speed or direction (or even both). This law also refers to the concept of momentum (or mass times velocity).

Newton's third law states that whenever a force acts upon an object, there is an equal and opposite reactive force. A heavy object, such as a dictionary, "presses" against the wooden stand its rests upon. At the same time, the stand presses up against the dictionary with a force that is equal and opposite. Thus, the dictionary remains on the stand. A person can also "feel" the force described in this law by pressing a thumb against a desk or table. Just as a person applies force by pressing down, he or she also feels the force pressing back against the thumb.

Equally important, Newton also studied the question of what force keeps the planets in motion around the sun. For this question, he applied three earlier laws of planetary motion outlined by Johannes Kepler (1571–1630), as well as his own laws of motion. One day in the country, Newton suddenly realized that the same force that makes objects fall straight to earth also secures the moon in its orbit. Using the mathematics of the laws, Newton explained the motion of the planets around the sun and the moon around Earth. He demonstrated that these orbits result from a force of gravity in proportion to 1 over the square of the distance between the sun and a planet or a planet and its moon. This famous law became known as Newton's inverse-square law. Newton had exposed the mysteries of heavenly bodies in a way no one else had ever imagined (though Galileo [1564–1642] had come close).

The law didn't stop there, however. Newton said that his explanation of gravity was not specific to the motion of heavenly bodies. Rather, he said it explains the effect of gravity on every object in the universe. Newton realized his law also explained everyday events, such as a pear falling off a tree. The same force that keeps the moon in orbit around the Earth also makes the pear fall toward the Earth. If the moon did not fall constantly toward the earth (like the pear), it would fly off at a tangent to its actual orbit. Attraction was not, therefore, a matter of specific objects. Rather, Newton showed that it depended on the amount of matter of the bodies being attracted (in addition to the distance between the bodies). For example, on Earth, an elephant weighs more than a mouse because it has more matter (and greater mass).

In 1704, Newton published a second book of terrific importance, *Optiks*. This book became the model for experimental physics for the rest of the century. One section detailed Newton's findings about the makeup and nature of light. Previously, most scientists thought that light traveled in waves, similar to those of sound. However, Newton reasoned that if light were made up of waves, it would travel around corners, like sound. Further, he was unable to explain some significant laboratory results with the wave theory. Newton proposed an alternate version of the makeup of light called the **corpuscular**, or particulate, **theory** of light. This model had light composed of tiny, discrete particles (or corpuscles) moving in straight lines at a finite velocity. Newton's corpuscular model was widely adopted for the next 100 years. (For more information, please see the entry on the Corpuscular Model of Light.)

In context of Newton's dual discoveries—the laws of motion and gravity, as well as the corpuscular theory of light—it is somewhat surprising that Newton himself did not actually predict the existence of black holes. These "dark stars," as they were often known, were objects in the universe with such fantastic gravitational pull that even particles of light could not escape. It took less than a century after the publication of the *Principa* for Newton's equations to be used to describe black holes. John Michell (1724–1793), an important English astronomer who also founded the science of seismology, suggested the existence of black holes in 1784. His theoretical discovery came to pass only when he applied Newton's ideas about gravity to Newton's corpuscular model of light.

Michell's work in astronomy was well regarded by his fellow scientists, and he had already made the first reasonable estimates of stellar distances. Michell also made reasonable guesses about the velocity needed for "particles" of light to escape various heavenly bodies. He knew the escape velocity of light from the Earth and, assuming for the moment that the sun had the same density as the Earth, Michell made an estimate of the sun's escape velocity. Because the sun is many times larger than the Earth, the required escape velocity would be much higher. Still, using various estimates, Michell found that the necessary velocity to escape from the surface of the sun is only 0.2 percent of the approximate speed of light (as figured by James Bradley [1693–1762]). As everyone well knows, light has no problem escaping from the sun.

But what if, asked Michell, objects existed in the universe with such mass that the escape velocity was greater than even the speed of light? Newton had long since shown that objects with greater mass exert greater gravitational pull, resulting in a much higher escape velocity. Michell proposed the astounding idea that these enormous objects would appear as "dark stars," completely invisible because the particles of light could not escape from their surfaces. He went on to calculate that a star with the same density as the sun and a radius about 500 times would exert just such a fantastically strong gravitational field. This type of star would therefore be invisible to the rest of the universe. Like cannonballs fired straight up from the surface of Earth, this star's particles of light would never escape its gravitational pull.

However, in the nineteenth century, scientists such as James Maxwell (1831–1879) revived a sophisticated version of the wave theory of light. Maxwell discovered equations demonstrating that waves in combined electric and magnetic fields, known as *electromagnetic waves*, travel at exactly the speed of light. Maxwell and most other scientists became convinced that, contrary to Newton's findings, light behaved similarly to the newly discovered radio waves.

Like the old wave theory and the corpuscular theory, the new wave theory of light would not last. In 1905, Albert Einstein (1879–1955) proposed his special theory of relativity. He also showed that light has properties that the wave model cannot explain—most especially, the act of knocking electrons out of metal (the photoelectric effect). Physicists now came to the tough conclusion that light is both a wave and a stream of particles (or photons). More accurately, light is *neither* truly a wave nor a particle in quantum theory. Sometimes light acts like one, and sometimes it acts like the other.

In 1916, Einstein proposed his general theory of relativity—a complete and universal theory of acceleration and gravity, akin to that presented by Newton's *Principia*. This theory proposed that, because light always travels at the same speed, black holes in the universe will be dark and prevent light from escaping. This theory also proposed the bending of the space–time continuum, which actually requires the existence of black holes. Because the gravitational force of a black hole is so powerful, all of its matter would be located at its central point. Quantum physicists call this point its *singularity*, the point where space and time are, quite literally, snuffed out.

More recently, British physicist Stephen Hawking (1942–) has made some of the most startling discoveries since those of Newton and Einstein. Hawking proposes that, even though light cannot escape, a black hole nonetheless spits out particles and radiation until it eventually explodes and disappears entirely. During the last 10 years, scientists using the Hubble Space Telescope have spotted evidence of more than ten monstrous black holes at the center of galaxies. In June 2000, astronomers snapped images of giant gas "bubbles" as they were spewed into space by an enormous black hole in the galaxy NGC 4438. Evidence of this sort has rekindled debates among physicists about what lies on the other side ("through the looking glass") of a black hole.

SEE ALSO Comets Follow Predictable Orbits, Cycles, and Returns; Corpuscular Model of Light and Color; Proper Motion of Stars and Sun; Stellar Aberration.

Selected Bibliography

Gribbon, John. *Unveiling the Edge of Time: Black Holes, White Holes, Wormholes.* New York: Harmony Books, 1992.

Krebs, Robert E. *Scientific Laws, Principles, and Theories: A Reference Guide.* Westport, CT: Greenwood Press, 2001.

Parkinson, Claire L. *Breakthroughs: A Chronology of Great Achievements in Science and Mathematics, 1200–1930.* Boston: G. K. Hall, 1985.

Sullivan, Walter. *Black Holes: The Edge of Space, The End of Time.* Garden City, NY: Anchor Press/Doubleday, 1979.

O

Optical and Mechanical Telegraphs
Richard Edgeworth, 1767
Claude Chappe, 1791

Along with the electric light, the telegraph is generally considered one of the most important inventions of the nineteenth century. It is also considered the forebear of the modern telephone, fax machine and modem. Credit for the electric telegraph is often shared among many inventors, as it had a long history of scientific development before American inventor Samuel Morse (1791–1872) won the first patent for it in 1837.

The telegraph eluded a great many of the eighteenth century's inventors, who (time and again) saw its theoretical potential, but not its actual creation. In fact, until the second half of the eighteenth century, communication was no faster than it had been in antiquity. Even the most urgent news (be it scientific, military, political or commercial) traveled only as fast as its human messengers aboard boat, ship or horse. Yet this rule of thumb is not without exception, as scientists of previous centuries experimented with various forms of telegraphic methods. Ancient Chinese scientists devised several elaborate systems of flags to relay numbers, letters and other codes. They also used kites (sometimes with human passengers) for both observation and communication. The ancient Greeks used complex systems of torches to signal wars and other calamities. Centuries later in Europe, rudimentary efforts involving beacons, trumpets, large bells, and other musical instruments met limited success.

By the late eighteenth century, the topic of electricity and its potential for communication occupied the minds of some of Europe's most important scientists. Francis Hauksbee (sometimes Hawksbee or Hawkesbee, ca. 1666–1713) greatly improved the **electric machine**, which produced current through friction. He also made the case for a relationship between electricity by friction and light. Stephen Gray (1695–1736) had discovered **conduction** (the actual flow of electricity) and **insulation**. Charles Dufay (1698–1739) and Benjamin Franklin (1706–1790) were publicly testing the nature of electricity, which they thought was a fluid. Luigi Galvani (1737–1798) discovered electricity in animals, though he incorrectly thought that it was somehow produced in the moist muscles of laboratory specimens. Finally, in possibly the century's most important work on electricity, Alessandro Volta (1745–1827) found that electricity was produced not in the tissue of animals, but in the moist combination of different metals. Through this research, he invented the first practical electrochemical cell, known as the **voltaic pile**. (For more information, please see the various entries on these topics.)

In light of these important advances in the science of electricity, it is not surprising that the technology for rudimentary electric telegraphs existed by the early part of the century. According to many historians, both Gray and Sir William Watson (1715–1787) could have simply used equipment in their possession to send crude "telegrams" in the form of coded electrical impulses over wires. Neither one ever did. In a 1753 issue of *Scots* magazine, an anonymous letter actually suggested setting up an electric telegraph employing twenty-six wires—one for each letter of the alphabet. Several scientists tried to make such devices, but none was particularly successful. Many similar letters were soon published.

The first successful telegraph of the century was not electric at all. It was a series of mechanical "machines" constructed by an Irish scientist of many talents, Richard Edgeworth (1744–1817). Historical records suggest that immediate necessity was the spark of invention for Edgeworth. One night in 1767, a noble friend of Edgeworth expressed regret over missing a horse race in Newmarket, England, because he had to stay in London. A third man bet he could have the results of the horse race to London by 9 P.M.—several hours after the race was to end. This third man proposed a kind of Pony Express precursor, by which he would station race horses along the path from Newmarket to London to carry the news. Edgeworth immediately bet that he could construct a system to return the news by 5 P.M.

Edgeworth proposed a series of stations with highly visible "machines" resembling wooden windmills no more than 16 miles apart—a distance easily seen by telescope. On seeing a visual message, a person at each station would relay it by contorting the long arms of his "machine" to an agreed-on position. This original proposal was never built because, on hearing the specifics, Edgeworth's competitor with the swift horses quickly withdrew his bet.

However, Edgeworth did build his noble friend a smaller version of the system, which effectively allowed the man to communicate between his

London house (at Downing Street) and Piccadilly. For this system, Edgeworth used large wooden letters, which he illuminated by lamps for viewing against the dark English sky. This system worked well enough, though the lamps rendered it too expensive for all but the most urgent of noble news.

Edgeworth made no other serious attempt until 1794, when widespread Irish fear of invasion by revolutionary France required rapid transfer of intelligence information over long distances. In this year, Edgeworth erected a series of stations consisting of groups of four long sticks holding one pointer each, like a single clock hand. Each pointer had 8 possible positions, with straight up signifying 0 and straight down signifying the number 4. The horizontally arranged pointers were read left to right, each signifying a unit (thousands, hundreds, tens and ones). In this fashion, the machines could relay numbers between 0 and 7,777. The big difference between this improved 1794 system and the less refined one of 1767 is that, in the intervening years, the semaphore or optical telegraph had been perfected by Claude Chappe (1763–1805) with the aid of his three brothers.

Previously, Chappe had experimented with aural, mechanical and electric telegraphs before deciding on the semaphore. He chose this optical method because it was faster and more discernible than sound-based systems. It was also less cumbersome than either the mechanical or electric telegraphs available to him.

During this time, the Frenchman Chappe actually coined the term *télégraphe* (or telegraph) literally from the Greek for "write at a distance." Though not coined until 1801, the term *sémaphore* (or semaphore) was also from the Greek, meaning literally to "bear a sign." The distinction is important. Mechanical and later electric telegraphs relayed an actual message through a theoretically transparent medium. Conversely, the success of the semaphore was that it was not limited by its ability to carry a message. Rather, the semaphore had only to signify a predetermined sign, analogous to a modern stoplight (where red means stop and green means go). There was no longer any need to pass along messages, which had thus far proven cumbersome and limiting.

In 1791, Chappe demonstrated two semaphore stations a quarter of a mile apart. Each station had a large wooden indicator that was arranged in a variety of positions. In the first public demonstration, two station operators traded signs with predetermined meanings. Later that year, Chappe demonstrated a 15-kilometer exchange of messages, using various stations between Brûlon and Parcé in France.

The local officials were impressed by the demonstrations—most especially for the military uses that the new invention readily offered the new republic. Chappe was allowed to build the world's first semaphore line of 16 stations for communication between Paris and an army installation at Lille (see

Figure 19. Semaphore towers. Previous to Claude Chappe's series of evenly spaced semaphore towers, communication was no faster than it was during antiquity. But with a little practice, Chappe's semaphore operators learned to transmit a simple sign over many miles in just a matter of minutes.

Figure 19). (This line was eventually lengthened to Dunkirk.) The line worked so well that, in 1798, Chappe built a second line from Paris to Strausbourg. The lines were series of stone towers capped with a long pole holding a cross-bar and long wooden arms protruding into the sky. In the rock tower below, an operator watched for the previous tower's sign through a telescope. On seeing it, the operator manipulated wires to pivot the long arms into position and was thereby able to signal in a code of around 10,000 words that Chappe developed.

Chappe's semaphore system was an overwhelming success, as lanterns even allowed its use at night and in light fog. It was also the first efficient, effective and affordable form of high-speed communication. Once the operators became highly skilled with the signs, messages were routinely transmitted to Paris from Lille (a distance of approximately 150 miles) in only about 2 minutes. Suddenly, French military units could transmit messages and wait for instructions in a reasonable amount of time. Eventually, similar systems were developed in countries such as England and Sweden. It was, however, Chappe's semaphore that aided the scattered armies of Napoleon in his conquests across Europe.

During the nineteenth century, the semaphore was widely replaced by the electric telegraph, which could transmit much more information. Once it was established, the telegraph also did not falter in poor weather conditions, such as heavy fog or strong wind.

Even today, both the telegraph and semaphore are still in limited use. Western Union Financial Services provides telegraph service in the United States. Huge numbers of fee-based messages known as *opiniongrams*—telegrams drafted by ordinary citizens—are sent to a Western Union office near the legislative offices of Washington, DC. From there, they are printed and delivered to members of congress. The U.S. Navy employs a system of semaphore flags for extremely short-range signaling. Also, until the latter half of the twentieth century, the Boy Scouts and similar organizations routinely taught how to signal with flags, as well as how to use Morse code.

SEE ALSO Chemical Cells; Electricity Can Produce Light; Electricity Is of Two Types.

Selected Bibliography

Canby, Edward Tatnall. *A History of Electricity*. New York: Hawthorn Books, 1968.

Crawley, Chetwode. *From Telegraphy to Television: The Story of Electrical Communications*. New York: Frederick Warne, 1931.

Fahie, John J. *A History of Electric Telegraphy to the Year 1837*. New York: Arno Press, 1974.

Meyer, Herbert W. *A History of Electricity and Magnetism*. Cambridge, MA: MIT Press, 1971.

Rowland, K. T. *Eighteenth Century Inventions*. New York: Harper & Row, 1974.
Wilson, Geoffrey. *The Old Telegraphs*. London: Phillimore, 1976.

Oxygen Is an Element That Supports Combustion
Carl Scheele, 1771–1772
Joseph Priestley, 1774
Antoine Lavoisier, 1779

Since antiquity, the nature, makeup, and properties of matter have spawned some of the most fundamental and contested theories in the entire history of science. In Greece during the 400s B.C., Empedocles (ca. 495–435 B.C.) taught that the four elements of air, earth, fire and water combined in varying proportions to create every object in the universe. Around 400 B.C., Democritus (ca. 460–370 B.C.) taught that tiny, indestructible units called atoms composed all objects. This theory claimed that differences in objects were caused by the size, shape and position of their constituent atoms. A little later, Aristotle (384–322 B.C.) claimed that each of Empedocles' four elements could be transmuted into one of the others by adding or removing heat and/or moisture. Following Aristotle's logic, Europeans from the 1100s to the 1600s (and beyond) were trying to transmute lead into gold through the practice of **alchemy**. Though it never produced gold, alchemy did produce the chemical knowledge that later formed the basis of modern chemistry.

Around 1700, German alchemist Georg Ernst Stahl (1660–1734) proposed the theory of **phlogiston**—alchemy's powerful final permutation. This theory said that **combustion** releases the hypothetical substance phlogiston from a burning object, such as a candle or a piece of charcoal. Combustion ceased only once an object had released all its phlogiston. The phlogiston theory swept across Europe and especially England. It provided a reasonable, rational explanation for most scientists' experimental results, as well as for everyday events. This was the theory that led to most of the eighteenth century's discoveries of new gases. It was also the theory to which many of the century's great scientists would subscribe for their entire lives.

One such scientist was the Swedish apothecary Carl Scheele (1742–1786). Though his achievements are often overlooked, many chemical historians consider Scheele the single greatest chemist of the eighteenth century. Books of Scheele's contributions and discoveries could fill libraries, given their importance and number. Scheele's chemical discoveries include arsenic acid, **chlorine**, glycerol, hydrogen cyanide, hydrogen fluoride, lactic acid, **nitrogen**, permanganates and uric acid. Several discoveries—like the isolation of chlorine gas—rank among the most useful in history. (For more on this topic, please see the entry on Chlorine.)

The reclusive Scheele conducted all of his research privately, working as an apothecary by day and researcher at night. Preferring to work nearly all of the time—and nearly always alone—Scheele turned down many prestigious offers from Europe's fabled academies.

In 1771, Scheele began a general study of the properties of air. A staunch phlogistonist, Scheele thought he was manipulating the elastic properties of a single, unified air. In one set of preparations, he produced a peculiar gas by heating saltpeter (potassium nitrate). Later, Scheele identified a strikingly similar gas when he heated manganese dioxide. This finding prompted Scheele to plan a series of separate heating trials. He heated heavy metal nitrates, then silver carbonate, and finally mercuric oxide. Scheele found that the gas they released had the same properties as before. Many tests later, Scheele concluded that heating each of these substances had produced the very same gas. Later, he also found this gas in the respiring air of plants and fish.

Scheele detailed each of these experiments in his only major publication, the 1777 book, *A Chemical Treatise on Air and Fire*. As the title suggests, Scheele staunchly believed in the phlogiston theory his entire life. Accordingly, he named his gas **fire air**. Scheele believed that he had isolated the substance that combined with phlogiston as it was released during reactions that released heat.

In 1772, English pneumatic chemist Joseph Priestley (1733–1804) documented the basic gaseous processes of plant **respiration** for the first time. He also explored the process now known as **photosynthesis**. In a series of trials, Priestley showed that the same gas that supports animal respiration is given off by green plants. He decided to investigate the matter much further. In a crucial investigation, Priestley enclosed a candle and the shoot of a green plant together in a laboratory container. Then, he burned the candle until the container's air would no longer support its flame. Priestley was careful not to burn the plant. He let several hours pass. When he was then able to burn the candle, Priestley concluded that the gas given off by the plant was the same gas that supported combustion.

Two years later in 1774, Priestley was surprised to rediscover his gas in the most unlikely of places. He had been conducting calcification (or oxidation) experiments in his laboratory for many months, with few useful results. One day, Priestley held his burning lens over a **calx** of mercury (or mercuric oxide). He collected the gas that this reaction released with an instrument known as a **pneumatic trough**. When he began tests on the gas, Priestley instantly recognized it as the same gas he had identified in plant respiration.

In subsequent experiments, Priestley determined that this intriguing gas supported a mouse's respiration much longer than an equal volume of ordinary air. Just as he thought, the gas also made a candle burn with a positively dazzling intensity. Priestley found the gas quite pleasant to breathe and not readily soluble in water. He broadly recommended its practical use in medical applications. A phlogistonist like Scheele, Priestley named his gas **dephlogisticated air**. He believed the gas had lost all its phlogiston and therefore hungrily grabbed it back from a burning object or a green plant. Several months later in 1774, Priestley traveled to Paris, where he was a popular figure due to his radical politics. Once there, he freely shared his discovery with the charismatic young chemist at the very center of the emerging chemical revolution, Antoine Lavoisier (1743–1794).

Unlike Priestley and Scheele, Lavoisier did not discover new gases. Rather, he reinterpreted the discoveries made by phlogiston chemists by fashioning a new theoretical framework. Lavoisier was the first scientist to pay close attention to the concept of mass in chemical reactions. He invented a laboratory balance capable of measuring tiny variations in mass. Stahl and his followers often cited calcification experiments as evidence of the phlogiston theory. When a substance was heated, calcification occurred and the substance lost mass. The phlogistonists said that this was due to the escape of phlogiston. However, Lavoisier knew of many calcifications (such as that of tin) in which the substance actually *gained* mass. Alchemists had actually puzzled over them for years. Stahl had tried to explain these circumstances as exceptions to the rule. He claimed that phlogiston sometimes has mass, sometimes has no mass, and sometimes has what amounted to a "negative" mass. Lavoisier was very skeptical of this account. (For more on this topic, please see the entry on Phlogiston.)

Ironically, it was the discoveries of phlogiston that would facilitate Lavoisier's destruction of the old theory. After studying Priestley's experiments, Lavoisier set off to repeat them with one key difference. This time, he would perform accurate measurements of changes in mass. After several experiments, Lavoisier turned his attention to the "calx" mercury (or mercuric oxide). Lavoisier carefully heated the substance, just as Priestley had done.

What he needed was a way to measure the mass of a calcification in a closed environment. Lavoisier enclosed the whole experiment in a container. When he heated the mercury, the "calx" formed, just as he expected. The entire container showed no change in weight. As he expected, no substance had either entered or left the glass container.

However, Lavoisier was surprised at what he found upon making a small opening in the glass. Air rushed in. Quickly, Lavoisier resealed the container. When he re-weighed the container, Lavoisier found that he had proved his conclusion. The weight gain of the container exactly equaled the weight gain of the mercury. Lavoisier repeated his experiment with many other "calxes," including tin.

"Something" from the air had united with the substance during heating. Lavoisier concluded that, as he suspected, the phlogistonists had it exactly backward. Phlogiston wasn't *separating from* a substance during heating or combustion. Something else from the air was *joining with* it. In strict Newtonian terms, phlogiston had no gravitational mass—it was purely hypothetical. But exactly what was this substance? Lavoisier demonstrated beyond doubt that it was the gas Priestley called "dephlogisticated air" and Scheele called "fire air." This gas supported combustion, animal respiration, and also "calcification."

Lavoisier mistakenly thought that the gas was present in all acids. In 1779, he gave the gas its modern name, **oxygen**, from the Greek for "acid producer." Even though his name was based in an erroneous conclusion, the discovery of oxygen changed everything. Scientists stopped talking about calcification and began studying oxidation. The simple formulation of *chemical + oxygen → chemical oxide* explained many chemical processes that phlogiston could not. It also explained that mass remains constant in chemical reactions—a formulation known as the **law of conservation of mass**. (For more on this topic, please see the entry on this law.)

In 1784, staunch English phlogistonist Henry Cavendish (1731–1810) demonstrated that water was not an **element**, as scientists had believed since antiquity. Rather, Cavendish showed that water is a **compound** of two elements, his **inflammable air** (or **hydrogen**) and oxygen. To the astonishment of many scientists, Cavendish successfully separated water into its constituent elements. (For more on this topic, please see the entry on Water Is a Compound.)

A year later, Lavoisier performed the opposite experiment. In a big public demonstration, Lavoisier carefully forced hydrogen and oxygen from two separate containers into a large flask. There, he united them with an electric spark. Lavoisier successfully "made" 5.5 ounces of water. Beyond doubt, water was a compound of two elements—hydrogen and Lavoisier's special gas, oxygen.

Once Lavoisier demonstrated that the special element oxygen was responsible for combustion and oxygenation, and was also an important constituent in many compounds, there was no turning back. The chemical revolution was in full swing. A new age had dawned by the time Lavoisier published the crown jewel of the chemical revolution, his 1789 book *Traité Élémentaire de Chimie*. More than any other, this book heralded the dawning age of modern chemistry. It disseminated its basic premises and empirical findings. For the first time, a book listed tables of modern elements and compounds. (For more on this topic, please see the entry on Substances: Elements and Compounds.) Many staunch proponents risked ridicule and professional humiliation by publicly abandoning phlogiston altogether.

To mark the book's success, Lavoisier's wife, translator and collaborator Marie-Anne Paulze-Lavoisier (1758–1836) dressed as a priestess and ceremoniously burned Stahl's old phlogiston treatises. One can almost hear them joking about releasing phlogiston. Also, the journal *Chemische Annalen* reported that a group of Lavoisier's friends also performed a celebratory play. In it, phlogiston was placed on trial, briefly defended by Stahl, and finally burned at the stake. Dramatically, this play exploited the fatal paradox of phlogiston theory. If combustion releases phlogiston, then what does burning phlogiston produce? By definition, it produces absolutely nothing. In the end, phlogiston had no where else to turn.

Today, the discovery of oxygen ranks among the most important and useful discoveries in history. Since Lavoisier's day, chemists have accepted—and built on—his theory of oxygen chemistry with near universality. For many years, Priestley received credit for the discovery of oxygen, because he was first to publish his results. Only recently have chemical historians discovered that Scheele's laboratory notes prove he discovered oxygen several years before Priestley. However, a series of publishing blunders and Scheele's self-effacing personality kept his discovery from being published until several months after Priestley published his findings. At the time, Priestley was completely unaware of Scheele's prior claim. For this reason, credit is generally given to both Scheele and Priestley for independently identifying this revolutionary gas, and to Lavoisier for explaining its importance.

SEE ALSO Chlorine; Law of Conservation of Mass; Phlogiston; Substances: Elements and Compounds; Water Is a Compound.

Selected Bibliography

Coulston, Charles Gillispie (ed.). *Dictionary of Scientific Biography*, volume XI. New York: Charles Scribner's Sons, 1972.

Gibbs, F. W. *Joseph Priestley: Revolutions of the Eighteenth Century*. Garden City, NY: Doubleday, 1967.

Greenberg, Arthur. *A Chemical History Tour: Picturing Chemistry from Alchemy to Modern Molecular Science*. New York: John Wiley & Sons, 2000.

Holt, Anne. *A Life of Joseph Priestley*. Westport, CT: Greenwood Press, 1970.

Lockemann, Georg. *The Story of Chemistry*. New York: Philosophical Library, 1959.

Millar, David, Ian Millar, John Millar, and Margaret Millar. *The Cambridge Dictionary of Scientists*. New York: Cambridge University Press, 1996.

Porter, Roy (ed.). *The Biographical Dictionary of Scientists*. New York: Oxford University Press, 1994.

Schwarts, A. Truman, and John G. McEvoy (eds.). *Motion Toward Perfection: The Achievement of Joseph Priestley*. Boston: Skinner House Books, 1990.

P

Phlogiston (or the First Reasonable, Rational Theory of Chemistry)
Georg Ernst Stahl, 1700

By the time **alchemy** reached Europe during the Middle Ages, the four elements of the ancient world had dominated chemical thought for nearly a millennium and a half. The alchemists mainly endeavored to turn "base metals" such as lead into "noble metals" such as gold. This idea, known as **transmutation**, came directly out of the proposal by Aristotle (384–322 B.C.) that each **element** contained two of the following qualities: heat, cold, moisture and dryness. Fire was hot and dry, air was hot and moist, water was cold and moist, and earth was cold and dry. A cornerstone of this ancient philosophy was that an object could burn because it contained within itself the element of fire. Under just the proper conditions, fire made its dismembering escape and burning ensued.

The alchemists had their own version of the element of fire, which they called the principle of "sulfur." This was not necessarily the element sulfur, though most alchemists also experimented with it. In 1669, a German alchemist, Johann Joachim Becher (1635–1682), attempted to expand the fire principle idea. Becher divided the four ancient elements into pairs: water and earth were the "important" elements, while he thought air and fire were simply agents of chemical change. Becher knew of the medical alchemist Paracelsus (also called Bombastus von Hohenheim, 1493–1541), who separated philosophical–chemical "matter" into three components: salt (the body), mercury (the spirit), and sulfur (the soul). Becher outlined his own three Earth Principles, which roughly corresponded to these. Salt became *terra lapida*, the essence of fixity and inertness. Mercury became *terra mercurialis*, the essence of fluidity. And most importantly, sulfur became **terra pinguis**, the essence of inflammability.

Becher said *terra pinguis*, or "fatty earth," was present in all combustible matter and was liberated upon **combustion**. When an object burned, light and especially heat were liberated. The object lost weight and turned to ash, because it had lost its matter of fire, its *terra pinguis*. Becher never tested his vague doctrine or nurtured it into a comprehensive theory. Like his fellow alchemists (dating back to Aristotle), Becher more or less tautologically explained that a substance burns because it is potentially combustible. Nonetheless, the first rational, unified theory of combustion grew directly out of Becher's work when German alchemist Georg Ernst Stahl (1660–1734) made another substitution. He said the essence of fire was not a philosophically indistinct "earthy" material like *terra pinguis*, but a definite, quantifiable, chemical component in all combustible materials. Stahl named this component **phlogiston**, from the Greek word *phlogistos*, or "burning."

Stahl's theory grew directly out of a series of observations. Central to phlogiston was Stahl's finding that metals frequently produced a **calx**—or crumbly, rusty residue—when he roasted them. In most cases, these calxes were less dense than the original metals. Stahl also had knowledge of many similarities between metals. This was a direct departure from the older alchemist distinction between "base" and "noble" metals. One of these similarities was that many calxes re-formed metals when they were heated with charcoal. Stahl correctly identified that the calxes he produced were the same substances most often found in nature. In other words, Stahl realized that metals mostly existed in calxes, or rocky ores in their natural state.

Stahl had thus far nurtured Becher's embryonic concept to a cohesive theory of combustion. Now, through a series of bold conclusions, Stahl shifted phlogiston to the very epicenter of his entire system of chemicals. With one stroke of genius, Stahl realized that his theory explained combustion in the conventional sense, as well as the calcification of metal. He said calxes were the "simple" form of metal because they were commonly found in nature. A calx was, therefore, a metal that is free of phlogiston. To form a metal from a calx, one need only add phlogiston—from burning charcoal—which the calx readily "absorbed": *calx + phlogiston → metal*. Alternately, a calx was formed by burning off the phlogiston: *metal → calx + phlogiston*. In these terms, phlogiston explained a chemical cycle, widely viewed as natural: Metal loses its phlogiston during combustion. When charcoal's phlogiston is added back, metal is restored from its calx. This cyclical insight was the first reasonable explanation of perhaps the world's oldest chemical practice: the separation of useful or precious metals from compound, rocky ores.

Stahl's theory made academic sense to scientists. It also just as effectively explained everyday chemical phenomena. Stahl explained that wood, charcoal and candles were especially rich in phlogiston because they leave only modest ash when they burn. Some scientists went so far as to equate charcoal with phlogiston itself, but Stahl never endorsed this comparison. Rather, he explained simple combustion in simple chemical terms. He said that when a piece of wood burns, flames escape from each pore as phlogiston begins passing from wood to the surrounding air. As burning progresses, the wood turns black and cracks wide open as phlogiston makes its escape. When all of the phlogiston is liberated, the flames die out. Then, as the residual phlogiston escapes, the remaining ash glows until, finally, all that is left behind is the wood's ash (see Figure 20)

It follows, therefore, that wood is made up of phlogiston and ash. Ash, not wood, is held to be the unique substance, the "thing" that exists independent

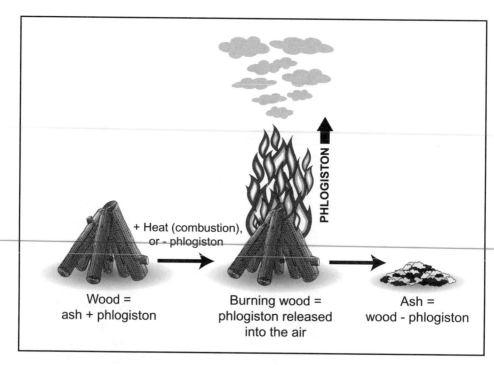

+ Heat (combustion),
or - phlogiston

PHLOGISTON

Wood =
ash + phlogiston

Burning wood =
phlogiston released
into the air

Ash =
wood - phlogiston

Figure 20. Phlogiston. Proposed in 1700 by George Ernst Stahl, phlogiston was the first rational, reasonable theory of chemical substances. This theory said that, when a substance burns, it releases the hypothetical substance phlogiston into the surrounding air. The process of combustion left behind a metal calx or other ashy matter. Though it was replaced by a better theory, phlogiston led to the first discoveries of independent gases.

of phlogiston. As for phlogiston, there was nothing hypothetical or elusive about it in Stahl's theory. Rather, he saw it as a definite, chemical component. In strictly Newtonian terms, it was conceived as a chemical with a specific gravitational mass.

Stahl widely expanded his durable theory when he correctly identified rusting as a kind of "combustion." When iron rusts, he said, it undergoes a very slow form of combustion. The metal iron, therefore, consisted of rust and phlogiston. The same was true for sulfur, which consisted of phlogiston and the choking fumes released during burning. Smelting was really just a matter of carbon giving its phlogiston to a calx, a process that resulted in metal. Stahl's theory applied equally well to "wet" reactions, such as those involving water, acids or any of the vitriols (or sulfates). Some physiologists even tried (unsuccessfully) to explain all bodily functions in terms of phlogiston. Yet, Stahl shrewdly identified only human **respiration** as a form of combustion.

Phlogiston succeeded because, like all good chemical theories, it provided an explanation for a variety of results and experiments. It explained why some objects burned and then later would not. It explained that combustion prematurely ceased in closed containers because the air became saturated with all the phlogiston it could hold. It also explained why perfectly healthy mice would suddenly die after only several hours in a closed container.

Phlogiston was also wildly successful in terms of praxis when, later in the century, it led to the first discoveries of unique "airs," or gases. In 1756, Joseph Black (1728–1799) discovered the first distinct gas in history when he isolated **carbon dioxide**. He named his gas **fixed air** because of its inability to absorb phlogiston (or support combustion). During the next decade, Henry Cavendish (1731–1810), one of the last scientists to believe the phlogiston theory, discovered the element **hydrogen**. He named his find **inflammable air**. Because of its violently combustive properties, Cavendish thought he had captured the elusive phlogiston itself. By this time, phlogiston chemistry was in full swing, with new gases being discovered on a regular basis. The great Swedish apothecary Carl Scheele (1742–1786) produced "dephlogisticated acid of salt" in his laboratory. It was not until the next century that Sir Humphry Davy (1778–1829) would name the gas **chlorine**.

Finally, Scheele and Joseph Priestley (1733–1804) independently discovered the century's most important gas, which Scheele called **fire air**. Priestley initially reported that this gas could sustain a mouse twice as long as ordinary air and that it made a candle burn twice as long and with twice the intensity. Just as Stahl thought charcoal was rich in phlogiston because it burned so cleanly, Priestley thought this gas was totally void of phlogiston, which it hungrily gobbled up during combustion. He named it **dephlogisti-cated air**.

It would take a revolution in chemistry for this gas to earn its modern name. All along, Stahl had been skirting a problem with his theory. Since the late 1490s, alchemists had steadily recorded specific metals whose calxes were actually *heavier*, not *lighter*, than the metals themselves. In 1630, the chemist Jean Rey (ca. 1582–1645) correctly recorded that the calx of tin is roughly 25% heavier than the original metal. If a substance loses phlogiston during combustion, how could this be? Stahl explained weight gain by postulating that, depending on the metals and the reaction, phlogiston can alternately have actual weight, have no weight at all, or have what amounted to a "negative" weight.

This explanation didn't sit well with several French chemists. It was especially troubling to Antoine Lavoisier (1743–1794), who would become the founder of modern chemistry at the century's end. Lavoisier did not discover new gases. Instead, he reinterpreted the discoveries made by the pneumatic chemists. Lavoisier was the first chemist to pay close attention to the concept of mass in chemical reactions. He made unprecedented studies of the mass of chemicals before and after a reaction. As he refashioned chemistry to a modern science, Lavoisier began thinking that something akin to phlogiston wasn't *separating from* the chemical during reaction. Rather, something in the air must be *joining with* the chemical. In a series of experiments, Lavoisier found that this "something" was none other than the newly discovered gas, "fire air" or "inflammable air."

Lavoisier gave this gas its modern name, **oxygen**, from the Greek for "acid producer." His choice of name was the result of the mistaken belief that oxygen is present in all acids. Nonetheless, his breakthrough theory of combustion changed everything. Chemical reactions became the basis of a new oxygen chemistry. No longer were scientists talking of *calxes*, but were instead concentrating on *oxidation*, as in: *chemical + oxygen → chemical oxide*. His careful studies of chemical mass (as in oxidation reactions) led to one of chemistry's great articulations, the **law of conservation of mass**. This law states that the weight of the products of combustion equals the weight of the reacting substances. This observation became foundational to the theory of oxygen chemistry. This theory has been accepted with virtual universality since Lavoisier's day.

Phlogiston was a powerful, fruitful theory that was eventually replaced by a better one. For this reason, a long-running debate continues among chemical historians. Some see phlogiston as a roughly 100-year distraction, a glitch in the procession toward modern chemistry. Others argue that it was a necessary link between the older concepts of alchemy and the new chemical framework, which was still only emerging at the century's end. Either way, phlogiston raised the right questions—and led to the right discoveries—for the later theory to supersede it.

SEE ALSO Carbon Dioxide; Chlorine; Hydrogen; Law of Conservation of Mass; Oxygen; Rational System of Chemical Nomenclature; Substances: Elements and Compounds.

Selected Bibliography

Asimov, Isaac. *A Short History of Chemistry: An Introduction to the Ideas and Concepts of Chemistry.* New York: Anchor Books, 1965.

Farber, Eduard. *The Evolution of Chemistry: A History of Its Idea, Methods, and Materials.* New York: Ronald Press, 1952.

Greenberg, Arthur. *A Chemical History Tour: Picturing Chemistry from Alchemy to Modern Molecular Science.* New York: John Wiley & Sons, 2000.

Krebs, Robert E. *Scientific Laws, Principles, and Theories: A Reference Guide.* Westport, CT: Greenwood Press, 2001.

Moore, F. J. *A History of Chemistry.* New York: McGraw-Hill, 1931.

Plant Physiology Is Independent of Animal Physiology
Stephen Hales, 1727

The idea that plant **physiology** is fundamentally similar to animal physiology is extremely old. This concept is known as the analogist approach because it draws comparisons between two different things, in this case animals and plants. The conceptual framework undergirding this approach can be seen in language, even today, in phrases like the "heart" of wood or the "veins" of a leaf.

This approach most likely developed when Aristotle (384–322 B.C.) began an intensive search for anatomical analogies between the conducting tissues of plants and the blood circulation of animals. Aristotle's vast metaphysical scale of being led him to the search because it theoretically organized all organisms in order of increasing functional and structural complexity. Some features, such as sexual reproduction, were specific only to "higher" level organisms. Aristotle held that others, such as absorption of food, were common to plants and animals alike. Aristotle's pupil, Theophrastus (372–288 B.C.), is often called the founder of **botany**. Theophrastus largely emulated his teacher's scale of being and classified plants in reference to their general form and appearance (trees, shrubs, herbs and so forth). These classifications dominated botany to the Middle Ages. However, aside from analogies, they make no mention of physiological or anatomical studies of plants.

The Italian naturalist Cesalpino (1519–1603) made some of the first pseudo-objective studies of plant physiology. Cesalpino studied the movement of sap, which had provided medieval botanists a great deal of confusion and mystery. With an emphasis on liquids, he postulated that plants use roots to draw the sap directly out of damp soil. He found that they convey the sap through tiny veins directly into the stem. Though he was largely mistaken, Cesalpino's experimental study marks a rupture with Aristotle's purely analogist approach. At the same time, however, Cesalpino reinforced the analogist approach when he announced that the "flesh" of fruits and vines "bleeds" exactly like that of an animal.

During the same century, Johann Baptist Van Helmont (1577–1644) performed his famous "tree experiment." In contrast with Aristotle's four elements, Van Helmont believed in only two: water and air. He thought trees were composed solely of water. To test his theory, Van Helmont carefully weighed 200 pounds of dry earth, added distilled water and planted a 5-pound tree sapling. Five years later, Van Helmont found that the tree weighed 169 pounds, but the soil (when dried) still weighed 200 pounds. Van Helmont concluded that the 164-pound difference was made purely of water that the tree had absorbed. Though the conclusion was completely errant, Van Helmont had broken with the analogist tradition in a major way. He had turned Aristotle's theory on its ear. Here was a study of plant mass and composition that related in no way to animal physiology. Here also was the infantile emergence of modern chemical methodology with Van Helmont's implicit assumption of the (as yet unformulated) **law of conservation of mass**.

Despite this early work, scientists of the seventeenth century turned more directly than ever to the analogist approach. This was spurred by the 1628 discovery of blood circulation in animals by William Harvey (1578–1657). In 1673, English botanist Martin Lister (1639–1712) publicly speculated that plants have a totally analogous system of internal fluid motion. Johann Daniel Major (1634–1693) had already published an intricate theory of plant circulation. Building on the work of Cesalpino, Major detailed how he thought sap circulated throughout a plant after being absorbed from the soil. Marcello Malpighi (1628–1694) reasoned that plants somehow transform simple materials into complex ones. He argued for a "downward" movement of sap—an uneven return to the soil. Robert Hooke (1635–1703) made the first detailed studies of plants using a microscope. In 1667, Hooke published the discovery of the cellular structure of cork and employed the term *cell* for the first time in this context. Finally, Nehemiah Grew (1641–1712) made detailed anatomical drawings of plant forms.

The seventeenth century produced many important botanical discoveries. However, each of these divergent studies suffered from strict adherence to the analogist tradition.

It was up to Stephen Hales (1677–1761), the originator of **pneumatic chemistry**, to replace the analogist tradition with a new experimentalist one in the early eighteenth century. Hales originally worked within the analogist tradition and became the first person to measure animal blood pressure. He had objectively studied the question by conducting experiments on horses' legs and his own arms. He began searching for the supposedly identical, elusive process in plants until one day, when he attempted to stop the "bleeding" of a broken grape vine. Hales said he suddenly envisioned a number of experiments to measure (among other things) the force of sap in plants. Hales's sleight-of-hand methodological switch was accomplished by measuring physiological activities objectively in plants, rather than stubbornly trying to graph concepts from animals. During experiments lasting the next 20 years, this new approach made all the difference for Hales's discoveries.

In 1727, Hales published his findings in *Vegetable Staticks*, which became the classical text for plant physiology. Hales had been in contact with Isaac Newton (1642–1727) at Cambridge in 1704. The opening of Newton's foundational text *Opticks*, published during that year, promises that the book is free of metaphor, hyperbole, conjecture, and **analogy**—approaches that he found hostile to science, when compared with objective knowledge. Hales was deeply influenced by Newton's work. His own book makes fifteen quotations or references to *Opticks*.

The first chapter of *Vegetable Staticks* jumped right into the question of water transport. It bore initial resemblance to Van Helmont's prior tree experiment. Hales watered a series of plants in sealed pots and measured their changes in weight. During short periods of 12 to 15 days, Hales found that the differences between the weight of added water and the weight of plant increase resulted from the varying amounts of water transpired by each plant. He proceeded to measure the surface area of each leaf and stalk and made a number of starling computations. Among these were rate of **transpiration**, rate of absorption of fluid, surface area of each root system, and the rate of flow from roots to leaves.

Hales also measured for such variables as climatic conditions and specimen variance. An early conclusion was that, weight for weight, plants pass substantially larger amounts of fluid than do humans. Circulation seemed less likely than ever. In fact, Hales found that sunflowers transpire fluid seventeen times faster than do humans (bulk for bulk). This meant that circulation would have to be improbably fast to move all the fluid. This experiment also established the importance of the "statical" (as well as Newtonian) method of inquiry to the whole endeavor.

Hales also detailed a deceptively ingenious experiment to measure the "force with which trees imbibe moisture." Hales filled a glass tube with water and cemented the cut end of a root to one end. He inserted the other end into a basin of mercury. Hales measured the "imbibing force" of the root by gauging how high the mercury rose into the tube to replace the absorbed water. Hales found that absorption (or imbibing) could proceed in all directions—laterally from the branch, from the root upward, or from a branch downward. Therefore, he concluded that "imbibing force" was a function of transpiration, rather than strict direction. He also concluded that sap motion resulted from both transpiration and a kind of "capillary" action.

Throughout his experiments, Hales had been seeing evidence that sap does not circulate in plants. In an exquisite, simple experiment (often repeated in modern botany courses), Hales demonstrated the direction of sap flow. He cut deep notches along the length and around the general circumference of a severed branch. Then he inserted the severed end in water. The branch adsorbed water with less success than an unnotched branch. Nonetheless, its leaves remained green. Hales correctly concluded that, because there was no line of passage between the branch's end and its leaves, sap had passed laterally between sap vessels. Subsequent experiments provided no evidence of either evaporation or transpiration at any of the notches. This experiment demonstrated the power of transpiration and its effect on sap flow in all directions. It also showed conclusively that plants exhibit a progressive motion, rather than the circulatory one found in animals.

All along, Hales had noticed pockets of air in the sap, particularly that of vines. This engendered a series of experiments of the mechanism by which plants draw air. These experiments were greatly overlooked at the time. It was only much later that scientists realized Hale had demonstrated that plants draw something specific from the air. It was even later that scientists identified this substance as **carbon dioxide**. (For more information, please see the entry on Plants Use Sunlight.)

Vegetable Staticks combined many disparate experiments into a coherent and cohesive system for the study of plants. Never before had rational experimentation, decisive analysis and functional equilibrium been applied to the study of plants. For these reasons, most scientists consider Hales the founder of plant physiology.

For more than a century, little was done to equal Hales's work in this area. In fact, it took the combined work of many nineteenth-century scientists to discover vascular tissue. This tissue is comprised of two types of specialized cells called *xylem* and *phloem*. In 1914, Irish botanist Henry Dixon (1869–1953) proposed a compromise model to understand *xylem*, the hotly debated cells responsible for carrying water and minerals from root to leaf. Similarly, several scientists gradually discovered that *phloem* tissues carry food produced by leaf **photosynthesis** to other parts of the plant.

SEE ALSO Animal Physiology; Animal Respiration; Carbon Dioxide; Plants Use Sunlight.

Selected Bibliography

Allan, D. G. C. and R. E. Schofield. *Stephen Hales: Scientist and Philanthropist.* London: Scholar Press, 1980.

Greenberg, Arthur. *A Chemical History Tour: Picturing Chemistry from Alchemy to Modern Molecular Science.* New York: Wiley-Interscience, 2000.

Knight, David M. *Natural Science Books in English: 1600–1900.* New York: Praeger, 1972.

Locy, William A. *The Growth of Biology.* New York: Henry Hold, 1925.

Plants Use Sunlight to Absorb Gas, and They Reproduce through Insect Pollination
Jan Ingenhousz, 1779
Joseph Koelreuter, 1761
Christian Sprengel, 1793

Ancient peoples understood certain basic facts about growing plants soon after the advent of agriculture about 12,000 years ago. They knew that fertilizing the soil with certain substances increased plant yield, and they saw that animal life depended, in large part, on eating plants. In time, they came to assume that plant life, therefore, depended on "eating" something in the soil. By the time of the ancient Greek philosopher Aristotle (384–322 B.C.), learned people considered the earth a vast "stomach" that nurtures the food for plants. Aristotle was not a scientist and did not conduct experiments or tests of his ideas. His student, Theophrastus (371–285 B.C.), founded the practice of **botany** through his study of plants in many environments and habitats. Though his ideas about plants depended on Aristotle's assertions, Theophrastus published two important texts that preserved his knowledge of many kinds of plants, *Enquiry into Plants* and *Cases of Plants*. These are the oldest known treatises devoted entirely to studying plants. They were authoritative in Europe for well over a thousand years.

Much later, Johann Baptista van Helmont (1578–1644) became the first person to experiment widely with plants. In his famous tree experiment, van Helmont planted a small sapling and weighed the sapling with its soil and pot. After several years, van Helmont found a tremendous increase in the weight of the tree. However, there was no change in the weight of the pot and very little in that of the soil. Van Helmont concluded that the tree's weight came entirely from the addition of water. This erroneous conclusion was due, in large part, to lack of knowledge about the gases in air and the fact that green plants utilize them. (For more on this point, please see the entry on Plant Physiology.) It was not until 1699 that English scientist John Woodward (1665–1728) first published results of a vague experiment in which he found that "something" from the earth or air, aside from water, seemed necessary for plant growth.

In the 1670s, Italian Marcello Malpighi (1628–1694) and Englishman Nehemiah Grew (1641–1721) made the first microscopic studies of plant structure. These studies first revealed the internal structures of plants and evidence that leaves are the nourishing organs of plants. In particular, Malpighi demonstrated that, when he removed the green leaves of plants, they ceased growth, even when the roots were intact. This directly refuted van Helmont's conclusion that plant growth was an effect of water.

These were important discoveries, but the revolutionary work of plant **physiology** would ultimately depend on advances in the **pneumatic chemistry** of airs. The founder of pneumatic chemistry, Stephen Hales (1677–1761), published what became the classical text in plant physiology in 1727, *Vegetable Staticks*. Hales conducted hundreds of experiments on plants that demonstrated the flow of sap in the roots and stalks of plants as well as leaf-based evaporation of water (a process known as **transpiration**). Hales investigated the absorption of air and showed that it was not analogous to animal **respiration** as most scientists believed. However, Hales and other scientists did not yet know that ordinary air was made up of several different gases. Hales thought plants were somehow able to modify the qualities of a single, highly elastic air.

In 1754, Joseph Black (1728–1799) discovered the first distinct gas ever identified, when he isolated **fixed air**, or **carbon dioxide**, in his laboratory. Though it took the later interpretive work of Antoine Lavoisier (1743–1794) to correctly understand this gas, Black's discovery eventually shattered the idea of a unified air. It also sent eighteenth-century pneumatic chemists on the fruitful journey of discovering many constituent gases in ordinary air. Later in the century, Joseph Priestley (1733–1804) became one of the discoverers of **dephlogisticated air**, or **oxygen**. In several experiments, Priestley noticed that a bird could not survive on the "spent" air left in a closed container after burning a candle to extinction. To his surprise, he found that a sprig of mint did not die. Rather, Priestley found that the plant grew without problems. He thought that the plant had somehow "purified" the air when he found it would again support a candle's flame or the bird's respiration. Many scientists were confused by this finding, and most were unable to reproduce Priestley's experiments. This led to a heated scientific debate. (For more information, please see the entry on Carbon Dioxide and also the entry on Oxygen.)

In 1770, Dutch physician Jan Ingenhousz (1730–1779) took up the issue in a series of over 500 well-planned experiments. These experiments authoritatively cleared up the issue when he published them in a 1779 book, *Experi-*

ments upon Vegetables. This book confirmed Priestley's findings about green plants and oxygen, and it also added light to the equation. For many years, scientists had observed that submerged plants give off tiny bubbles. The accepted explanation was that water simply "frees" trapped air from the leaves. Ingenhousz suspected a much more sophisticated process. After isolating many variables, Ingenhousz finally found that when he heavily shaded the plants in his laboratory, the bubbles stopped. He thought that the plants therefore required heat in order to produce the bubbles. But putting a flame next to them made no difference. Finally, Ingenhousz demonstrated that the plants gave off bubbles only on receiving sunlight.

Delving further, Ingenhousz collected the gas that his plants gave off in sunlight. He found that it supported **combustion** brilliantly, and he correctly concluded that it was the same gas as Priestley's oxygen. However, when Ingenhousz collected the gas given while he shaded the plants, he was surprised to learn that plants "injure" the air by making it unfit for combustion. He correctly concluded that this gas was the same as Black's carbon dioxide— long known to shut down combustion and respiration. Ingenhousz had discovered the process by which green plants absorb carbon dioxide and emit oxygen when exposed to sunlight. He had also discovered that they reverse this process in heavy shade or darkness.

Today, this process is known as **photosynthesis** (see Figure 21), a word that means "putting together with light." Later in the century, Jean Senebier (1742–1809), a Swiss scientist, confirmed the discoveries of many of Ingenhousz's experiments and as well as the general efficacy of his photosynthesis theory. Today, scientists widely accept Ingenhousz's theory as valid. Curiously, none to date has been able to artificially replicate photosynthesis in a laboratory setting. As a modern equation, photosynthesis appears as $6CO_2 + 6H_2O + light\ energy \rightarrow C_6H_{12}O_6 + 6O_2$. Scientists also consider Ingenhousz the first to draw attention to the concept of interdependence between living creatures, such as plants and animals. Animals depend on plants for food and oxygen. Plants depend on animals for carbon dioxide, soil-enriching waste, and decomposed carcasses.

While Ingenhousz was discovering photosynthesis in plants, other scientists were beginning to understand the complex mechanism of reproduction in plants. As early as 1500 B.C., people in Israel knew how to pollinate date palms. Previous cultures, such as the Egyptian and the Assyrian, may have perfected the process thousands of years before. Ancient Japanese and Chinese gardeners widely produced species of flowers of which the majority result from dusting one species with the pollen of another species (such as roses, camellias, and chrysanthemums). In the seventh century, the Arabic writer Kazwini (died ca. 682) lamented that the only tree he knew how to artificially fertilize was the date palm. He thus provided evidence of continued widespread

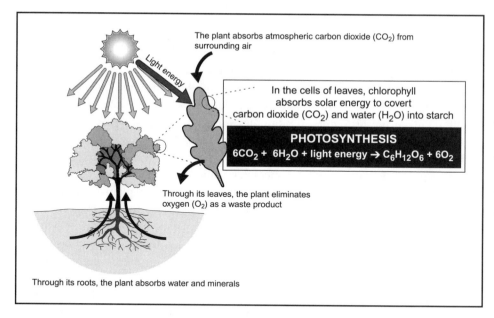

The plant absorbs atmospheric carbon dioxide (CO_2) from surrounding air

In the cells of leaves, chlorophyll absorbs solar energy to covert carbon dioxide (CO_2) and water (H_2O) into starch

PHOTOSYNTHESIS

$$6CO_2 + 6H_2O + \text{light energy} \rightarrow C_6H_{12}O_6 + 6O_2$$

Through its leaves, the plant eliminates oxygen (O_2) as a waste product

Through its roots, the plant absorbs water and minerals

Figure 21. Photosynthesis. Photosynthesis is one of the most important chemical reactions in nature. It is also the process by which plants make carbohydrates from carbon dioxide and water, in the presence of chlorophyll and light. A by-product of this reaction, oxygen, is fundamental for animal survival. Jan Ingenhousz first experimented with photosynthesis in the early 1770s.

practice. The ancient Greeks, Aristotle and Theophrastus, first delineated the metaphor of "union" between "male" and "female" flower parts.

It was not until 1682, however, that Nehemias Grew found that pollen must reach the stigma to ensure proper seed development. He found that this could happen artificially—through human intervention—or naturally, though he did not elaborate on the "natural" source of **pollination** between individual plants. Twelve years later, Rudolph Camerarius (1665–1721) published experiments in which he found that two different parts of the flower, the pistil and the stamen, worked together to produce ripe seeds. Following the residual practice of projecting animal physiology onto plants (a practice known as the analogist approach), Camerarius announced that these parts were "true sexual organs." In 1750, Arthur Dobbs (1689–1765) realized that this metaphor was only valuable when kept in nonliteral perspective. The "male" element, or pollen, must fall on the "female" stigma for fertilization to take place.

In 1761, Joseph Koelreuter (1733–1806) conducted a series of systematic studies on the newly understood concept of pollination. A prevailing theory held that the wind transferred pollen between flowers for pollination.

Koelreuter exposed showy flowers such as irises and water lilies to the wind, but excluded as many other possible agents as he could. The vast majority of the flowers did not produce seeds. When he drew nectar from his flowers with a glass tube, Koelreuter found that the nectar attracted insects. Though he did not prove it experimentally, Koelreuter speculated that insects are primarily responsible for pollination.

In 1793, Christian Sprengel (1750–1816) began an intensive study of plant physiology that was heavily influenced by the work of Hales. He found a direct relationship between flower form and pollinating insects. Building on this knowledge, Sprengel discovered that the "male" and "female" sexual parts of many flowers do not mature at the same time. He therefore correctly concluded that cross-pollination between individual flowers was much more crucial than scientists had previously thought.

He also demonstrated that pollination is primarily carried out by insects, such as bees, that carry pollen between flowers. Scientists were slow to embrace this discovery. In fact, they did not generally accept pollination by insects until Charles Darwin (1809–1882) cited Sprengel in his 1859 *Origin of Species*. Darwin championed Sprengel's work as evidence of co-dependence between flowers and insects, a significant component in his theory of evolution.

SEE ALSO Animal Respiration; Carbon Dioxide; Oxygen; Plant Physiology.

Selected Bibliography

Allan, D. G. C. and R. E. Schofield. *Stephen Hales: Scientist and Philanthropist*. London: Scholar Press, 1980.

Buchmann, Stephen L., and Gary Paul Nabhan. *The Forgotten Pollinators*. Washington, DC: Island Press, 1996.

Greenberg, Arthur. *A Chemical History Tour: Picturing Chemistry from Alchemy to Modern Molecular Science*. New York: Wiley-Interscience, 2000.

Knight, David M. *Natural Science Books in English: 1600–1900*. New York: Praeger, 1972.

Krebs, Robert E. *Scientific Laws, Principles, and Theories: A Reference Guide*. Westport, CT: Greenwood Press, 2001.

Proctor, Michael, Peter Yeo, and Andrew Lack. *The Natural History of Pollination*. Portland, OR: Timber Press, 1996.

Smith, Ray F. (ed.). *History of Entomology*. Palo Alto, CA: Annual Reviews, 1973.

Priestley's Chemical Experiments, Discoveries and Inventions
Joseph Priestley, 1760s–1770s

Many chemical historians consider Joseph Priestley (1733–1804) the single most prolific chemist of the late eighteenth century. They cite his many discoveries, his important experiments, and his laboratory inventions that greatly added to the sum total of growing chemical knowledge. Priestley lived during an exciting time for chemistry—one in which new discoveries were frequent and contested theories were on the tongues of virtually all scientists. Nonetheless, Priestley took a while to recognize his devoted passion for science.

Priestley was already 33 years old, a respected teacher and ordained minister, before he performed any scientific experiments. On a 1766 trip to London, Priestley met the man who would impact the rest of his life. In a series of conversations lasting far beyond the confines of the trip, Benjamin Franklin (1706–1790) sparked Priestley's life-long interest in science, especially the emerging science of chemistry. From this point on, Priestley consciously combined his many religious and civic responsibilities with his growing commitment to advanced scientific research.

Once realized, Priestley's scientific ability took off very quickly. Less than a year after meeting the elder scientist, Priestley wrote a successful compendium of electrical knowledge up to that point. In this text, Priestley also made a startling observation. He found that both charcoal and coke were good conductors of electricity. Previously, scientists thought that metal and water were the only true conductors. This first discovery was impressive and useful, but it was just the beginning. The field of **pneumatic chemistry** was where Priestley would garner lasting, international respect through revolutionary discoveries.

In the mid-eighteenth century, scientists were still delineating the foundations of modern scientific methods and inquiries. Chemistry was just beginning to emerge from the messy hodgepodge and centuries-old practice of **alchemy**. Most scientists accepted the **phlogiston** theory of substances. This theory said that **combustion** releases the hypothetical substance phlogiston from an object, such as a piece of charcoal. Combustion continued until the object had rendered its reserve of phlogiston. Phlogiston was the first rational theory of chemistry. It also led to most of the important chemical discoveries of the eighteenth century. (For more on this topic, please see the entry on Phlogiston.)

Like many of his discoveries, Priestley's gas inquires enjoyed a genesis of partial serendipity. Priestley's house was very near a brewery, and he became intrigued by the different gases that the chemical process of fermentation regularly released. When he began experimenting with the gases released from the brewery's vats, Priestley had a startling discovery. He was surprised to find a familiar gas. In 1756, Joseph Black (1728–1799) identified **carbon dioxide** as a distinct gas that he called **fixed air** because it would support neither combustion nor animal **respiration**. This was the first discovery of a distinct gas. (For more on this topic, please see the entry on Carbon Dioxide.)

Priestley's tests proved that this gas was present as a by-product in the process of chemical fermentation. Priestley designed a method to capture the gas in his laboratory. This method won Priestley a prestigious award from the Royal Society. Also, when he used pressure to force the "fixed air" into water, Priestley set off a European craze for a tasty new product, soda water. Priestley soon presented some of his most important work before the Royal Society.

In identifying "fixed air," Black had implicitly assumed he had manipulated the elastic qualities of a single, ordinary air. Priestley proposed that he had really freed some kind of separate "air" (or a unique gas). Though he was far from the modern understanding of chemically distinct gases, Priestley's proposal was startlingly advanced. It was also startlingly fruitful, as it led him to examine the qualities of different "airs" and to perfect the techniques for liberating and collecting them.

The established pneumatic method involved bubbling a gas into a jar containing water. In effect, the water was forced out and the gas was collected. This method was effective for many gases, but presented particular problems for water-soluble gases. Priestley designed a way around the problem by substituting dense mercury for water in an apparatus he built, the **pneumatic trough**. Nearly every eighteenth-century chemist soon came to rely heavily on this apparatus. Priestley was no exception. He used the pneumatic trough in conjunction with a powerful magnifying glass that he called a burning lens. The burning lens turned the sun's rays into a powerful heat source for his experiments. Though he did not realize it, one of his first experiments with the trough and lens produced an important gas. When he heated sodium chloride and sulfuric acid together, Priestley isolated the water-soluble gas hydrogen chloride.

As Priestley became more prolific with the properties of different gases, he meticulously recorded and described the different "airs" he produced. Priestley quickly became known for discovering more gases than anyone prior to him. In one set of inquiries, he also became the first scientist to empirically study the support of combustion in different gases. He elaborately recorded flame size of different gases, many of which had only been partially recognized as distinct. Priestley also recorded the different spectra of colors produced by burning different gases in his experiments.

In 1772, one of the most important of these experiments produced a gas that Priestley called "nitrous air" (or nitric oxide). Priestley found that this second gas reacted with iron and sulfur and gave off a second gas. Priestley found that this gas was combustible. It was only later that Humphry Davy

(1778–1829) identified the second gas as **nitrous oxide**, colloquially known as "laughing gas." In 1798, Davy experimented with using this important gas as an anesthetic. Many years later, laughing gas became one of the first widely used surgical anesthetics.

At this time, many scientists had prepared solutions of **ammonia** and water. However, no scientist had successfully isolated ammonia gas on its own. Priestley began researching the matter. After several inquiries, he heated water and ammonia, and carefully collected the gas through his new method of mercury displacement. Almost immediately, Priestley recognized that he had successfully isolated the gas. Priestley's subsequent tests demonstrated that solutions of ammonia were basic. Priestley knew that solutions of hydrogen chloride were alkaline. He began wondering what "air" or gas would form through the simple combination of these substances. When he united them, Priestley recorded that a white cloud formed. Gradually, the cloud gave way to a fine powder which Priestley called "sal ammoniac," or ammonium chloride.

Several years later, in 1774, Priestley learned how to replicate the obnoxious odor of burning sulfur by heating "oil of vitriol" (concentrated sulfuric acid) with mercury. Priestley also pioneered a method of producing the gas by heating the acid with copper turnings. This same method is still used to make sulfur dioxide.

Over and above these important discoveries, Priestley's single most lasting contribution is the isolation of one of the most basic gases of the atmosphere. In 1772, Priestley had successfully documented the basic gas processes of plant respiration for the first time. In a series of innovative trials, Priestley found that the same gas that supports animal respiration is given off by green plants. To investigate, Priestley enclosed a green plant and a candle together in a glass container. He burned the candle until the air in the container would no longer support its flame. Then, he waited patiently as several hours passed. When he tried again, the candle burned—even more brightly than it had before. Priestley had proven his theory that the gas that supported combustion was the same one released by green plants.

In 1774, Priestley rediscovered his gas in an unlikely place. Priestley was deep into calcification (or oxidation) experiments through burning. His trials had yielded few results, and he was growing frustrated. Then one day, Priestley happened to hold his burning lens over a **calx** of mercury (or mercuric oxide). Carefully, he collected the gas that was released in his pneumatic trough. When Priestley began tests on the gas, he instantly recognized it as the same gas from green plants and candle combustion. In subsequent experiments, Priestley determined that this gas could support a mouse's respiration much longer than would an equal volume of ordinary air. He found it pleasant to breathe and thereby generally assumed it was the "respirable" portion of ordinary air.

A staunch phlogistonist, Priestley named his gas **dephlogisticated air**. He believed that the air had lost all of its phlogiston and therefore grabbed it back from a burning object or respiring animal. For many years, Priestley was universally credited with the discovery of this important gas. However, chemical historians uncovered evidence that Swedish apothecary Carl Scheele (1742–1786) first isolated the gas he called **fire air** in 1771. Because of a publishing delay, Scheele published his findings several months *after* Priestley in 1777. Neither scientist knew of the other's findings. For this reason, Priestley and Scheele share credit for this important discovery. (For more on this topic, please see the entry on Oxygen.)

In 1774, Priestley showed his discovery to a colleague in Paris, Antoine Lavoisier (1743–1794). Lavoisier was the young French chemist at the very center of the chemical revolution. He was also first to name this gas **oxygen** from the Greek for "acid producer." This name resulted from Lavoisier's mistaken belief that oxygen was present in all acids. Nonetheless, over the next several years, Lavoisier realized that "calcification" was really oxidation—a process that involved oxygen *joining with* a substance, rather than phlogiston *separating from* it. Practical impacts of this realization led to the theoretical dismantling of phlogiston. (For more on this topic, please see the entry on Phlogiston.)

Throughout his professional life, Priestley preferred a kind of physical chemistry of mechanical explanations over a system of rigorous elemental classification. Nonetheless, Priestley's experiments and discoveries were absolutely critical to the development of Lavoisier's revolutionary theory of oxygen chemistry.

SEE ALSO Carbon Dioxide; Oxygen; Phlogiston; Rational System of Chemical Nomenclature.

Selected Bibliography

Gibbs, F.W. *Joseph Priestley: Revolutions of the Eighteenth Century*. Garden City, NY: Doubleday, 1967.

Gillispie, Charles Coulston (ed.). *Dictionary of Scientific Biography*, volume XI. New York: Charles Scribner's Sons, 1972.

Holt, Anne. *A Life of Joseph Priestley*. Westport, CT: Greenwood Press, 1970.

Meldrum, Andrew Norman. *The Eighteenth Century Revolution in Science—The First Phase*. New York: Longmans, Green, 1930.

Porter, Roy (ed.). *The Biographical Dictionary of Scientists*. New York: Oxford University Press, 1994.

Schwarts, A. Truman, and John G. McEvoy (eds.). *Motion Toward Perfection: The Achievement of Joseph Priestley*. Boston: Skinner House Books, 1990.

Principle of Least Action
Pierre-Louis de Maupertuis, 1744

The ancient Greeks first conceived of the notion that nature takes the shortest, easiest and most direct route in moving objects between two points. The orator Isocrates (436–338 B.C.) noted generally that the smallest forces in the universe produce the motion of the large masses. A little later, the philosopher Hero (ca. first century, also known as Heron) theorized that a ray of light always travels the shortest distance between an object (such as a mirror) and a person's eye. Ptolemy (ca. 100–165) greatly expanded the concept, claiming that nature never does anything without designing a direct method for its accomplishment. He also found that rays of light move in straight lines, take the least amount of time possible to travel between points, and follow a uniform velocity before and after reflection.

These philosophical positions were articulations of the minimal principle— the earliest studies of teleology. Teleology is the ancient belief that the natural movement of an object is governed by its telos, or final destination. Teleology stated that the movement of a ray of light was governed by its final destination, the human eye. Though some enlightenment thinkers of the eighteenth century replaced antiquity's teleology with efficient cause explanations, teleology nonetheless led to the general study of the behavior of light and, eventually, to the eighteenth-century's **principle of least action**.

During the Middle Ages, philosopher–theologian William of Ockham (sometimes known as Occam, ca. 1284–1347) stated the principle known as **Ockham's razor**. This principle states that a problem should be stated in its most basic, simplest terms. For enlightenment scientists, this meant that the simplest theory that fully explains a phenomenon or problem should be the one selected. Scientists such as Leonardo da Vinci (1452–1519) also commonly accepted the idea that nature acts in the mathematically shortest way possible. During the seventeenth century, Isaac Newton (1642–1727) proposed his laws of motion. The idea that nature tends to economize its actions is key to Newton's first law of motion. Known as the principle of inertia, this law states that an object that is moving in a straight line continues moving in a straight line until or unless it is acted on by an outside force. The same holds true for objects at rest.

Perhaps, thought many enlightenment scientists, there is some larger, more grandiose principle at work in the universe. Many, such as the French mathematician Pierre de Fermat (1601–1665), began searching for just such a principle. Fermat founded modern number theory and discovered much of the mathematical foundation for **calculus**. He also mathematically studied the law for bending (or refraction) of light. Fermat found that light follows the path that takes the shortest time. He wrote that a ray of light traveling between an object and a human eye follows the path with the least travel time out of all the possible paths. This correctly accounted for the principle of refraction when it showed that light is bent when it passes (for example) from air to water because refraction is the shortest path. However, like the ancients, Fermat only predicted this behavior. It would be another two centuries before it was experimentally proven.

During the next century, French mathematician Pierre-Louis de Maupertuis (1698–1759) reformulated the work of Fermat and others into the principle of least action. The physical laws, Maupertuis believed, indicated the teleological influence of nature itself. In a paper "Accord of different laws of nature which hitherto had appeared incompatible," Maupertuis defined his principle of least action. He said action was the product of mass (m), velocity (v) and distance (s) (m × v × s = action). The principle states that any natural motion of a particle between two points tends to minimize the product, action. Originally, Maupertuis applied this concept only to the path of light rays between two points. However, he later expanded it to cover virtually all types of particles or bodies in motion. For example, most people know that water running downhill follows the steepest, fastest, most direct descent—just as this expanded principle expects. Though some natural philosophers objected to the characterization of nature as "lazy" (in taking the simplest path), Maupertuis' principle was eventually adopted by scientists in a range of disciplines.

Maupertuis's principle can be philosophically simplified as the idea that nature is thrifty in all its actions. Its most basic form—seen in the principle of refraction of light—can also be understood by way of metaphor. A person standing on a beach at point A hears someone in the water cry for help. One hundred yards to the person's left and a little way out in the surf, a swimmer is drowning at point C. If the person on the beach decides to help, he or she likely will run to the point at the surf closest to the person (point B) before going into the water. This is because the person, in all likelihood, moves faster on land than in water. The fastest path to the drowning swimmer is, therefore, not a straight line from point A to point C. Instead, it is two segments of straight lines, from point A to B and from point B to C.

The same is true for light—it moves faster in air than in water. Between point A and C on the hypothetical beach, then, a ray of light will not travel in a straight line, because point C is in the water. Rather, the ray will take a straight path from point A to point B at the edge of the water, then turn a little bit, and follow another straight path to point C in the water (as did the person performing the rescue). This process is called *refraction,* and it was one of the bases of Maupertuis' principle of least action.

However, refraction was not the only basis of Maupertuis's principle. From the start, Maupertuis thought he could somehow discern intentionality from the very principles that guide the universe, and therefore scientifically prove the existence of a deity or higher power. A enlightenment intellectual, Maupertuis reasoned that because a ray of light somehow knows where it is going before it gets there, and because it follows the simplest route in getting there, then the guiding wisdom of a higher power must be helping it along. However, these teleological claims failed to impact either French culture or his fellow scientists. They did, however, earn Maupertuis the ire, suspicion and even disdain of many fellow enlightenment scientists, most notably Voltaire (1694–1778).

Although Maupertuis lacks the reputation of his greatest contemporaries, his formulation of the principle of least action has become fundamental to modern physics and the fields of relativity, optics and quantum mechanics. It was a forerunner of virtually all laws and theories dealing with conservation, energy, and mass. Soon after Maupertuis, the Swiss mathematician Leonhard Euler (1707–1783) turned the principle of least action into a theorem by providing an accepted mathematical rationale for dealing with variations in calculus. Euler's formulation laid the foundation for future studies by showing that the path of a point under influence of a "central force" minimizes the action taken. Building on this work, the French mathematician Joseph-Louis Lagrange (1736–1813) published his 1788 book *Analytical Mechanics*. This book reconstituted major parts of the Newtonian system in terms of the principle of least action.

Yet, as big an impact as the principle had on enlightenment physics, it was not until the twentieth-century advent of quantum mechanics that the principle was fully recognized as one of the greatest generalizations in the history of science. As Albert Einstein (1879–1955) formulated his theories of relativity, which both built on and replaced Newton's laws of motion, the principle of least action took on new scientific meaning. According to the general theory of relativity, for example, a planet orbiting a star follows a geodesic curve. This is the shortest distance in the curved geometry of space-time. Tellingly, Max Planck (1858–1947), the chief founder of quantum theory, often noted that the principle of least action comes closest to the final aim of his life-long theoretical research.

Today, scientists have amended the principle of least action to the law of the whole. This law states that the specific behavior of single objects (particles and waves) cannot be understood except in aggregate form. This is a departure from classical physics, which studied specific parts of the universe as a method of understanding the whole. In quantum physics, the whole web of relationships and the interaction of many "parts" forms an inseparable unit. The whole forms the part. In this way, even today's advanced quantum theory is indebted to Maupertuis's original formulation of the principle of least action.

SEE ALSO Corpuscular Model of Light and Color; Heat Is a Form of Energy; Stellar Aberration.

Selected Bibliography

Beiser, Arthur. *Modern Physics: An Introductory Survey*. Reading, MA: Addison-Wesley, 1968.

Karplus, Robert. *Physics and Man*. New York: W. A. Benjamin, 1970

Krebs, Robert E. *Scientific Laws, Principles, and Theories: A Reference Guide*. Westport, CT: Greenwood Press, 2001.

Lockett, Keith. *Physics in the Real World*. New York: Cambridge University Press, 1990.

Parkinson, Claire L. *Breakthroughs: A Chronology of Great Achievements in Science and Mathematics, 1200–1930*. Boston: G. K. Hall, 1985.

Taylor, John G. *The New Physics*. New York: Basic Books, 1972.

Proper Motion of the Stars and Sun through Space
Edmond Halley, 1718
William Herschel, 1782

It is often said that astronomy is the oldest science. It is true that several prominent civilizations gave much thought to the stars and much respect to those who interpreted them. Long before the great stargazers of antiquity, however, prehistoric humans had made the most basic observations about the heavens. One was the alternation between day and night that more or less corresponded with the sun's daily journey across the sky. To prehistoric peoples, the sun no doubt made a regular, reassuring journey across the sky in an easily predictable pattern. So did the moon. The moon's shifting shape during its lunar cycle gave rise to the first months. These, in turn, provided lunar calendars, which date back as far as ancient Babylonia.

As civilizations became more sophisticated and interdependent, they came to need more sophisticated methods of marking time. Ancient peoples developed solar time, with its 24-hour days, by studying the movement of the sun across the sky. Either the ancient Egyptians or Babylonians first conceived of a circle as having 360 degrees. The Babylonian base number system consisted of 10s and 60s (rather than modern 10s and 100s). Therefore, they divided each degree into 60 parts and each of these parts into 60 parts. Later, the Romans called the first division *partes minutae primae*, or "first small parts." They called the second division *partes munutae secundae*, or "second small parts." Eventually, these terms were shortened to the modern *minute* and *second*. This hybrid time scheme, of course, is still used today.

A majority of the astronomical observations of antiquity came together around the year 150, when the astronomer Ptolemy of Alexandria (ca.

100–165) published his monumental thirteen-part text, *Mathematik Syntaxis*. In this text, Ptolemy refuted the concept that the Earth moves. Instead, he said, everything in the universe moves to the center of the Earth. Two parts of his text were dedicated to a thorough catalog of the stars. Ptolemy claimed that the stars were stationary points of light far away from the Earth. Against this tapestry of light, Ptolemy said that each of the other five known planets (Mercury, **Venus**, Mars, Jupiter and Saturn) were in orbit around Earth. Most importantly, so was the sun. Ptolemy's conception of the universe was heavily endorsed by the Church and other powerful authorities for more than 1,400 years. Many European astronomers spent their time trying to discern the motion of the sun and planets around the Earth.

It was not until 1593 that the Polish astronomer Nicolaus Copernicus (1473–1593) challenged the Ptolemaic conception of the universe. For this, he is usually considered the founder of modern astronomy. Copernicus's masterpiece, *On the Revolutions of the Heavenly Spheres,* stated that Ptolemy's concepts of heavenly movements were mistaken perceptions of movement, rather than actual movement. The sun, he said, did not revolve around the Earth. Instead, it only appeared to do so. This is because the Earth, like the other planets of the solar system, orbits the sun.

Copernicus's explanation was much less complex and had fewer mathematical problems than Ptolemy's. Its impact on astronomy was catastrophic. The entire schema and understanding of the universe was changed. Though many European authorities (especially the Church) did not embrace the Copernican view, many astronomers did. Within 20 years, Johannes Kepler (1571–1630) and Galileo Galilei (1564–1642) were already developing the math and physics to prove Copernicus's theory.

Copernicus would eventually change the most basic foundations of humanity's beliefs about its own existence. The universe was not as it seemed. The sun did not move across the sky at all. The Earth and other planets orbited the sun. However, astronomers who accepted Copernicus's reorganization of the universe soon began asking some very tough questions about the sun and stars. Even though Ptolemy was wrong, what if the sun is somehow moving, after all? Does the Earth's orbit necessarily preclude some kind of solar movement, albeit vastly different from that of Ptolemy's conception?

In 1612, Galileo began studying sunspots, the relatively dark areas on the surface of the sun. He noticed that the spots appeared to move. He knew of all kinds of outlandish theories of tiny planets near the sun and Mercurial conjunctions. Rejecting these tall tales, Galileo surmised that the sun rotates on its axis in a period easily determined by the apparent motion of the sunspots, which rotate with the sun. The sun made some movement, after all. On its axis, at least, it was certainly not stationary.

Then, in the early part of the eighteenth century, the great astronomer Edmond Halley (1656–1742) renewed widespread interest in antiquity. Halley made several important archaeological findings, as well as several key translations of ancient astronomy texts. As a young man, Halley became a favorite collaborator of many renowned scientists. Early on, he had established himself by developing the astronomical basis for a reliable system of **celestial navigation**. Then, he traveled to Danzig (now Gdansk, Poland) to meet with Johannes Hevelius (1611–1687), 45 years his senior. The young Halley also collaborated with the great Giovanni Cassini (1625–1712) while traveling through Europe.

Halley was also an outstanding editor. From 1685 to 1693, he edited the *Philosophical Transactions of the Royal Society* during one of its most important and fruitful periods. One of Halley's greatest collaborations took place with Sir Isaac Newton (1642–1727). By virtually all measures, it was at Halley's insistence that Newton published one of the most influential books in history, the *Principia*. The *Principia* contained Newton's famous laws of motion. Halley worked as Newton's editor and became so important to the project, in fact, that Newton took to referring to the *Principia* as "Halley's book."

After this important work, Halley was primed and ready for another "collaboration" of sorts—this time with a fellow astronomer who had lived 1,500 years before. In 1718, Halley began a series of inquires into Ptolemy's catalogue of stars from *Mathematik Syntaxis*. Like Ptolemy, Halley had also made a catalog of stars—the first one ever compiled of stars visible from the southern hemisphere. Since antiquity and right up to Halley's day, astronomers assumed the stars were fixed in place. Halley had never had reason to challenge this assumption.

However, in his inquiry, Halley began noticing discrepancies between Ptolemy's stellar positions and those compiled by Halley and his contemporaries some 15 centuries later. Halley assumed that Ptolemy had been wrong. Certainly, Ptolemy had made many well-documented mistakes—his conception of an Earth-centered universe had led scientists astray for centuries. Also, the ancient catalog was not entirely Ptolemy's original catalog. Parts were based on one compiled by Hipparchus (ca. 190–120 B.C.), who lived several centuries before Ptolemy. Perhaps these other astronomers had made mistakes that Ptolemy had reprinted in his catalog.

Yet, the discrepancies Halley was identifying were so enormous they were hard to believe. Ptolemy's chart was way off. In some cases, the placement of stars looked like rough estimations. Could Ptolemy have been this grossly mistaken? Could he have been so negligent with his positions of stars? Halley could not believe so.

If Ptolemy had not made mistakes, then Halley had to reexamine his own carefully compiled catalogs. Could he and his contemporary collaborators have made such careless, even outlandish mistakes? He rechecked his measurements and performed his calculations again. If he had made mistakes, he could not find them.

When he was assured that his calculations were impeccable, Halley arrived at a startling, risky hypothesis. If both catalogues were correct, then the stars were not fixed and stationary, as scientists had assumed for all those centuries. If both catalogues were correct, the only conclusion to draw was that the stars were moving.

Halley began a long series of inquiries into his hypothesis of stellar motion. Specifically, he studied three very bright stars: Sirius, Arcturus, and Procyon. When he studied the relational position of the stars, Halley detected changes in their position. These changes were not attributable to any known or theoretical stellar factors such as **parallax**. Halley carefully made sophisticated records of the stars' exact nighttime positions. From these records, Halley published the first irrefutable proof that the stars have a proper motion.

Halley also postulated that other stars possessed motions, but were dimmer and therefore further away. He said that detection of their proper motion was, because of location, not possible. It was only with the invention of much more powerful telescopes that detection of the proper motion of these stars was finally determined—a century and a half later. Instruments finally caught up with Halley's discovery of stellar motion. Curiously, however, Halley did not speculate about the motion of the closest star, the sun. These speculations were left to other astronomers, who began intensifying their solar studies later in the century.

Just a year after discovering **Uranus**—the first planet discovered since the advent of history—the English astronomer William Herschel (1738–1822) took up the question of the sun's motion through space. He knew about theories that all the stars of the **Milky Way** are in motion around a common center. Herschel observed that the galaxy's stars seem to converge toward a specific part of the sky. He claimed that this is because the galaxy—including the sun and its solar system—rotates in the opposite direction.

Herschel's 1783 paper to the Royal Society outlined the proper motions of thirteen stars. This paper was based in no small part on Halley's findings about stellar movement. Based on the motions of these stars, Herschel estimated a point in space toward which the sun is moving, which he called a **stellar apex**. He also found evidence that the sun orbits the center of the Milky Way, just as planets orbit the sun. Central to Herschel's finding was his claim that the sun was moving in the direction of the constellation Hercules at a speed roughly equal to the Earth's own orbit.

With his discovery of the sun's movement through space, Herschel had correctly applied Copernicus's idea of revolution to the sun. He had also turned it on its ear. Just as the Earth moves, so does the sun. Because the sun is now understood to be on a never-ending voyage through the galaxy, it is hardly the center of the entire universe, as it was with Copernicus's theory. This was the first step toward the field of stellar astronomy, the study of the region beyond the Earth's own solar system. For this reason, Herschel is credited with founding this field. Stellar astronomy opened astronomers' eyes to the vastness of space beyond the reaches of sun, the solar system and even the Milky Way.

Copernicus had set the world in motion in two ways. First, the Earth was in daily rotation on its axis. Second, the Earth followed an annual orbit around the sun (not the other way around). Now, Herschel had added a third kind of motion—that of the sun and its relatively tiny solar system on its vast journey toward the star Lambda in the constellation Hercules.

Herschel was correct about the sun's orbit and also about its direction. However, his estimate of speed was flawed. Modern astronomy has shown that the sun makes a revolution as part of the Milky Way at about 250 kilometers (km, or about 155 miles [m]) per second. At this speed, the rotation period of the Milky Way and its stars is roughly 225 million years. (For more on this point, please see the entry on the Milky Way.) During its rotation, the sun travels roughly 10 billion times the distance between it and the Earth. The sun is moving in the direction of Hercules at roughly 20 km (12 m) per second. During its motion through space, the sun is spinning on its axis, as is the Earth. Just as the earth takes a day to rotate on its axis, the sun takes about a month.

Many twentieth-century discoveries about the motion of the sun and Milky Way were made by the Dutch astronomer Jan Hendrik Oort (1900–1992). Oort identified the **Oort cloud**, a vastly distant collection of comets. He also found that rotating stars in a galaxy follow Newtonian laws of motion. This meant that the inner stars of a galaxy have a demonstrably faster motion than do the outer stars. Stellar rotation is, therefore, not uniform, but differential (e.g., stars rotate at different speeds). Oort went on to calculate the sun's revolution in the Milky Way at around 225 million years.

SEE ALSO Milky Way; Newtonian Black Holes; Stellar Aberration; Uranus Is a Planet.

Selected Bibliography

Abetti, Giorgio. *The History of Astronomy*. Translated by Betty Burr Abetti. New York: Henry Schuman, 1952.

Calder, Nigel. *The Comet is Coming!, The Feverish Legacy of Mr. Halley*. New York: Viking Press, 1980.

Cook, Sir Alan. *Edmond Halley: Charting the Heavens and the Seas*. Oxford: Clarendon Press, 1998.

Doig, Peter. *A Concise History of Astronomy*. London: Chapman & Hall, 1950.

Krebs, Robert E. *Scientific Laws, Principles, and Theories: A Reference Guide*. Westport, CT: Greenwood Press, 2001.

Lancaser-Brown, Peter. *Halley and His Comet*. New York: Sterline, 1985.

Motz, Lloyd, and Jefferson Hane Weaver. *The Story of Astronomy*. New York: Plenum Press, 1995.

Pannekoek, A. *A History of Astronomy*. New York: Interscience, 1961.

Ronan, Colin. *The Astronomers*. New York: Hill and Wang, 1964.

Thiel, Rudolf. *The Discovery of the Universe*. Translated by Richard Winston and Clara Winston. New York: Alfred A. Knopf, 1967.

Yeomans, Donald K. *Comets: A Chronological History of Observation, Science, Myth, and Folklore*. New York: John Wiley & Sons, 1991.

Proust's Law of Constant Proportions
Joseph Louis Proust, 1799

In antiquity, Greek philosophers and scientists believed in four distinct elements: air, earth, fire and water. They believed that these elements combined in varying proportions to constitute all matter in the universe. Later, they proposed tiny units called atoms that contained the very kernel of an object or **element**. Atoms of water, thought the Greeks, were slippery and wet, while atoms of iron were hooked, with sharp edges. In this ancient view of matter, atoms of one element could be changed to atoms of another by manipulating proportions of heat and water. The name given this hypothetical process was **transmutation**. The belief in transmutation led to the practice of **alchemy**, which became the ancestral practice of chemistry in Europe. Among other things, alchemy was mainly interested in changing ordinary metals to gold.

In the eighteenth century, French chemists led by Antoine Lavoisier (1743–1794) began to challenge the old beliefs about elements. Early on, Lavoisier proved that the so-called elements (the alchemists listed many more than did the ancients) could not be transmuted to one another. Along with others, Lavoisier proved that, far from an element, water was actually a **compound** of two different elements. In one famous public experiment, Lavoisier "made" water by first decomposing it into **oxygen** and **hydrogen**, which he collected separately using a **pneumatic trough**. He then combined the elements back into five and half ounces of water.

At the same time, English phlogiston chemists were challenging the idea of air as an element. Ordinary air was beginning to look more and more like a mixture of gases. In fact, by the late eighteenth century, new gases were being discovered almost overnight. Joseph Black (1728–1799) had long since discovered **fixed air**, or **carbon dioxide**. More recently, Joseph Priestley (1733–1804) and Carl Scheele (1742–1786) had independently discovered oxygen, which would eventually form the basis of Lavoisier's new, modern chemistry. After extensive experiments on oxygen, Lavoisier thought that all acids require high proportions of oxygen. Though this specific finding turned out to be incorrect, it led to some very important discoveries about both oxygen and acids.

As other scientists turned their attention toward acids, they identified certain properties of acids. These substances tended to be highly reactive with metals and to cause severe burns to human skin. Other acids caused pigments to change colors. Scientists also identified chemicals that became known as bases. These substances also had a distinct set of properties. Most importantly, scientists found that when they mixed acids and bases, the resulting mixture exhibited properties of neither group. In fact, when they mixed these chemicals in the proper and specific proportions, scientists were intrigued to find they had made a solution of salt. The acids and bases had been neutralized.

One of these scientists was the German chemist Jeremias Richter (1762–1807), who found that neutralization reactions required specific, definite, and exact amounts of both acids and bases. Recipes for everything from dessert pastries to building mortar called for specific ingredients in specific proportions. However, nothing approached the minute laboratory proportions required by the neutralization reactions. Richter called this specific proportion an *equivalent weight*, and he published this finding in 1792.

Soon after, two prominent French chemists, Claude L. Berthollet (1748–1822) and Joseph Proust (1754–1826), began a prominent scientific debate. This public argument centered on the question of whether exact, specific proportions existed only in acid-base neutralizations or whether they were found throughout the chemical world. Accompanying Napoleon on his 1798 expedition to Egypt, Berthollet had found, much to his surprise, huge deposits of soda (sodium carbonate) on the shores of many salt lakes. Chemical tests found that salt and limestone (calcium carbonate) in the bottom of lakes had reacted to form the soda. This completely contradicted the well-documented chemical reaction, in which soda and calcium chloride react to form salt and limestone. Berthollet thought that the unusually high concentration of salt had forced the usual reaction to "reverse."

In modern terms, Berthollet had discovered an equilibrium reaction:

$$CaCl_2 + Na_2CO_3 \leftrightarrow CaCO_3 + 2NaCl$$

Berthollet had also discovered the reversibility of chemical reactions. However, all of this would only be understood later. Instead, Berthollet concluded that chemicals compounded together in variable and shifting proportions. In other words, he concluded that composition was indefinite and fickle, dependent on the method of preparing a particular compound. This conclusion marked a huge rupture with the exacting direction that chemistry had been taking.

Proust still disagreed with this conclusion. Many of his findings suggested that Berthollet had either misinterpreted his results or, at best, isolated several (as yet unexplained) anomalies. Proust undertook a number of reactions, in which he ensured painstakingly accurate analyses. Early on, Proust demonstrated that two oxides of tin and two sulfides of iron existed. His tests revealed that each oxide or sulfide had its own specific, definite composition, in contrast with earlier findings by other chemists. The previous findings, said Proust, were due to uncertainty about the specific mixture of each compound.

Proust had therefore demonstrated that at least several metals can form multiple oxides and sulfides. Directly refuting Berthollet's findings, Proust showed that each oxide or sulfide had a specific, definite composition. Proust persuasively demonstrated in 1799 that copper carbonate contains definite weight proportions of copper, carbon, and oxygen. He showed that this was true for both the copper carbonate produced in the laboratory and the copper carbonate isolated from mined earthen extractions. For this compound, the proportions were always 5.3 parts copper to 4 parts oxygen to 1 part carbon—regardless of how it was produced.

In tests that followed, Proust showed that the compositions of other compounds were equally constant and specific. In 1799, Proust stated the **law of constant proportions**, which is sometimes called Proust's law. This law states that an "invisible hand" holds the composition of compounds so that it is always the same proportion, regardless of how it is made in nature or in a laboratory. A modern statement of this law is: A chemical compound always contains the same proportion (mass) of its constituent elements, regardless of its source or preparation. A simple, distilled version of Proust's law could be stated thus: *Compounds always contain fixed proportions of elements.*

Berthollet continued to challenge Proust's law. In fact, he cited numerous examples in which combinations of two elements formed more than one compound. Some of Berthollet's examples were shown to result from either inaccurate analysis or insufficiently purified reactants. However, others would have to wait for explanation. Even though Proust had published his law much earlier, the famous Swedish chemist Jöns Jacob Berzelius (1779–1848) credited Proust with the law in 1811. After this highly visible publicity, Proust's law fell into general acceptance in nearly all areas of chemistry.

Early in the nineteenth century, Irish scientist William Higgins (1763–1825) claimed that when elements combine to form more than one type of molecule in a compound, they do so in small, whole numbers of a ratio. Higgins's claim was purely speculative and based almost entirely on Proust's findings of constant proportions. In 1803, English chemist John Dalton (1766–1844) proposed his atomic theory of matter. It both depended on and explained Proust's law of constant proportions. Dalton's theory held that each chemical element is composed of its own unique atoms, and that each atom has the same specific, relative weight. His theory therefore explains why a fixed weight of a substance always combines with a fixed weight of another substance when forming a compound.

Dalton's theory explained many of the leftover anomalies established earlier by Berthollet and unexplained by Proust's law alone. Dalton found that when two (or more) elements combined in different proportions under different conditions, the results could be multiple compounds. For example, when he combined 3 parts carbon and 8 parts oxygen (strictly by weight), he got carbon *di*oxide. But when he mixed 3 parts carbon with 4 parts oxygen, he got carbon *mono*xide. Dalton also noticed that all the combinations were in the form of small whole numbers.

Dalton had therefore discovered what was to become one of the basic laws of chemistry, the *law of multiple proportions*. This law is based on three assumptions. First, Lavoisier's **law of conservation of mass** states that matter can neither be created nor destroyed. Dalton said that *atoms* are neither created nor destroyed. (Today, this law holds true in all ordinary [e.g., non-nuclear] chemical reactions.) Second, Proust's law of definite proportion established that compounds contain fixed proportions of elements. Dalton said that *atom* ratio is fixed and that therefore mass is constant. Third, Dalton's law of multiple proportions resulted where two elements form more than one specific compound. Dalton's statement of Lavoisier's theory and Proust's law became foundational to modern chemistry.

Like Proust's, Dalton's work was not accepted right away or without the help of other scientists. In 1808, English physicist William Wollaston (1766–1828) and Scottish physicist Thomas Thomson (1773–1852) conducted a series of experiments offering indisputable proof of Dalton's law—and, therefore, of Proust's law. Today, like Dalton's atomic theory, Proust's conception of definite composition is an essential characteristic of chemical compounds.

SEE ALSO Carbon Dioxide; Law of Conservation of Mass; Oxygen; Substances: Elements or Compounds.

Selected Bibliography

Asimov, Isaac. *A Short History of Chemistry*. Garden City, NY: Anchor Books, 1965.
Bensaude-Vincent, Bernadette, and Isabelle Stengers. *A History of Chemistry*. Translated by Deborah van Dam. Cambridge, MA: Harvard University Press, 1996.

Brock, William H. *The Norton History of Chemistry*. New York: W. W. Norton, 1992.

Greenberg, Arthur. *A Chemical History Tour: Picturing Chemistry from Alchemy to Modern Molecular Science*. New York: John Wiley & Sons, 2000.

Krebs, Robert E. *Scientific Laws, Principles, and Theories: A Reference Guide*. Westport, CT: Greenwood Press, 2001.

Partington, J. R. *A Short History of Chemistry*. New York: St. Martin's Press, 1960.

R

Rational System of Chemical Nomenclature
Antoine Lavoisier, Louis-Barnard Guyton de Morveau,
Marie-Anne Paulze-Lavoisier, Claude Louis Bertholette
and Antoine Francois de Fourcroy, 1787

As the eighteenth-century dawn of modern chemistry emerged, it inherited a confusing babble of inarticulate chemical names. The **alchemy** names of substances had evolved slowly and unevenly over many centuries and in many places. In this derelict pastiche, alchemists had named substances haphazardly, based on any number of varying characteristics. Often, chemical names varied according to country, with no regard to scientific paradigms. Beginning in 1700, the **phlogiston** theory stated that the hypothetical substance phlogiston was liberated from a substance during **combustion**. Phlogiston was alchemy's final gasp before being shattered by modern **oxygen** chemistry. Just as phlogiston was failing to accurately describe the makeup and property of matter, its inherited language was failing to describe the foundational substances of chemistry, which were theoretically emerging in France.

In alchemy's jumbled mess, some chemicals were named for their discoverers, while others were named for their subjective resemblance to familiar objects. Some were named for their place of discovery, while others were named for planets. Some of the known metals were named for ancient polytheistic gods. Tin was "Jupiter," iron was "Mars," and iron oxide was the monstrosity "astringent Mars saffron." Sulfuric acid was "oil of vitriol" or "Nordhausen oil of vitriol," and copper sulfate (apparently a hybrid of name strategies) was "vitriol of **Venus**." An oxide of a metal was its **calx**. Magnesium sulfate was "Epsom salt"—apparently after the mineral springs in Epsom (Surrey), England. Perhaps most unflattering, zinc oxide was "philosophic wool" and calcium acetate was "shrimp eyes salt." A mixture of potassium hydroxide and potassium carbonate was known by the rather unfortunate moniker "oil of tartar per deliquium." These names may have served the needs of alchemy just fine, but they were miserably failing the more sophisticated needs of precise chemistry.

Even more infuriating than the lack of rational nomenclature, though, was the lack of standardized names that everyone agreed to use. French chemists—at the forefront of the chemical revolution—had to borrow from English, German and Swedish technical terms to describe their findings. Worse yet, widespread and varied practical applications of chemistry developed an entirely new, unforeseen disparity between professions. Merchants rarely used the same terminology as scientists, preferring "green blotches" to "vitriol of Mars" and "white blotches" to "zinc vitriol." Most fantastically, the main substances of alchemy's focus were nearly impossible to speak about with any degree of clarity. Lead had no less than sixteen different names, while mercury boasted a staggering total of thirty-five.

Despite the tremendous frustration of these names, chemistry was somehow still exploding with new discoveries. The number of known substances was growing every year, as was the amount of demonstrable knowledge about each one. Previously, seven metals had been documented. Suddenly there were seventeen. Where earth was previously considered an **element**, five different types were classified in a very short span of time. Stone was being subdivided again and again into different types. Even masterful chemists most certainly had a hard time keeping up with the revolution in gases. Scientists were making discoveries about gases almost overnight.

All of this led French chemist Louis-Barnard Guyton de Morveau (1737–1816) to shrewdly ask, "Does not the understanding of such a nomenclature require more effort than the understanding of the science itself?" (quoted in Poirier 1996, 183). Indeed, the fact that there was no language to describe this young science shows just how revolutionary and new chemistry really was. It seemed as though modern chemistry was born paradoxically ahead of itself, idiosyncratic to its own moment, with such furious speed of achievement that the language of its birth needed time to catch up. Substances were changing fast. Soon, so were their names.

In 1787, after much frustrated discussion, a group of chemists began meeting frequently. The location was the laboratory of Antoine Lavoisier (1743–1794)—as close a place as any to the birthplace of chemistry. This location was fitting, given the group's daunting project of creating a fitting, practical language for chemistry. The linguistic powerhouse included Lavoisier, de Morveau, and three others: Lavoisier's wife, translator, and collaborator,

Marie-Anne Paulze-Lavoisier (1758–1836); famous chemist Claude Louis Bertholette (1748–1822); and French politician, doctor and chemist Antoine Francois de Fourcroy (1755–1809). The act of collaboration was itself evidence of enormous philosophical change. Where alchemy relied on carefully guarded knowledge, chemistry was quickly becoming open to cooperation and careful dissemination of freely shared knowledge.

In his important work as a writer and editor, de Morveau had long been bothered by an inadequate chemical lexicon. Collaborating with Lavoisier was an obvious choice. Lavoisier had already rescued his own special gas from the sobriquets **fire air** and **dephlogisticated air**. Whether or not Shakespeare's perennial posy is still a rose by any other name may be open to debate. But these two scientists knew good and well that "Libavius smoking liquor" is simply not the same as tin chloride.

In many ways, de Morveau was the first to really think long and hard about the problems of naming in the new science. He was first to suggest that publishing a book might eliminate the confusion of the old names. As early as 1782, he wrote about the need for a table of chemical nomenclature that should respect five fundamental principles. First, de Morveau said, a phrase is not a name. A name should apply in all circumstances regardless of context—just like a person's name. Second, a name should conform to the nature of a substance itself. An elemental name should derive only from the simple substance in question. The **compound** name should refer to—and only to—the nature of the compound's constituent substances. Third, if little is yet known about a substance, the name should imply nothing, rather than something that could convey a false impression about it. Fourth, new designations ought to be formed from Greek or Latin roots for easy memorization. And finally, designations should be carefully mated to the genus of the language for which they are being formed.

De Morveau's second principle especially bears close examination, for it would emerge as the most revolutionary rupture with the old names. This principle stated that the system for naming substances should be anything but haphazard. Rather, a name should entirely reflect a substance's unique composition. Today, this rule of chemical nomenclature is so universally accepted that it seems like common sense. Many chemical historians consider de Morveau's principle as equally important with any discovery of a new substance. Making the names of chemicals readily understandable helped advance the new science of chemistry.

The exciting meetings in Lavoisier's laboratory coalesced in 1787 with the magnificently influential book *Méthode de Nomenclature Chimique*. This book disseminated the new system of nomenclature. In doing so, it forcefully made the case for modern chemistry. In the process, it persuaded many prominent

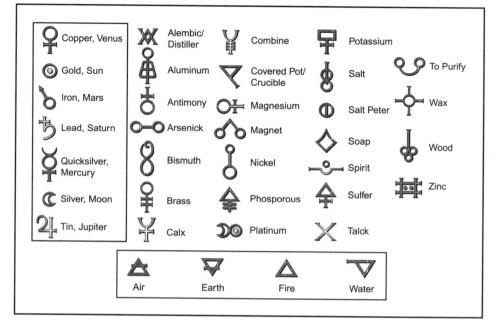

Figure 22. Table of common alchemy symbols.

chemists away from phlogiston. At its heart, *Nomenclature* was a table arranged like a dictionary, with the old alchemic names next to the new chemical ones (see Figure 22).

In its system of nomenclature, the chemicals named first were considered "simple." They roughly corresponded with modern elements (with a few erroneous inclusions, such as heat, light, and so-called heat fluid, **caloric**). Names were mainly derived from Greek and they expressed well-known, easily identified characteristics. The next group contained the principal elements of acids (phosphorus, sulfur and carbon were named here). The third group comprised the metals. For the first time, they were considered "simple" substances (phlogistonists had considered many of them compounds). After metals came a group the chemists called "simple earthy substances" (lime, magnesia, silica). These were followed by the alkalis or bases (soda, potash, ammonia). Next on the table were the compounds. These were defined as substances that contain two or more "simple" substances. Next, instead of the old category "salts," was the new category of nitrites, sulfates and sulfites, phosphates, and carbonates.

Finally, the table showcased a crucial substitution. Instead of alchemic "calces," *Nomenclature* listed chemical oxides. There were simple combinations of any one of a number of elements with the first element that Lavoisier ever named, oxygen. Chemical "calces" had been fundamental to the theory of phlogiston. Lavoisier's destruction of phlogiston was centered on the discovery that this reaction caused a gain in mass. He showed this was due to the action of oxygen joining with the substance, and therefore renamed the reaction "oxygenation." Many of these classifications may sound familiar, because they form the basis of chemical nomenclature today—more than two centuries later.

The renowned group of chemists went on to publish a journal and to publish several updated editions of important chemistry books. In the latter capacity, in 1784, Marie-Anne Paulze-Lavoisier translated Irish chemist Richard Kirwan's *Essay on Phlogiston* from English to French. This edition included critical notes by members of the group. Because it stubbornly defended the phlogiston theory, Kirwan's book became a rallying point, a focus of French criticism. This critique was such a coup de grace to the old theory that, in 1792, Kirwan publicly abandoned phlogiston altogether.

After *Nomenclature*, chemistry really took off. The language had finally caught up with the science. Also, many other scientists adopted chemistry during the next several years. Since this time, Lavoisier's chemistry has been accepted with near universality. Like French subjects storming the Bastille, the linguistic task force tore down the tower of alchemic babble. The former eventually emerged as citizens of a powerful new republic. The latter emerged as founders of a powerful new science.

SEE ALSO Heat Is a Form of Energy; Law of Conservation of Mass; Oxygen; Phlogiston; Substances: Elements and Compounds.

Selected Bibliography

Greenberg, Arthur. *A Chemical History Tour: Picturing Chemistry from Alchemy to Modern Molecular Science*. New York: John Wiley & Sons, 2000.

Guerlac, Henry. *Antoine-Laurent Lavoisier: Chemist and Revolutionary*. New York: Charles Scribner's Sons, 1975.

Lockemann, Georg. *The Story of Chemistry*. New York: Philosophical Library, 1959.

Poirier, Jean-Pierre. *Lavoisier: Chemist, Biologist, Economist*. Translated by Rebecca Balinski. Philadelphia: University of Pennsylvania Press, 1996.

Riedman, Sarah R. *Antoine Lavoisier, Scientist and Citizen*. New York: Abelard-Schuman, 1967.

S

Scientific Morphology and Taxonomy of Animals and Plants

Carolus Linnaeus

Plant and animal classification in *Systema Naturae*, 1735 and 1758 editions

Plant classification in *Species Plantarum*, 1753

Taxonomy, or the practice of classification, is a very old field of biology. The ancient Egyptians, Assyrians and Chinese may have arranged gardens according to an esoteric classification order. The word *taxonomy* comes from the Greek for "naming in order," and indeed the ancient Greeks largely began the practice. Theophrastus (372–288 B.C.) became known as the founder of biology after he classified many of the plants originally collected by his teacher, Aristotle (384–322 B.C.). Aristotle made some of the earliest studies of the natural world. In fact, his elaborate taxonomy impacted the supposed order of living things for two thousand years, and continues to do so today.

Aristotle based his classification system on four postulates. These were drawn from philosophical doctrine, not empirical findings. The first postulate stated that each species never changes and is fixed. Aristotle assumed that species change was not possible and adhered to the idea of eternal or "real" species. His second postulate stated that each species reflected a true and perfect ideal, or *idos*. Each species had an "essence," which made variations insignificant. Aristotle's third postulate, *Scala Naturae*, described his great ladder of being or chain of nature. The great ladder organized all organisms of the natural world from tiny and simple creatures to large and complex ones. As vastly different creatures from all others, humans were the very embodiment of biological complexity and were comfortably perched on the highest rung. Humans shared little with the "inanimate" matter at the bottom;

the plants just above it; the organisms of the water; or the insects, fish, reptiles and mammals of the Earth. Finally, the fourth postulate stated that the natural world is always held in perpetual, perfect equilibrium. Because nature was the very model of balance, Aristotle had no use for variations caused by ecological or environmental factors.

In Aristotle's system, there was little change in the world or its creatures, or in their relations with one another. Aristotle and his students first used the word *genus* to describe a group of similar organisms. They then tried to describe the *differentio specifica*—the specific difference—that made each type of organism unique within the group. Aristotle's basic ideas about the order of things would not be seriously challenged until the enlightenment in Europe, during the seventeenth and eighteenth centuries.

Another Greek scientist, Pedanius Dioscorides (ca. 40–90), classified plants by their known medical properties. He published his findings in *De Materia Medica*. A Roman naturalist, Pliny the Elder (23–79), recorded many of the known species of plants in a 37-volume text, *Historis Naturalis*. Like Aristotle's metaphysical scale, each of these books was authoritative in Europe for more than one thousand years.

In the sixteenth century, the European botanists known as herbalists began to appear. Though largely dependent on the work of Dioscorides and Pliny, these scientists reinvigorated the study of plants, mostly for medicinal use. During this period, many naturalists relied on arbitrary criteria for classification of animals. This gave rise to subjective, culturally specific, slippery categories such as "water creatures" and "domesticated animals." These categories still owed their genesis to Aristotle. However, there would be a shift in thinking during the industrial revolution, which demonstrated that applied science could drastically alter the natural world within a span of a few short decades. John Ray (1627–1705) first began studying all parts of an organism during different phases of its development. He also first suggested broadly applying the concept of different species of animals to their classification. Later, Johann Baptist van Helmont (1577–1644) performed his famous "tree experiment." Though it was greatly flawed, this experiment was one of the first to study the structure of organisms in great detail. (For more on van Helmont's tree experiment, please see the entry on Plant Physiology.)

During the next century, Stephen Hales (1677–1761) founded a new experimentalist tradition for plant **physiology**. Also, Baron Cuvier (1769–1832) founded comparative anatomy, the branch of zoology dealing with comparisons and contrasts in body structures of animals. Cuvier also proposed a four-phyla schema for animal classification. Aristotelian classification was on the verge of complete overhaul. (For more information, please see the topic on Plant Physiology.)

Beginning in 1735, Swedish scientist Carl von Linné (1707–1778)—better known by the Latin version of his name, Carolus Linnaeus—published his classification of living things, *Systema Naturae*. At first, *Systema Naturae* was a slim pamphlet of mostly plants. In 1753, Linnaeus published an expanded two-volume study of plants, *Species Plantarum*. This book first elaborated a system of botanical taxonomy. Then, subsequent editions of *Systema Naturae* grew to multivolume, weighty tomes. Several editions, such as the ones published in 1758 and 1767, were especially important because they included classifications of animals as well as plants. In these texts, Linnaeus expanded classification by species into a full-blown system of scientific taxonomy. This was a major step beyond Aristotle's broad grouping of animals within genera.

Linnaeus also classified creatures according to the differences between species. However, when he wanted to broaden his taxonomic classification beyond even genera, Linnaeus ran into problems. He found that scientists had yet to invent applicable nomenclature. The practice was so modern and so new that it lacked an appropriate technical language. Scientists currently described species in long, inconsistent names that they altered nearly at will. Generally, plant names consisted of two elements. The first was a word or series of words common to some denomination of plants. The second was most often a phrase stating the marks that characterized some part of the plant. Two names might apply to a species, or several species might share a name. Worse yet, scientists in different areas had different names for the same plants and therefore lacked a common vocabulary. This problem was deeply compounded by the huge numbers of "new" plants and animals being brought from Asia, America, and Africa by explorers.

Linnaeus outlined an ingenious terminology of paired Latin names—a kind of scientific shorthand known as systematic **binomial nomenclature**. This naming strategy was analogous to that used for human beings in many Western cultures (though it reversed the order). The first name was common to all species of one large group (or genus). The second name was unique to a single kind of organism (or species within the genus). Working first from humans, Linnaeus combined *Homo* and *diurnis* for "man of the day." Later editions of *Systema Naturae* refined humans to the genus *Homo* and species *sapiens* for the familiar species name meaning "wise man" (or human). Linnaeus then expanded his classifications beyond species. He grouped similar species into genera, similar genera into families, similar families into orders, similar orders into classes, similar classes into phyla and phyla into one of two kingdoms—plants or animals. With some modifications, this is basically the taxonomic system that scientists use today.

Linnaeus and later scientists explained this new system through the metaphor of a tree in which large branches divide into smaller branches again and again until the final limbs, representing the species. It was only later, with the advent of biological evolution, that scientists fully understood the concept of taxonomy as itself a metaphorical kind of adaptable "living creature."

Much of this work in classification had come through *Species Plantarum*, in which Linnaeus classified plants according to a "sexual" system of stamens and pistils. In this book's 1,200 pages, Linnaeus classified 5,900 species in 1,098 genera—every plant he had ever known. Today, botanists generally accept the 1753 publication of *Species Plantarum* as the starting point of modern botanical nomenclature. Zoologists similarly refer to the 1758 publication of the first volume of the tenth edition of *Systema Naturae* as their field's taxonomic founding. These two texts contain, respectively, the oldest plant and animal names that are currently accepted as scientifically valid.

Today, scientists accept the seven categories of Linnaeus's system with some modifications. Most significantly, the two-kingdom model was replaced in the 1960s, when knowledge of microscopic structure and biochemistry finally made it obsolete. Scientists now accept a five-kingdom model consisting of Animalia, Plantae, Fungi (such as mushrooms, molds and lichens), Protista (many kinds of algae, sporozoans and flagellates), and Prokaryotae (bacteria and cyanobacteria). With over a million documented species, Animalia is the largest kingdom. Like Linnaeus, scientists base modern classification on the concept of the absolutely unique species—even for similar organisms. For example, both lions and tigers are members of the kingdom Animalia, the phylum Chordata, the class Mammalia, the order Carnivora, the family Felidae and the genus Panthera. But each has a specific species. The lion is classified in the species *Panthera leo*, the tiger in *Panthera tigris*.

SEE ALSO Animal Physiology; Modern Entomology; Plants Use Sunlight; Plant Physiology; Rational System of Chemical Nomenclature.

Selected Bibliography

Asimov, Isaac. *Asimov's Chronology of Science and Discovery*. New York: Harper & Row, 1989.

Blunt, Wilfrid. *The Compleat Naturalist: A Life of Linnaeus*. New York: Viking Press, 1971.

Broberg, Gunnar (ed.). *Linnaeus: Progress and Prospects in Linnaean Research*. Pittsburgh: Hunt Institute for Botanical Documentation, 1980.

Frängsmyr, Tore (ed.). *Linnaeus: The Man and His Work*. Los Angeles: University of California Press, 1983.

Krebs, Robert E. *Scientific Laws, Principles, and Theories: A Reference Guide*. Westport, CT: Greenwood Press, 2001.

Larson, James L. *Reason and Experience: The Representation of Natural Order in the Work of Carl von Linné*. Los Angeles: University of California Press, 1971.

Spinning Machines
James Hargreaves, Spinning Jenny, 1764
Sir Richard Arkwright, Water Frame Spinning Machine, 1769
Samuel Crompton, Spinning Mule, 1779

Many historians believe that, after the **steam engine**, the invention with the single greatest impact on the eighteenth century was the **spinning machine**. Through a series of short-spaced machines, the entire process of spinning was revolutionized from an outdated and home-based process to a remarkably modern and mechanized one. By the end of the eighteenth century, spinners had become a distinct class of worker and their new practices laid the foundation for the factory model of efficient production.

Spinning is an ancient art form that changed little for many hundreds of years. A person used turning spindles to twist animal or plant fiber into thread. These primitive machines were little more than rotating sticks with a notch at one end to gather thread. In the fourteenth century, spinning wheels from India made their way to Europe. They consisted of a large wheel, which turned a spindle to draw thread. Later wheels, such as the Saxony wheel, were powered by a foot pedal. These wheels were a big improvement over spinning by hand, but they still depended on the skill and sheer numbers of individual people spinning individual threads.

In 1733, John Kay (1704–1764) invented the flying shuttle loom. (For more on this topic, please see the entry on Weaving Machines.) Soon, **weaving**—the ancient art of making cloth by crossing threads—was completely transformed. The traditional method of weaving required a laborious hand-driven process carried out by one or two weavers on each and every loom. With Kay's flying shuttle, the process was mechanized and the speed was greatly enhanced. The time it took to produce cloth was drastically decreased. By some estimates, Kay's flying shuttle doubled the production of cloth throughout Europe. The insatiable market for cotton products drove these levels of production. As a result, weavers needed more yarn than ever if they were to meet demand for cloth. This put enormous pressure on spinners to produce cotton. However, even the best spinning wheels allowed each spinner to produce only one thread on one spindle at one time. Every factor pointed to an improvement in spinning that would allow multiple thread production at once.

The first spinning machine was built in 1738 by Lewis Paul (ca. 1759) and John Wyatt (1700–1766), two English inventors. This machine introduced roller spinning, a technique that would become vital to later inventions. One roller fed a piece of cotton or wool at a steady speed, while a second roller pulled the fiber at a slightly greater speed. The result was a variance in the tightness of thread. When combined, these non-uniform threads were found to be much stronger than ones of uniform tightness. This machine failed to catch on for a variety of reasons—both technical and economic. Nonetheless, it was an important first step that led the way for some of the most important machines ever invented.

The first of these machines was the **spinning jenny** invented by James Hargreaves (ca. 1722–1778) in 1764. Hargreaves took inspiration for his spinning jenny when he saw a spinning wheel continue to turn after it was knocked on its side. Indeed, Hargreaves's spinning jenny did resemble a number of spindles placed vertically and next to one another.

To operate the spinning jenny, a spinner turned a large wheel, which turned the row of spindles—at first 8 then 16, and then as many as 120 per machine. A clasp held the spindles. On the opposite side, Hargreaves mounted a moveable carriage to hold the bits of cotton. A single spinner could therefore operate the entire machine. With one hand, he or she turned the large wheel, often by means of a crank. With the other, he or she could slide the carriage frame back and forth, as necessary, to draw the thread on each and every spindle at once. In a separate operation, the drawn thread was collected and wound onto bobbins.

Hargreaves's machine had the enormous advantage of producing multiple threads at once (see Figure 23). As important as it was, another machine had an even greater effect. This was the **water frame spinning machine** invented by Sir Richard Arkwright (1732–1792). Arkwright held many professions before finally turning his attention to the problem of mechanized spinning. He knew of several small machines that were successfully driven by horse power. He even knew of one that was driven by the power of running water from a river. But the problem was only partly with harnessing power. Until now, even the greatly improved jenny could not meet the demands for cotton yarn. Also, the jenny successfully mechanized the actual spinning operation, but not the tedious operations that preceded it.

Arkwright's machine perfected the technique of roller spinning, for which it employed two pairs of rollers. The top rollers were covered in leather to tightly grip the cotton, and the bottom ones were fluted to allow it to pass through. One pair of rollers turned significantly faster than the rest, with each pair turning successively slower. This is how the machine drew out the cotton to the required tension, achieving varying levels of coarseness. For the most part, the water frame made very coarse, strong yarn. Twisting the cotton was carried out by spindles in front of the pairs of rollers.

Arkwright may have borrowed parts of his technique from Wyatt and Paul's earlier use of roller spinning, but he perfected the machinery in two key ways.

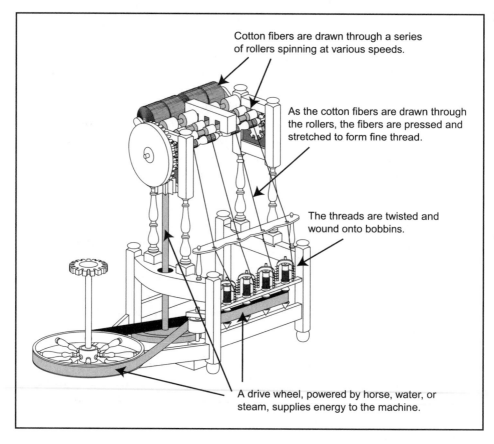

Cotton fibers are drawn through a series of rollers spinning at various speeds.

As the cotton fibers are drawn through the rollers, the fibers are pressed and stretched to form fine thread.

The threads are twisted and wound onto bobbins.

A drive wheel, powered by horse, water, or steam, supplies energy to the machine.

Figure 23. Spinning machine. The many eighteenth century spinning machines revolutionized the spinning of cotton into yarn. Since antiquity, an individual had been able to spin only a single thread of yarn at once. Spinning machines allowed for spinning multiple threads at one time, thereby greatly increasing production. Later models, like the one shown here, were powered by horse, water, or steam energy.

First, he placed the pairs of rollers an effective distance apart for drawing. Previously, they were generally too close or too far apart. The former mistake caused the yarn to snap, and the latter caused it to become unbalanced or even lumpy. Second, Arkwright also applied weights to the top rollers, which made them push firmly against the lower ones. This prevented unbalanced yarn, because the cotton could not run between the rollers and into the drawing zone.

Together, Hargreaves's jenny and Arkwright's water frame began the transformation in spinning. Rather than conflicting, they complemented one another. While the jenny made fine, fragile yarn, the water frame made strong, coarse yarn. A decade after Arkwright's machine appeared, a third machine joined the two others. In 1779, an English inventor named Samuel Crompton (1753–1827) introduced his **spinning mule**. This machine was aptly named. Just as a mule is a cross between two animals (a horse and a donkey), the spinning mule is a cross between the two machines. Like its animal namesake, the spinning mule combined the best of both worlds. It could produce whichever yarn one desired, strong or fine. A secondary advantage was that the quality of yarn produced by the spinning mule was much easier to control than with the jenny or water frame.

Many parts of Crompton's machine were incorporated from the jenny and water frame, but the movable carriage of forty-eight spindles was the machine's most ingenious innovation. It accomplished the weaving of both fine and strong cotton, as well as simple quality control through the movable carriage. By sliding the carriage, the cotton could be stretched gently or firmly to produce either a fine or a strong yarn. It could even produce yarn of varying degrees between the two extremes. Equally important, the movable carriage made it possible—with a little practice—for the spinner to quickly correct production problems as they surfaced. Because forty-eight spindles were mounted on the moveable carriage, the tension employed in roller spinning was achieved. This was done by simply moving the carriage away from the rollers at a slightly faster rate than the rate of cotton delivery to the last pair of rollers. Within several years, Crompton's spinning mules were in widespread use, superseding even the two machines it combined.

The spinning machines changed everything in the eighteenth-century world of spinning cotton (and, to a lesser degree, wool). Their effect on the economies of England and the New World was unprecedented. Just as Kay's weaving machines created an insatiable market for spun cotton, spinning machines created a nearly limitless demand for cleaned cotton. The **cotton gin** of Eli Whitney (1765–1825) would answer that latter call. Where spinning (and, to a lesser degree, weaving) machines transformed industry, this machine revolutionized agriculture. (For more on this point, please see the entry on the Cotton Gin.)

The changes to industrial manufacture went even deeper, however. More than anyone else, Arkwright combined different processes of yarn manufacture (carding, roving, drawing and spinning) into one smooth operation. In this manner, the spinning machines contributed enormously to the development of modern factories with their effective, efficient processes. By some estimates, the spinning machines announced a fundamental shift in basic human organization as well as sophisticated means of production.

SEE ALSO Cotton Gin; Weaving Machines.

Selected Bibliography

Britton, Karen Gerhardt. *Bale o' Cotton: The Mechanical Art of Cotton Ginning.* College Station: Texas A&M Press, 1992.

Burton, Anthony. *The Rise and Fall of King Cotton.* London: British Broadcasting Corporation, 1984.

Fitton, R. S. *The Arkwrights: Spinners of Fortune.* Manchester, UK: Manchester University Press, 1989.

Hills, Richard L. *Power in the Industrial Revolution.* Manchester, UK: Manchester University Press, 1970.

Mirsky, Jeannette, and Allan Nevins. *The World of Eli Whitney.* New York: Macmillan, 1952.

Rowland, K. T. *Eighteenth Century Inventions.* New York: David & Charles Books, 1974.

Steamboat
Claude de Jouffroy, 1783
John Fitch, 1787

By the 1780s, enlightenment scientists had made many important discoveries and inventions. Pneumatic chemists had discovered the world's first independent gases **carbon dioxide**, **chlorine**, **hydrogen**, **oxygen**, and **nitrogen**. In France, Antoine Lavoisier (1743–1794) was furiously reworking their discoveries into a theory of modern oxygen chemistry. During the 1780s, English inventor James Watt (1736–1819) and scientist Joseph Black applied the concept of **latent heat** to the **steam engine** in the form of the separate condenser. As much as any other single event, Watt's steam engine forged the foundation for the industrial revolution. Around the same time, French inventors Joseph (1740–1810) and Étienne (1745–1799) **Montgolfier**, as well as Jacques Alexandre Cesar Charles (1746–1823), were, respectively, inventing **hot-air** and **gas balloons**. Benjamin Franklin (1706–1790) and Charles Dufay (1698–1739) were discovering applications of electricity. In 1781, William Herschel (1738–1822) discovered the first planet, **Uranus**, since the advent of history. The world was changing faster than it ever had before. (For more information, see the various entries on these topics.)

However, despite the furious pace of change during the late eighteenth century, transportation over long distances was little better than it had been in antiquity. Balloons offered some improvement, but they certainly could not move large numbers of people over large distances on a regular basis. For that, people still relied (as they always had) on old forms of energy, such as animal or wind power. The average person could not journey far from where he or she lived. Transportation was a slow, inefficient, and often dangerous proposition.

Since the early eighteenth century and the invention of the first practical steam engine, inventors had hoped to harness the power to propel their boats. In fact, the question of effective boat propulsion had vexed inventors since antiquity. The ancient Egyptians had the first really usable paddle wheel, which drove small boats for centuries. During the Italian Renaissance, Leonardo da Vinci (1452–1519) drew a diagram for a much more effective paddle wheel boat, though the craft was never built. In the mid-sixteenth century, Spanish inventor Blasco de Garay (ca. 1543) unsuccessfully tried to run a boat with paddle wheels and a cryptic "vessel of boiling water." Scientists are unsure about this reference, though they believe de Garay's boat never achieved self-propulsion because reliable steam engines did not yet exist.

In 1707, Denis Papin (ca. 1647–1712) designed a craft to run off the energy supplied by his early steam-powered water pump. Papin set a pair of waterwheels to drive a shaft from water supplied by one of his pumps. Papin's boat was built, but never moved. His pump was not powerful enough. Still, Papin's boat represented a vast improvement over older models. Papin understood the need to efficiently transfer energy from his pump or engine to his waterwheel. The main problem was not in the design of his boat, but the lack of power from his engine.

In 1712, Thomas Newcomen (1663–1729) built the first practical, usable steam engine. Its impact on English industry was enormous. Newcomen's huge engines effectively cleared flooded mines by driving pumps, and also enhanced older forms of power, especially water. The world's second self-acting machine (after clocks), Newcomen's engine was entirely driven by atmospheric pressure. In a **Newcomen steam engine**, a blast of steam at greater than atmospheric pressure was forced to the bottom of a brass cylinder, which held the engine's piston. Then, a jet of cold water was injected to quickly condense the steam. This created a vacuum, which caused the pressure of the atmosphere to slam the piston downward, creating a power stroke. For the next power stroke, the same brass cylinder had to be reheated by steam, and the process began again. At its prime, a Newcomen engine was capable of pumping over 3.5 million gallons (13.2 million liters) of water a day. This offered enormous potential for **transportation**.

One inventor who saw this potential was English military inventor Jonathan Hulls (1699–1755). In 1736, Hulls outfitted a small paddle wheel tugboat with a scaled-down Newcomen steam engine. Hulls mounted his engine at the stern and transferred its energy power with a counterweight driven by the engine's piston. Hulls hoped to use his boat to rescue becalmed Royal Navy ships in harbor. Hulls's tugboat was, by far, the most scientifically sound attempt at building a steamboat to date. But in the end, its method of crude power transmission lost too much energy and the boat never moved.

The first inventor to move a boat with the power of steam was a French nobleman, the Marquee Claude de Jouffroy d'Abbans (1751–1832). In the late 1770s, Jouffroy began a study of the natural power of animals in different circumstances. Jouffroy spent a long exile in Provence for dueling as a young man. There, he began an extensive study of ornithology, the study of birds. In fact, Jouffroy became so impressed with the paddling mechanisms of aquatic birds, that he became one of the world's foremost experts. On returning from exile, Jouffroy decided to build a boat from an idea he had considered for several years. His boat would have a steam-driven, mechanical web paddle fastened to the end of a rod. The boat would be driven by a webbed "foot," just like aquatic birds.

Jouffroy hired a local coppersmith to build a small version of a Newcomen steam engine. In this ingenious engine, a single sheet of copper was bent into shape and smoothed into the shape of a barrel with iron rings. An extremely heavy counterweight pulled the engine's piston to the top of its brass cylinder before each downward power stroke of its piston. On the boat, Jouffroy designed a series of eight-foot-long rods that ended with frames fitted with hinged flaps like aquatic paddles. The engine's power stroke moved a system of ratchets and pulleys, which effectively transferred energy to each aquatic paddle. This forced each paddle open (like a hand-held fan) and back, so that it cut through the water and propelled the boat forward. Because Newcomen engines had only a single (rather than a reciprocating) stroke, the counterweight took hold at the end of every cycle. It pulled each rod back toward the bow of the boat, shut the webbed paddle, and raised the piston for the next power stroke.

Newcomen engines could not produce rotary, even motion. Rather, they produced uneven, irregular power strokes. The webbed paddles were perfectly suited to this delivery of energy. However, Jouffroy's trials demonstrated that his engine did not produce enough power strokes to achieve an acceptable level of forward thrust for the boat. To obtain a much more continuous forward motion, Jouffroy designed the first twin-cylinder engine, which he timed to deliver alternating power strokes. At the moment that one piston entered its recovery stroke, the other piston produced its power stroke. In this manner, Jouffroy produced an effective (if uneven) forward motion for his boat.

In 1778, Jouffroy's boat was ready for a public trial on the river Doubs. Jouffroy designed his boat—from the start—to paddle upstream, like an enormous bird. He had anchored his boiler onto a vessel of about 40 feet in length, with a 6-foot beam. He inclined his engine's cylinders about 50 degrees and used a sliding tile beneath the valve chest to admit steam and water to the underside of each piston. Controlled trials had gone well, and Jouffroy was confident in his invention. However, the day did not go well. The river's gentle current caused the webbed paddles tremendous trouble in opening. After only several power strokes, Jouffroy realized he had made a mistake. He had not properly calculated for the resistance of even a gentle current. His webbed paddles were doomed. Nonetheless, Jouffroy learned a tremendous amount from this trial and went back to his designs.

During the next several years, Jouffroy learned of the revolutionary improvements to the steam engine of James Watt (1736–1819). Even though these machines were strictly controlled by the English government (export to France was especially forbidden), Jouffroy was able to study their design. Watt designed a separate condenser for the steam engine to run much more efficiently. He also invented the first parallel-motion, double-acting steam engine. This engine provided much more dependable power by forcing the piston both down *and* up. Watt had freed the steam engine from total reliance on atmospheric pressure and counterweight contraptions. His steam engine now provided much better efficiency and much more uniform power. Jouffroy realized that this would overcome many of the problems of his earlier, webbed-paddle failure. (For more on steam engines, please see the entry on this topic.)

With a reciprocating engine built specifically for him, Jouffroy realized he could not use his webbed paddles. In 1783, Jouffroy debuted his new ship, the *Pyroscaphe*, with a double-acting steam engine, which actuated a pair of large paddle wheels. Because of its reciprocating engine, the *Pyroscaphe* required only one 25.6-inch-diameter cylinder with a 77-inch stroke.

On July 15, 1783 (little more than a month after the Montgolfiers's first balloon trial), several thousand people gathered at the river Saone for the *Pyroscaphe*'s trial run. Jouffroy lit his boiler and waited for pressure to build. After a long, anxious pause, the heavy piston began to rise and fall. Jouffroy cut his boat loose from its dock. The *Pyroscaphe*'s enormous paddle wheels began to slap the water, slowly at first, and then with greater intensity until the boat began to move. For a full 15 minutes, the *Pyroscaphe* steamed against the current of the river. This 15-minute journey marked the first time in history that a vessel of any kind moved under its own power.

The first really usable steamboat was independently invented several years later. Across the ocean, one of the biggest questions facing the agrarian, newly formed United States of America was how to navigate its vast wilderness. Many inventors felt that riverboats were the only option. But the problem was that the rivers did not flow west into the heart of the resource-laden wilderness. In 1786, an unhappy American inventor named John Fitch (1743–1798) invented a boat that achieved a top speed of 3 miles per hour (mph) against the current. Fitch's craft was driven by a set of six paddles on each side, which were actuated by a crude steam engine. Before designing this craft, Fitch had only read about Newcomen steam engines. He had never seen one, so he hired a German immigrant mechanic, Johann Voigt, to build one. By many historical estimates, Voigt was one of the most skilled mechanics in early America.

With Voigt's technical assistance, Fitch set about redesigning the 1786 craft with the goal of dependable, regular steam service on the Delaware River. On August 22, 1787, he unveiled a 45-foot craft carrying an improved engine with a 12-inch bore and a 36-inch stroke. Fitch and Voigt did not have the benefit of many of Watt's innovations, which had not yet made it to America. Instead, they invented mechanisms to solve many of the problems that Watt had already addressed. Instead of the weight that was generally used to raise an engine's pistons, Fitch cleverly used a set of springs, which would not risk capsizing the boat in turbulent weather.

On this boat, three rectangular paddles at the stern replaced the six on either side, which greatly reduced resistance in the water. However, Fitch realized that these paddles also worked much better with even, steady strokes. To convert the engine's reciprocating motion to a circular motion, Fitch invented a ratchet wheel rotated by ropes on pulleys from the piston. Effectively, this independently rediscovered the rotary flywheel, which was James Watt's method of converting up-and-down piston motion to rotary action through the principle of inertia.

On August 22, Fitch's steamboat made a long, successful trip up the Delaware River. On this day, Fitch and several others became the world's first steamboat passengers. Less than 3 years later, Fitch completed his dream of creating the world's first commercial steamboat service on the Delaware. Through several more innovations, Fitch's vessel achieved a top speed of 7 knots with an efficient single-cylinder engine with separate condenser. During the summer of 1790, Fitch's steamboat service offered passengers one-way or round-trip service between Philadelphia and Trenton, New Jersey. Throughout the summer, Fitch's steamboat proved reliable and suffered few breakdowns in service. It marked the first time in history that regular transportation was provided by a vessel operating under its own power.

Unfortunately, during this same time, the relatively low-technology land stagecoach had come into widespread use. Because it relied on animal energy, the stagecoach proved to be significantly less expensive. Fitch's steamboat never turned a profit, his funding dried up, and the steamboat's service ended after only several months.

A few years later, Fitch's main competitor, American inventor James Rumsey (1743–1792), struck a deal with James Watt's firm to power his steamboat. Rumsey built a 101-ton vessel in England called the *Columbian Maid*. This vessel was to be propelled through a different kind of mechanism, a water-jet (suction) system. Ultimately, water-jet propulsion never really got off the ground, though modern diesel-powered catamarans and gasoline-powered personal water crafts employ a version of water-jet propulsion.

During the nineteenth century, steamboat service became a widely used form of transportation. In the United States, it was so successful that steamboats became a symbol of the country itself, steaming ever westward against the backward flow of ancient rivers. In 1807, American inventor Robert Fulton (1765–1815) designed the first commercially successful steamboat, the *Clermont*, in New York. The Clermont made its first successful trip up the Hudson River, from New York City to Albany, on August 17 of that year. This steamboat's steady trip ushered in a new era in the history of safe, affordable, and efficient mass transportation. During the following months, the *Clermont* became the first practical, financially successful steamboat. In 1812, Fulton also designed one of the earliest steamships built expressly for military purposes, the *Fulton the First*. This ship was to defend New York Harbor during the war of 1812, but it was not completed before Fulton died in 1815.

To this point, even successful boats like the *Clermont* were never built with oceangoing aspirations in mind. In 1809, the *Phoenix* became the first steamboat to successfully make a voyage on the Atlantic Ocean. After American engineer John Stevens (1749–1838) built the boat, the *Phoenix* traveled from New York City down the Delaware River and along the Atlantic coast to Philadelphia. The trip took a whopping 13 days. Sailboats routinely made the trip in 2 to 3 days. But Stevens had shown that steam vessels could navigate the choppy waters of the Atlantic.

In 1819, another American vessel, the steamship *Savannah*, crossed the Atlantic, traveling from New York City to Liverpool in 29 days. This trip was made partly on steam power and partly through the use of conventional sails. In 1838, the British side-wheeler *Sirius* offered the first regular, steam-powered service across the Atlantic. The trip took just 18.5 days, ushering in a new, unprecedented era of mobility across the Atlantic Ocean.

Great steamships and steamboats dominated water transportation during the nineteenth century. Their use only began to fade in the 1890s, when English engineer Charles Parsons (1854–1931) designed the marine steam turbine, a much more powerful kind of engine. In steam turbine engines, steam produced in a ship's boiler spins enormous bladed wheels of a turbine. In turn, the turbine drives a propeller shaft, making the ship's propeller revolve, thereby propelling the ship forward. Originally, these ships burned coal to produce steam. In the early twentieth century, most ships began burning gas or oil to produce steam to drive their turbines. In the mid-twentieth century, a few began using nuclear reactors.

SEE ALSO Balloon Flight; Latent and Specific Heat; Steam Engine.

Selected Bibliography

Briggs, Asa. *The Power of Steam: An Illustrated History of the World's Steam Age*. Chicago: University of Chicago Press, 1982.

Cox, Bernard. *Paddle Steamers*. Dorset, UK: Blandford Press, 1979.

Dayton, Fred Erving. *Steamboat Days*. New York: Frederick A. Stokes, 1925.

Flexner, James Thomas. *Steamboats Come True: American Inventors in Action*. Boston: Little, Brown, 1978.

Griffiths, Denis. *Steam at Sea: Two Centuries of Steam-Powered Ships*. London: Conway Maritime Press, 1997.

Hindle, Brooke. *Emulation and Invention*. New York: New York University Press, 1981.

Rowland, K. T. *Eighteenth Century Inventions*. New York: David & Charles Books, 1974.

Rowland, K. T. *Steam at Sea: A History of Steam Navigation*. New York: Praeger, 1970.

Virginskii, V. S. *Robert Fulton, 1765–1815*. Translated by Vijay Pandit. Washington, DC: Smithsonian Institution and National Science Foundation, 1976.

Steam Engine
Thomas Newcomen, 1712
James Watt, multiple ongoing innovations, 1763–1799

When the Manchester Museum of Science and Technology opened in 1968, a team of engineers from the University of Manchester agreed to build a working model of a **steam engine**. They decided that the engine would a be a one-third scale model based entirely on a 1719 drawing depicting a steam engine of the type invented by Thomas Newcomen (1663–1729). Throughout the process, none of the engineers was prepared for the endlessly frustrating pitfalls and trials of making the machine work. Building the engine had proved much more difficult than they ever imagined. What's more, as several engineers pointed out, Newcomen did not enjoy the advantages of modern engineering knowledge nor even the steady assurance that the machine could be made to work. Rather, Newcomen's invention was the result of highly concentrated work in an area that scientists and engineers knew little about.

During the seventeenth century, scientists had only begun to examine the properties of vacuums and pumps. Early on, German scientist Otto von Guericke (1602–1686) designed a hand pump that could remove air from sealed containers. Von Guericke proved that he had created a vacuum by showing that candles would not burn and bells would not ring inside his container. In 1654, Von Guericke demonstrated before Emperor Ferdinand III that a team of eight horses was unable to separate two half-spheres being held together by internal vacuum.

In 1698, English military engineer Thomas Savery (ca. 1650–1715) invented one of the earliest steam-powered machines, which he used to power a water pump, also of his invention, the *Miner's Friend*. The *Miner's Friend* used a steam-and-condensation chamber to create a vacuum for pumping water out of flooded coal mines. Savery's pump had no moving parts. Rather, it relied on expansion and suction effects generated by producing and then condensing steam in a single chamber. This formed a vacuum inside the chamber. When a valve was opened, water was forced up from the mine. However, the *Miner's Friend* only had a range of about 25 feet. It never lived up to its name.

In 1707, British inventor Denis Papin (1647–1712) improved on Savery's design with a pump that employed moving parts. Papin's pump had a moving piston, two receiving chambers and a crude safety valve. Though it was more efficient than Savery's pump had been, Papin's invention lacked the pressure needed to clear badly flooded mines.

Around the year 1700, British blacksmith and inventor Thomas Newcomen entered into a business partnership with a skilled mechanic and plumber named John Calley (1685–1729). During the next decade, Newcomen and Calley studied various water pumps based largely on Savery's original design. They studied the machines' central failure, which was its inability to pump water farther than a few feet. Newcomen and Calley also invented several mechanisms to overcome this lack of a powerful piston stroke before entirely

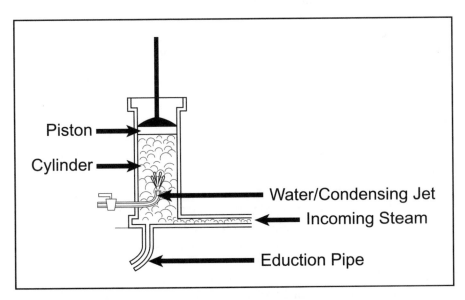

Figure 24. Newcomen steam engine. Around 1712, Thomas Newcomen introduced the first workable steam engine capable of pumping over 3 1/2 million gallons of water each day. Newcomen's engine used only one chamber for both heating (steam) and cooling (condensation). Many years later, James Watt and Joseph Black discovered that this meant the engine had to overcome latent heat for each and every stroke. This meant that it was very inefficient and required huge quantities of fuel.

redesigning a new steam engine. After many working models and some legal problems with patents, the two unveiled a new model in 1712 (see Figure 24). After this year, Newcomen continued working on his steam engines.

The **Newcomen steam engine** was so technologically advanced that it required parts not yet invented. The size and complexity of each part presented previously unknown manufacturing challenges to Newcomen. From the start, he had to adapt many parts that had never been intended for use in his steam engine. Newcomen's engine required a boiler of gigantic proportions. It also required a water-tight brass cylinder. Brass cylinders were certainly available, but no seals had ever been produced. Newcomen crudely outfitted his cylinder with leather sealing rings of his own production. He also constructed a huge wooden post to support a pivoting, horizontal "rocking beam," which connected to his brass piston at one end and to a large mechanical water pump at the other.

When each of these parts worked correctly (a feat in itself), they made Newcomen's steam engine only the second self-acting machine in history (clocks were the first). Newcomen's steam engine was entirely driven by atmospheric pressure. To begin with, inside a brass cylinder, steam at greater than atmospheric pressure was forced to the underside of the engine's single piston. Here, a jet of cold water was injected to condense the steam. In turn, this created a vacuum, which caused atmospheric pressure at the opposite end of the open cylinder to force the piston back down, thereby producing a downward power stroke. This single stroke was generally used to power a water pump or similar device. For the next power stroke, the same brass cylinder, which had just been cooled for the power stroke, had to be heated to begin the process (of heating and cooling) all over again.

This was a laborious, time-consuming and fuel-hungry process. Nonetheless, Newcomen's engine was the first useful engine in places such as mines and deep wells, where transporting water was a problem. Additionally, for the first time, steam power began to enhance older forms of power, such as that of water. In textile mills especially, Newcomen's engine was used to pump water through a mill time and time again. This made the mill more efficient and more cost effective, even though it still depended on the natural power of water. Where every other inventor had failed, Newcomen had harnessed the power of steam for the first time. Though Newcomen engines needed frequent repair, they were put into use in many mines, wells, and mills across England. At its prime, a Newcomen engine was capable of pumping over 3.5 million gallons (13.2 million liters) of water a day.

Newcomen's steam engine was one of the most useful inventions in history. Before this machine was crafted, usable forms of energy were little better than those available in antiquity. Even highly evolved industries totally depended on the work of animals, water, or wind. Despite their usefulness, Newcomen steam engines were constantly and consistently haunted by difficult, proprietary, expensive repairs. Few good mechanics had the skill to fix a broken engine. Prominent inventors began to acknowledge a disconnect between the usefulness of Newcomen's engine and the limits of its outdated parts. Among these inventors was Scottish scientist and inventor James Watt (1736–1819). In 1763, the University of Glasgow purchased a broken down Newcomen engine and charged Watt, a maker of instruments, with repairing it.

As Watt toiled away at the frustrating task, he began to envision design improvements to the machine's old parts. After several months, Watt began drawing up plans to completely redesign Newcomen's problematic steam engine. During the past half-century or so, industrial machine parts had made some major advances. Watt began adapting many of these precision parts for use in the Newcomen steam engine. One major improvement was a self-sealing brass cylinder made with rubberized taffeta.

In addition to being an instrument maker, Watt had studied mathematics, and he knew many scientists. In fact, one of England's greatest scientists was at the University of Glasgow during this time. In 1756, Joseph Black (1728–1799) had successfully isolated the first independent gas, **fixed air** (or **carbon dioxide**). In 1761, Black had also conducted a series of inquires on a problem he had discovered. In his laboratory tests, Black had found an "extra" amount of heat that was necessary to make water boil or freeze. Further investigation showed Black that the amount of this "extra" heat was tremendous when compared with that required to bring water to the boiling point. He named this phenomenon **latent heat**, or the heat required to make a substance change states. (For more on this topic, please see the entry on Latent and Specific Heat.)

During this same period, Watt was independently discovering that Newcomen's engine was extremely inefficient. To create the necessary atmospheric pressure for a power stroke, its single cylinder had to be heated and cooled for every cycle of condensation and vacuum. By using the machine, Watt realized that this required tremendous amounts of excess fuel. In fact, through his own laboratory trials, Watt found that heat from just 1 pound of steam would bring 6 pounds of water to its boiling point. In other words, he found that steam contains a lot of latent heat. But the cycle of heating and cooling was the very process that made Newcomen's machine work. Watt realized that he could run the engine on a fraction of the fuel—if only he could somehow stop having to overcome latent heat with each and every stroke.

Watt and Black had many meetings during this time. In fact, they became close friends and partners, often discussing scientific questions. They realized that each had independently discovered the process that Black named *latent heat*. Where Watt identified latent heat as a problem in his technological innovations, Black had found it through empirical tests. Once Black explained the scientific process of latent heat, Watt understood better than ever why Newcomen's machine needed so much fuel: He was practically throwing it away to change water to steam and back again. Watt understood the challenge, but didn't know what he could do about it.

After he became famous throughout England, Watt liked to tell the story of how he overcame this problem of latent heat. One Spring evening in 1765, Watt said, he was walking through Glasgow Green, as was his custom. Suddenly, he felt an idea hit him—as powerful as a bolt of lightning: What if he divided the steam engine's labor between the existing hot cylinder for producing steam, and a separate, cold chamber for condensing water? Successfully doing this could save a lot of time and fuel (or energy). Almost instantly, Watt said, he saw a clear visual arrangement of his steam engine in his mind.

Building the new steam engine turned out to be a much different story, however. It took Watt over 10 years of diligent and frustrating experiments to develop a steam engine entirely based on his new innovation. During these years, Watt perfected a boring machine that drilled brass cylinders with near uniformity—an incalculable advantage over the ones he had used before. This improved cylinder freed Watt from total dependence on atmospheric pressure. He redesigned his dependable engine to force the piston both down and up, providing much better efficiency and much more uniform power.

During this time, Watt also entered into a seminal partnership with Matthew Boulton (1728–1809), a wealthy engineer who ran a metals factory renowned for its precise machine parts. In 1775, the pair first began selling a double-action cylinder. They also kept refining its design based in large part on Watt's legendary aptitude for applied science and Boulton's impeccably engineered parts. Finally, in 1784, Watt and Boulton patented a steam jet condenser as part of an impressive parallel-motion, double-acting steam engine. With one ingenious innovation—the separate condenser—Watt had completely sidestepped the problem of inefficiently heating and cooling the same chamber (see Figure 25). For the first time, the cylinder stayed hot and the condenser stayed cold. Latent heat was no longer a problem—it was not even a significant part of the equation.

Many industrialists were floored at how far the steam engine had come since Newcomen's original model. Yet, every one of Watt's innovations more or less sprung from the process of improving Newcomen's original invention (which was itself an improvement over older models). It was all there to be built upon—either through invention or perfection.

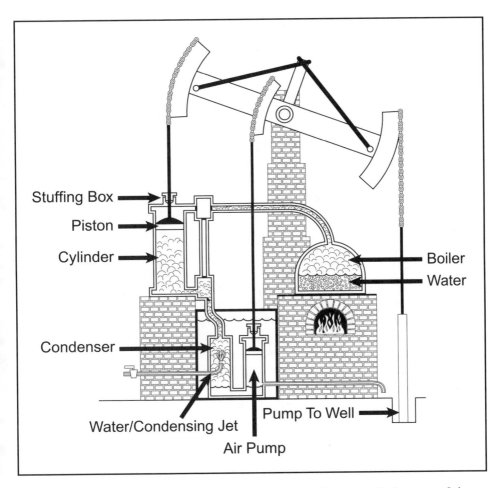

Figure 25. Watt steam engine with separate condenser. During the final quarter of the eighteenth century, James Watt introduced a revolutionary new steam engine. Though Watt manufactured many different models, his steam engines generally shared the important innovation of a separate condenser. Through Watt's work with thermal efficiency and Joseph Black's discovery of latent heat, Watt realized that a tremendous amount of heat was lost in heating and cooling the same cylinder. When he employed a separate condenser, the steam engine became much more efficient. Watt could eventually run his engine on about a quarter of the fuel required by previous models.

Even though James Watt did not properly "invent" the steam engine, he did make the necessary improvements for its universal use during the industrial revolution. After the invention of the separate condensing chamber, Watt's engine really began to take off. Watt continued to make important improvements, such as the rotary flywheel. The flywheel operated on the principle of inertia to convert the up-and-down stroke of a piston to a circular motion that ran more smoothly and much more constantly. Along with the separate condenser, the rotary flywheel made the widespread use of steam power a reality.

With the combination of these innovations, the popularity of steam power began to increase. Steam power spread across England at an unprecedented rate. Soon, pump bellows and huge hammers powered industry. Factories were no longer dependent on animal, water, or wind power. The location of factories was now almost entirely dictated by economics, rather than location of land-energy resources. During the next decade (and despite the English government's efforts to the contrary), steam power spread to France and across Europe and even to the newly formed United States.

During the next few years, Watt also rose to a celebrity, even legendary, status in England. Historically, Watt represents a link between empirical science and technical application, especially in context of his various partnerships with Black (scientifically) and Boulton (technologically). Watt continued to make major improvements to his steam engine, including a governor to control its speed, meters to count its strokes (and therefore measure its efficiency compared with fuel consumption), and a scale to measure its power in horses (which is still in popular use). Virtually all scientists rank the steam engine as one of the single largest contributions to the industrial revolution and also as one of the most important inventions in history.

In 1824, French scientist Sadi Carnot (1796–1832) published a major study about the efficiency of steam engines. For this work, scientists often consider him the founder of **thermodynamics**, the study of various forms of energy (especially the transfer of heat). In this book *Réflections sur la puissance motrice du feu,* he described the familiar **Carnot cycle**, which demonstrated that the efficiency of a reversible engine depends on the highest and lowest temperatures it reaches in a single cycle. As an equation, the Carnot cycle could be expressed as $E = (T1 - T2)/T1$, where E is efficiency and T is temperature.

In purely scientific terms, Carnot demonstrated that the difference between the two temperatures forms a direct proportion of a heat engine's thermal efficiency. It also forms part of the second law of thermodynamics, stating that a limit less than 100 percent exists for the efficiency of all heat engines. In entirely practical terms, Carnot demonstrated the relationship between an engine's work and the difference of its high and low temperatures. This formed an important discovery for improving efficiency of steam engines through temperature extremes.

Also, many people in the late eighteenth century and early nineteenth century had become seduced by the power of steam. A popular myth stated that the power of steam was "limitless." Many people even believed that steam would one day carry them (condenser and all) to the moon. The Carnot cycle popularly demonstrated that all forms of heat engines were far from limitless and that the "perfectly efficient heat engine" did not exist.

Beginning with the advent of the steamboat and later with the steam locomotive, steam power revolutionized nineteenth-century transportation as it had eighteenth-century factories. (For more on this topic, please see the entry on the Steamboat.) Before the application of steam in these inventions, transportation had been virtually unchanged since antiquity. Traveling by land or river was nearly as slow and often as dangerous as it always had been. After the invention of steam engines small enough for transportation, steam power created the first glimpse of truly mobile societies. In fact, steam power was not challenged until the early twentieth century, when internal combustion engines began to replace steam-powered ones.

SEE ALSO Heat Is a Form of Energy; Latent and Specific Heat; Steamboat; Water Is a Compound.

Selected Bibliography

Briggs, Asa. *The Power of Steam: An Illustrated History of the World's Steam Age.* Chicago: University of Chicago Press, 1982.

Constable, George (ed.). *Time Frame AD 1700–1800, Winds of Revolution.* Alexandria, VA: Time-Life Books, 1990.

Dickinson, H. W. *A Short History of the Steam Engine.* London: Cambridge University Press, 1938.

Dickinson, H. W., and H. P. Vowles. *James Watt and the Industrial Revolution.* New York: Longmans Green, 1943.

Hart, Ivor B. *James Watt and the History of Steam Power.* New York: Collier Books, 1961.

Hills, Richard L. *Power from Steam: A History of the Stationary Steam Engine.* New York: Cambridge University Press, 1989.

Rolt, L. T. C., and J. S. Allen. *The Steam Engine of Thomas Newcomen.* New York: Science History Publications, 1977.

Rutland, Jonathan. *The Age of Steam.* New York: Random House, 1987.

Siegel, Beatrice. *The Steam Engine.* New York: Walker and Company, 1986.

Stellar Aberration and Speed of Light
James Bradley, 1728

In the mid-1720s, English astronomer James Bradley (1698–1762) mounted a 212-foot-tall telescope to the chimney of his house. If this gigantic feat at first seems less than practical, the same cannot be said of the astronomer. Bradley was a true eighteenth-century man of fantastic intelligence and enlightenment skill. His formal education was largely religious, and he became a vicar in the Church of England in 1717. However, through his uncle, James Pound (a highly skilled amateur astronomer and friend of the great Edmond Halley [1656–1742]), Bradley discovered that his true fascination was with astronomy. By 1721, he had returned to Oxford University as a professor of astronomy. In 1742, Bradley was named England's third Astronomer Royal, an office he held until his death 20 years later.

Many astronomers in Bradley's day were concerned with trying to measure the phenomenon known as stellar **parallax**, the observed change in a star's position due to the motion of the Earth around the sun. As the Earth orbits the sun, astronomers noticed that stars appear at varying distances. The theory was that, as the Earth orbits the sun, an observational displacement is produced. Viewing a nearby star from one side of the sun and then from the other should produce a parallax displacement equal to the distance of the Earth's orbit. In other words, when astronomers made measurements, stars did not appear in the same position. Scientists thought this was the effect of parallax.

Beginning in the early 1720s, Bradley had become intrigued with the problem. In fact, he was hoping to chart stellar parallax through the telescope mounted on his chimney. Proving the existence of parallax would solve the problem of displacement. It would also offer the first indisputable, empirical evidence of the Copernican hypothesis that the Earth orbits the sun. Though they had accepted this theory for many years, scientists had not yet devised a successful form of empirical proof.

Bradley had little luck in measuring stellar parallax. For several years, he studied the star known as *Gamma Draconis* with fellow astronomer Samuel Molyneux (1689–1728). Throughout their study, Bradley became increasingly puzzled with one finding. To his surprise, Bradley found that the star's greatest displacement occurred not between December and June, as the parallax scenario had led him to expect. Curiously, the largest displacement occurred between the months of September and March.

Several centuries earlier, Polish astronomer Nicolaus Copernicus (1473–1543) had suggested that parallax was practically undetectable because of the relatively small distance of the Earth's orbit. He thought that the Earth's orbital distance was far too small to have an impact on the enormous distance that light travels between a star and Earth. In effect, Bradley's measurements suggested the same thing, that the stars were too far for parallax to have any measurable effect. But Bradley still had to account for the displacement he consistently recorded over the course of several years. If not parallax, then what?

Bradley began looking for an alternate explanation. After many trials and jettisoned theories, he arrived at one in 1728. This theory was so beautifully simple that it had somehow proved elusive. It dawned on Bradley that perhaps the displacement arose due to the motion of the Earth in contrast to the straight line of observed starlight. In other words, Bradley realized he was observing a "stationary" object (a star) from a moving one (the Earth). (For more information on this topic, please see the entry on Proper Motion of the Stars and Sun.)

Bradley conducted several months of tests on his theory. Each one lent validity to his theory of a moving Earth affecting the comparatively "fixed" light of a star. This motion results from the diurnal rotation of the Earth, the Earth's orbit around the sun and the entire solar system's motion through space. Therefore, because Bradley's telescope was rotating (along with everything else on Earth), it needed to be tipped slightly in order to catch the starlight. In effect, the Earth's own motion accounted for the displacement so many astronomers had observed and mistaken for stellar parallax. This is analogous to the way in which an umbrella must be tipped when a person walks briskly through a rainstorm in which the drops are falling vertically (see Figure 26).

Bradley had explained the phenomenon behind displacement. He named his discovery **stellar aberration**. Bradley was extremely reluctant to publish his findings without intense scrutiny (often lasting many years). However, he used these data about stellar aberration to prepare an accurate chart of the positions of over sixty thousand stars. The correct nature of aberration was an enormous discovery. Scientists could now accurately calculate the positions of the stars. Bradley's own chart adjusted for displacement, as Bradley himself had explained it. Nearly three centuries later, this chart is still useful to scientists.

Bradley's discovery also formed the basis of several other revolutionary conclusions. He realized that the degree to which he needed to angle his telescope allowed him to determine the ratio of the speed of light to the speed at which the Earth moves. The method used by Olaus Romer (1644–1710) in 1675 was only an approximation. No one had yet discovered a more accurate method.

Because the speed of the Earth in orbit was well known, aberration formed the basis of a more accurate way of calculating the speed of light. Through this new method, Bradley was able to calculate that light moves 10,000 times faster than the Earth. Because the speed of the earth in orbit was known to be 18.5 miles per second (mps), Bradley put the speed of light at 185,000 mps. For many years, Bradley's figure was accepted as accurate. Indeed, it was only about 1,000 mps shy of the currently accepted figure of approximately 186,000 mps.

The existence of aberration meant that the Earth must necessarily rotate in the opposite direction of the apparent daily movement of the heavenly

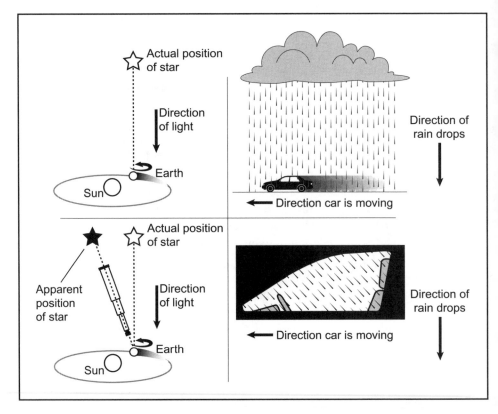

Figure 26. Stellar aberration of light. Stellar aberration is the observational displacement of a heavenly body from its actual position. This effect is caused by the combined effect of the Earth's motion and the time for a ray of light to travel the length of a telescope. James Bradley showed that an astronomer must tilt his or her telescope to properly "catch" a ray of light. Aberration is analogous to a person briskly walking in a rain shower in which the drops fall straight down. A person has to tip his or her umbrella to "catch" the drops of rain. In the illustration, aberration is also analogous to a car moving in a similar shower.

bodies. Because the stars appear to move from east to west, the earth therefore rotates from west to east. This simple deduction was the first empirical evidence of the Earth's direction of rotation. Before Bradley's discovery, scientists were divided on its true direction of rotation.

The discovery of aberration considerably advanced a long, fruitful series of light inquires that eventually culminated in the twentieth century's theory of relativity. First proposed by Albert Einstein (1879–1955) in 1905, this theory revolutionized basic scientific thought with new conceptions of space, mass, time, motion, and gravitation. It also laid the theoretical groundwork for the atomic age.

Bradley never found the parallax he set out to discover. In the end, however, aberration provided scientists with the first empirical evidence of the Copernican hypothesis. Just as effectively as parallax would have done, aberration demonstrated that the Earth is necessarily in orbit around the sun.

SEE ALSO Comets Follow Predictable Orbits, Cycles, and Returns; Newtonian Black Holes; Proper Motion of the Stars and Sun; Uranus.

Selected Bibliography

Crew, Henry. *The Rise of Modern Physics*. Baltimore: Williams & Wilkins, 1933.

Englebert, Phillis. *Astronomy & Space: From the Big Bang to the Big Crunch*. New York: UXL, 1997.

Engelbert, Phillis and Diane L. Dupuis. *The Handy Space Answer Book*. Detroit: Visible Ink Press, 1998.

McGraw-Hill Encyclopedia of Science and Technology. Volume 1. Washington, DC: McGraw-Hill, 1997.

Millar, David, Ian Millar, John Millar, and Margaret Millar. *The Cambridge Dictionary of Scientists*. New York: Cambridge University Press, 1996.

Panek, Richard. *Seeing and Believing: How the Telescope Opened Our Eyes and Minds to the Heavens*. New York: Viking Penguin, 1998.

Porter, Roy (ed.). *The Biographical Dictionary of Scientists*. New York: Oxford University Press, 1994.

Trefil, James S. *Space Time Infinity*. Washington, DC: Smithsonian Books, 1985.

Substances Fall into One of Two Distinct Chemical Categories: Elements or Compounds
Antoine Lavoisier, culminating in *Traité Élémentaire de Chimie*, 1789

Ancient Greek philosophers proposed that the four elements of air, earth, fire, and water make up all matter in the universe in differing proportions. They also thought that tiny constituent units called atoms contained the essence of an object or **element**. Atoms of water were slippery, but atoms of iron were jagged and had hooks. A little later, Aristotle (384–322 B.C.) proposed the theory of **transmutation**, which held that each element could be changed into one of the others by adding or removing heat and/or moisture. Europeans spent much of the 1100s to the 1600s following a modified version of transmutation in which they tried to turn lead into gold through the practice of **alchemy**.

Alchemists never developed an adequate understanding of elemental chemistry. They did manage to identify a few substances that occurred naturally in pure or nearly pure form. Among these were gold, copper, carbon, and sulfur. Later alchemists learned how to extract pure metals such as iron, lead, silver and tin from **compounds**. (Please see Figure 22 in the entry Rational System of Chemical Nomenclature, which shows a chart of alchemical symbols.) But the theory of four basic elements was more or less still widely accepted.

In retrospect, one can trace the theoretical destruction of the four-element theory to the year 1661 and the publication of *The Sceptical Chymist* by Robert Boyle (1627–1691). In this book, Boyle forcefully argued that there is no valid basis for limiting the number of elements to four—or indeed to any predetermined number. Instead, scientists should determine, through rigorous empirical investigation, how many elements exist. This was a startling proposal. In effect, Boyle was saying that theories of chemical elements should be adapted to fit scientific findings, not the other way around. Boyle's proposal was so divergent and new that it would take a chemical revolution in the late eighteenth century to fully realize its significance.

Boyle never outlined a comprehensive scientific method of classifying elements. He also never published a table of specific chemical elements. Over a century later, the task of creating both the method and the table was picked up by Antoine Lavoisier (1743–1794), the prolific young chemist at the very epicenter of the chemical revolution in France.

In early inquiries into elemental chemistry, Lavoisier began investigating the old alchemical claim that the so-called element water could be transmuted to earth by adding heat. In past experiments, Lavoisier had found that when he repeatedly distilled rain water, it left behind solid residue. However, Lavoisier thought that the residue was more likely a by-product of the experiment than the transmutation of elements.

In 1768, Lavoisier designed a decisive experiment. First, he weighed an empty glass container, filled it with distilled water, and weighed it again. By subtracting, he had quickly obtained the weight of the water. Lavoisier then sealed the container and proceeded to boil the water for over 100 days. After just several weeks, Lavoisier indeed recorded the formation of solid particles inside the container. At the end of his experiment, however, Lavoisier again weighed the closed container and found no change in weight. He emptied the container and carefully saved the water and the solid particles. Then, Lavoisier weighed the container yet again. He found a slight drop in its weight. The water showed no change in weight. However, when Lavoisier meticulously weighed the solid particles, he recorded their aggregate weight as identical to the weight loss of the glass container. This experiment was so simple that Lavoisier expressed public surprise that no one had thought of it before. Regardless, Lavoisier was the first scientist to pay close attention to the concept of mass in chemical reactions—and it had proven decisive here.

The solid particles were hardly the result of some mysterious transmutation. They were merely partially dissolved solid bits from the glass. This experiment led Lavoisier to doubt the entire idea of transmutation. He also began to seriously doubt that water was an element. Water experiments would prove a constant source of knowledge for Lavoisier—one to which he would return many times. For several years after this experiment, though, his focus shifted to one of the other so-called elements of antiquity, air.

By 1774, Lavoisier had adopted the opinion that ordinary air is no more an element than water. In his **oxygen** experiments, Lavoisier had shown that a specific gas in the air combined with the metal during calcinations (or oxygenation). For many years, Lavoisier had come to believe "that the air of the atmosphere is not a single thing" (quoted in Meldrum 1930, 39). He believed that ordinary air was made up of at least several distinct gases. By 1777, Lavoisier was able to isolate and name one of these gases in the air. This was the gas responsible for both **combustion** and calcinations (or oxygenation). In fact, this gas became the basis of Lavoisier's entire theory of chemistry— the one that shattered the old theories of the alchemists. This substance was (of course) oxygen. (For more on this point, please see the entry on Oxygen.) Air was not an element. Rather, it was *composed of* several independent elemental gases, including oxygen.

Oxygen was the element responsible for many of ordinary air's qualities. Equally important was another gas, which had been discovered many years earlier by Joseph Black (1728–1799). In 1756, Black isolated **fixed air** (or **carbon dioxide**), the first independent gas ever discovered. Lavoisier identified this gas in the "leftover" air of combustion and animal **respiration**. (For more information, please see the entry on Carbon Dioxide and the one on Animal Respiration.) Lavoisier was certain—and he was quickly convincing many other scientists—that ordinary air was a mixture of many different gases. In other words, Lavoisier was certain that air was a compound.

If ordinary air was definitely not an element, then what of the three other renegade elements from antiquity—the ones that had led scientists astray for centuries? In 1785, Lavoisier designed a second big experiment on water— in the form of a big, public demonstration. Recently, the composition of water had been definitely established by Henry Cavendish (1731–1810). (For more on this topic, please see the entry on Water Is a Compound.) Armed with this knowledge and in front of more than thirty scientists, Lavoisier dripped water through an incandescent iron gun barrel. This effectively decomposed it into oxygen (which Lavoisier discarded) and **hydrogen** (which he collected in a **pneumatic trough**). Lavoisier proceeded to produce oxygen by heating calcined mercury and collecting the gas in a gasometer. Then, Lavoisier carefully forced both chemicals into a large flask. There, combustion from an electric spark united them.

On this day, Lavoisier created five and a half ounces of water. More importantly, he had demonstrated, once and for all, that "water is not an element, that it is on the contrary composed of two very distinct principles, the base of vital air [oxygen] and that of hydrogen gas; and that these two principles enter into an approximate relationship of 85 to 15, respectively (quoted in Poirier 1996, 151). In other words, water was a compound, not an element. For the first time in history, scientists understood the composition of water. The old four-element transmutation theory was effectively shattered. Lavoisier's experiment also had a practical note. He had publicly demonstrated a method for producing hydrogen gas. His method formed the basis for collecting hydrogen for **gas balloon** flights like those pioneered by fellow French chemist Jacques Alexandre Cesar Charles (1746–1823). (For more on this topic, please see the entry on Balloon Flight.)

In 1789, Lavoisier cracked the whole issue wide open with his momentous book, *Traité Élémentaire de Chimie*. This book ushered in a new age of chemistry and completed its transformation by providing detailed tables of chemical elements. It shattered previous theories like transmutation and **phlogiston** that had either misjudged or completely ignored elements. *Traité* defined an element as a "substance not decomposed—the simple and indivisible molecules that compose bodies" (quoted in Riedman 1967, 140).

Traité also reproduced Lavoisier's tables of the "substances not decomposed" that he had first published in 1787 in *Méthode de Nomenclature Chimique*, another important book. *Traité* greatly expanded the included substances. Where *Nomenclature* had proposed a system for chemical nomenclature, *Traité* saw it to practical fruition when it actually named a wide variety of elements. (For more on this topic, please see the entry on Rational System of Chemical Nomenclature.) *Traité*'s tables listed the new elemental names of chemistry next to the old, vague names of alchemy. In this way, *Traité* formalized the chemical revolution through visual arrangement on the page. Chemistry had slipped through the looking glass now. With elemental categories, tables and names, there was no turning back.

Tellingly, *Traité* contained only thirty-three substances in its list of elements. Nineteen organic radicals were eliminated since the writing of *Nomenclature*, as were the three alkalis (potash, soda and ammonia). This evidenced a refining process on the part of Lavoisier, who readily adapted his theories to his findings (not the other way around). In fact, Lavoisier publicly expressed skepticism about whether several elements on his list were correctly categorized. He thought a few of his listed elements might actually turn out to be chemical compounds. His scientific honesty (instead of dogma) was proved correct early in the next century. Sir Humphry Davy (1778–1829) proved beyond a doubt that potassium—and not potash as Lavoisier had originally thought—

belonged on the list of elements. Accordingly (and as one might expect from the shrewd Lavoisier), *Traité* also listed a table of common compounds. These were common substances formed by chemical combination of two or more different substances.

When Lavoisier published *Traité*, chemistry was only just emerging as a distinct, modern science. The chemical revolution still had a long way to go. *Traité* was so far ahead of its own discipline that it included a number of mistaken elements. It listed the hypothetical substance **caloric** as a kind of elemental heat. With the publication of *Traité*, Lavoisier had not only demolished transmutation, but also shattered alchemy's sophisticated final gasp, phlogiston. In this theory, combustion releases the hypothetical substance phlogiston into the air. In many ways, Lavoisier's "heat fluid" caloric became a substitute for phlogiston. After Lavoisier was guillotined during the Reign of Terror, his widow and collaborator married the American-born (but Tory-loyal) chemist Count Benjamin Thomson Rumford (1753–1814). In 1798, Rumford, with the help of his student Humphry Davy, demonstrated that heat was a form of energy, not the so-called element caloric. (For more information, please see the entry on this topic.)

Today, much of Lavoisier's work is familiar because it founded the modern science of chemistry. In this context, caloric is a useful reminder that Lavoisier did not have the benefit of even the simplest chemistry laws or books from which to work. Rather, Lavoisier all but created the laws and wrote chemistry's first textbook. Where phlogiston was a sophisticated final permutation of alchemy, chemistry was a new discipline. Lavoisier's theory of oxygen chemistry has been accepted with virtual universality since the end of the eighteenth century.

Lavoisier's special designation of elements was absolutely foundational to modern chemistry. The very idea of elements made modern chemistry possible, fostered its creation, and nurtured its scientific growth. In the nineteenth century, Russian chemist Dmitri Mendeleev (1834–1907) recognized that the properties of chemical elements recur in regular patterns. This theory formed the basis of his periodic table of elements—a very fruitful arrangement of chemical elements. The periodic table has been used to predict the existence of elements, based on similarities between known elements. Today (and for the past 60 years or so), most newly discovered elements have been made by scientists working with particle accelerators or nuclear reactors. In a nuclear reaction, the structure of a nucleus is changed when it gains or loses one or more neutrons or protons, thereby changing it into the nucleus of a different isotope of element.

There seems to be transmutation of elements after all. However, this is only in special cases born far beyond the gold-laden dreams of alchemists, as well as the rational organization of late eighteenth century chemical scientists.

SEE ALSO Heat Is a Form of Energy; Oxygen; Phlogiston; Rational System of Chemical Nomenclature; Water Is a Compound.

Selected Bibliography

Greenberg, Arthur. *A Chemical History Tour: Picturing Chemistry from Alchemy to Modern Molecular Science*. New York: John Wiley & Sons, 2000.

Guerlac, Henry. *Antoine-Laurent Lavoisier: Chemist and Revolutionary*. New York: Charles Scribner's Sons, 1975.

Holmyard, Eric John. *Makers of Chemistry*. Oxford: Clarendon Press, 1931.

Lockemann, Georg. *The Story of Chemistry*. New York: Philosophical Library, 1959.

Meldrum, Andrew Norman. *The Eighteenth Century Revolution in Science—The First Phase*. New York: Longmans, Green, 1930.

Poirier, Jean-Pierre. *Lavoisier: Chemist, Biologist, Economist*. Translated by Rebecca Balinski. Philadelphia: University of Pennsylvania Press, 1996.

Riedman, Sarah R. *Antoine Lavoisier, Scientist and Citizen*. New York: Abelard-Schuman, 1967.

Sootin, Harry. *12 Pioneers of Science*. New York: Vanguard Press, 1960.

T

Thermometry Instruments and Scales
Guillaume Amontons, 1709
Gabriel Fahrenheit, 1709 and 1714
Rene Antoine Ferchault de Réaumur, 1730
Anders Celsius, 1742

A **thermometer** is an instrument that measures temperature through the use of materials that change in ways reflecting temperature changes. Some thermometers use a liquid, such as alcohol or mercury, which expands as its temperature rises. As the temperature rises and falls, so does the length of the liquid trapped in the column. Most liquid thermometers have numerical markings along their sides. Converting the number to a temperature is as simple as calibrating the thermometer to a standardized scale. The most widely used scales are **Fahrenheit, Celsius**, and **Kelvin** scales.

Until the early seventeenth century, very little was known about heat and temperature. The chief source of knowledge was the medical literature of Galen (130–200) from antiquity. Galen based his clinical knowledge of thermometry on the teachings of Aristotle (384–322 B.C.), who thought that heat and cold existed in different quantities in various parts of nature (including individual people). By the mid-sixteenth century—more than a millennium later—medical writers such as Hasler of Berne still depended on Galen's textual advice in the prescribing of remedies for illness based on patients' temperatures. Doctors understood so little about heat that they devised a complicated method of dispensing drugs based on the general climate in which a patient lived.

In early Renaissance Europe, there were no instruments that functioned as thermometers. Several ancient civilizations of the Middle East probably understood that air expanded on heating, but failed to measure the expansion in any meaningful way. The first instrument for measuring temperature, called the **thermoscope**, was probably invented during the early part of the sixteenth century. One of the few known creators was an Italian physiologist named Santorio of Padua (1561–1636, sometimes known as Sanctorius), whom many believe was the first to furnish his instrument with a numerical scale.

During this time, Galileo (1564–1642) also devised an early thermoscope consisting of a thin tube of glass with a closed bulb at one end and an opening at the other. This was one of the first working air thermometers. Galileo heated the bulb of his instrument and then inverted the open tube into a pan of water. When the bulb began cooling, water was drawn into the narrow tube, toward the bulb end. Galileo could crudely measure the temperature of the outside air by gauging how far the water traveled up the tube. Later, Galileo realized he could enclose the water in the large glass cylinder of his thermometer. In the water, he floated hollow glass balls keyed to varying water densities. Once the device reached thermodynamic equilibrium with the air around it, the balls would rise or fall to varying degrees, thereby indicating the air temperature. Today, working Galileo thermometers are widely sold in stores and catalogs specializing in science educational products.

From this point on, these more accurate liquid-in-glass thermometers began slowly replacing air thermometers. However, Guillaume Amontons (1663–1705) further developed the air thermometer in the early part of the eighteenth century. Amontons realized that the air thermometer demonstrates a fundamental process of gases. Working with his thermometer, Amontons demonstrated that air volume decreased when air cooled and increased when it was heated. Amontons found that this happened constantly and consistently upon heating and cooling of air—not just in certain cases. Only later in the century, after development of much more accurate alcohol thermometers as well as a quantitative scale of temperature, was Amontons's discovery reformulated as **Charles's law** of gases. In this law, V = volume, and T = temperature. If pressure remains constant, a gas expands by the same fraction of its original volume, as its temperature rises. Therefore, Charles's law is written V/T = constant.

Early liquid-in-glass thermometers were invented in the seventeenth century by the astronomer Olas Rømer of Copenhagen (1644–1710) and the grand duke of Tuscany, Ferdinand II. These thermometers actually used a mixture of water and alcohol (such as wine) in attempt to expand the thermometers' boiling and freezing points beyond those of water. This was especially important to prevent the common problem of freezing. Rømer also created one of the first quantitative scales of temperature based on two universal **fiducial points**. Rømer labeled the boiling point of water at 60°C and the melting point of ice at 7.5°C. This latter number was not as purely arbitrary as it might seem. Rømer intentionally reserved one-eighth of his scale to stand below the freezing point, so he could base 0 degrees at what scientists erroneously thought was the coldest possible temperature, an endothermic slurry of melting ice and salt. However, Rømer did not publicize his thermometer or publish his scale.

In 1708 or 1709, Rømer was visited in Denmark by the German instrument maker Daniel Fahrenheit (1686–1736). Fahrenheit watched Rømer calibrate his thermometers to his scale. Fahrenheit originally attempted to reproduce the scale on his own thermometers. However, he based his scale on what he thought was the primary fiducial point of the temperature of human blood (found by placing his improved, enclosed alcohol thermometer's bulb in the mouth or armpit of a healthy adult male). Fahrenheit divided each degree into four parts (no one knows entirely why), so that the upper point registered approximately 90°F and the lower point (for the melting point of ice) registered 30°F. (Zero degrees was still the lowest point, for the salt and ice mixture.) Later, he refined these points to 96°F and 32°F to eliminate what he called "inconvenient and awkward fractions" (see Figure 27).

Much later, after Fahrenheit's death, scientists began setting Fahrenheit thermometers with the universal boiling point of water, which registered 212°F. (They also refined body temperature to 98.6°F.) Also, in 1714, Fahrenheit revolutionized the design of thermometers when he began using quicksilver (mercury) in place of alcohol. The use of this liquid metal eliminated the effects of atmospheric pressure and greatly expanded the scale of freezing and boiling, because mercury melts at approximately −101.96°F (−38.87°C) and boils at 673.84°F (356.58°C).

In 1731, French physicist and naturalist Rene Antoine Ferchault de Réaumur (1683–1757) constructed an alcohol thermometer with a new temperature scale. This scale was special because it was based entirely on objective, universal fiducial points. For these, Réaumur chose the freezing point of water, which he set at 0°, and the boiling point of water, which he set at 80°. One degree Réaumur is therefore one-eightieth the difference of the melting and boiling points of water. Scientists quickly found they could navigate Réaumur's 80 positive degrees with much greater efficiency than they previously could with Fahrenheit's 212°. Use of the Réaumur scale became widespread (especially in France), though far from universal.

The first scientist to perform and widely publish experiments aimed at constructing an international temperature scale was Swedish astronomer Anders Celsius (1701–1744). Celsius conducted a long, detailed series of meticulous experiments aimed at establishing that the freezing point is independent of latitude and even atmospheric pressure. Using a Réaumur thermometer, Celsius established the same freezing point in Uppsala and Torne (at different latitudes) that Réaumur had found in Paris.

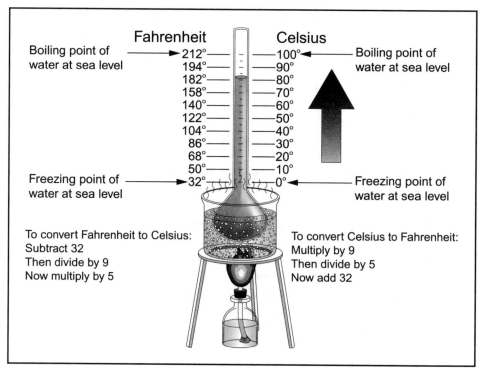

Figure 27. Thermometry instruments and scales. Early thermometers and thermoscopes existed before the eighteenth century. Eighteenth century scientists developed much more accurate instruments, such as mercury and alcohol thermometers. They also developed the first scales employing universal fiducial points, such as boiling and freezing points of water. The most important scales were Celsius, Réaumur, and Fahrenheit. Later, Lord Kelvin developed the Kelvin scale, an absolute temperature scale.

In 1742, Celsius published a paper in the Annals of the Royal Swedish Academy of Science entitled "Observations of two persistent degrees on a thermometer." These two degrees, 0 and 100, became the basis of his new temperature scale. However, Celsius designated 100° as the freezing point of water and 0° as the boiling point. For modern people habituated to things "heating up" and "cooling down," this thermometer seems reversed and awkward.

As Celsius's experiments taught, though, temperature scales are as arbitrary as the temperatures themselves are universal. Just as inverting a map has no effect on its accuracy (north is no more naturally "up" than it is "down"), Celsius's "reverse" thermometer seems to have served its historical purpose, as evidenced by its eventual adoption by scientists in much of Europe. One hundred degrees of temperature (a nice, round, simple number) apparently lent itself well to many divergent and varied measurements. More important than the scale itself, however, was Celsius's demonstration of universal fiducial points of temperature.

In 1743, French scientist Jean Pierre Christin (1683–1755) inverted the Celsius scale, placing the freezing point of water at 0° and the boiling point at 100°. This is the Celsius scale used today. In 1948, the ninth General Conference of Weights and Measures officially named it the Celsius (rather than centigrade) scale as part of the **SI** standard metric scale for temperature. As part of the **metric system**, this scale is used in nearly all countries, except the United States, which still measures temperature in Fahrenheit. Also, the Réaumur scale is still used in a few parts of France (often in conjunction with degrees Celsius).

Today, however, neither the Celsius nor the Fahrenheit is the international standard for scientific temperature measurement. The standard scale for this is the Kelvin scale, an absolute temperature scale. (Many scientists also measure in Celsius.) In 1862, Lord Kelvin (1824–1907) proposed a scale based on the hypothetical temperature of absolute zero. Though physicists theorize that it is impossible to attain a temperature of precisely absolute zero, this temperature equals −273.15° on the Celsius scale, −459.67° on the Fahrenheit scale, and 0° on the Kelvin scale. The coldest temperature ever recorded, less than one-billionth of a degree Kelvin, was attained by cooling and demagnetizing the nuclei of silver foil.

SEE ALSO Charles's Law; Heat Is a Form of Energy; Latent and Specific Heat; Metric System.

Selected Bibliography

Krebs, Robert E. *Scientific Laws, Principles, and Theories: A Reference Guide*. Westport, CT: Greenwood Press, 2001.

Middleton, W. E. Knowles. *A History of the Thermometer and Its Use in Meteorology*. Baltimore: Johns Hopkins Press, 1966.

Parkinson, Claire L. *Breakthroughs: A Chronology of Great Achievements in Science and Mathematics, 1200–1930*. Boston: G. K. Hall, 1985.

Quinn, T. J. *Temperature*. New York: Academic Press, 1983.

Three "Primary" Colors Are Sufficient to Produce All Others
Jakob Christof Le Blon, 1710
Moses Harris, 1766

Long before scientific experiments confirmed some theories and disproved others, many ancient philosophers developed theories about the nature and makeup of color. In Greece during the fifth century B.C., Empedocles (ca. 495–435 B.C.) theorized that minute particles traveled between a person's eyes and the object of his or her gaze. He said that each particle was a particular color, which the eyes recognized. During the next century, Plato (ca. 427–347 B.C.) said that color wasn't due to particles, but "rays" that shot from a person's eyes to an object.

A little later, Aristotle (384–322 B.C.) drew the first connection between color and light. This was a significant achievement that would become especially crucial thousands of years later. Aristotle thought that something transparent—akin to a kind of screen—between a person's eyes and an object caused perceptions of color. In a treatise, *De coloribus* (*On colors*), Aristotle outlined a theory of seven colors (white, yellow, red, purple, green, blue, black) and arranged them on a light-to-dark color scale.

In this first color scale, Aristotle said that the most pleasing mixes of five of his colors matched gray-scale tones (tones constructed solely of black and white) in precise, whole number ratios of $3:2$ or $4:3$ in terms of their brightness. He outlined three methods of ancient color mixing: adjacency (of constituent patches), layers (of translucent coats) and mixtures (of the pigments of substances). One of Aristotle's concepts—mystical color associations—influenced European color theory for well over a thousand years. In this assumption, he aligned four of his colors with the supposedly "true" colors of the four ancient elements. Red stood for fire, blue for air, green for water, and gray (or black and white) for earth.

In China during this time, a similar matching of elements and colors also took place. The philosopher Dzou Yen (ca. 300 B.C.) aligned each of five colors with an **element** and a direction. At the center was earth, aligned with the color yellow and the metal gold. To the east was wood, which was azure and aligned with lead. To the south was the red copper of fire. To the west was the white silver of metal. Finally, to the north was the black iron of water. These concepts were combined with the principle of opposites, yin and yang, which intersected many color theories.

The most important physician of ancient Rome, Galen (ca. 129–210), thought that rays from the eyes caused the air to carry a tiny, colored image of an object to the eyes. Then, he thought the image itself was carried by a vague "spirit" from the eyes (where it was received) to the brain (where it was analyzed).

During the Italian Renaissance, Leonardo da Vinci (1452–1519) began a long tradition of color theories relating specifically to painting. Da Vinci thought there were six colors that corresponded to the four elements. His was a theory of complementary colors in which red "enhanced" green and blue "enhanced" yellow. Da Vinci lived during a time in which European science was only just beginning to take hold. As such, he made many astute observations that he was not able to explain, such as the ability to see "tones" but not colors in a poorly lit environment.

After Da Vinci, investigators such as Giovanni Lomazzo (1538–1600), as well as the Flemish painter–scholar Peter Paul Rubens (1577–1640), made similar color perceptions. However, they never achieved an understanding of the basis for different colors. The color scale was also still completely arbitrary. Even painters with the most masterful command of artistic techniques were tormented by the need for scientific understanding of colors.

In 1704, Sir Isaac Newton (1642–1727) published *Opticks*, which contained his **corpuscular theory** of light. Prior to this book, most scientists accepted the wave model of light. Instead, Newton said that light is made of tiny, discrete particles moving in straight lines at a finite velocity. In arriving at this conclusion, Newton had discovered the relationship between white and colored light. Since antiquity, scientists had believed that white light was "pure," untainted and homogenous. They thought that glass prisms somehow "tainted" white light, so that colored light was white light that had been modified. In *Opticks*, Newton published his **experimentum cruces** (crucial experiment), in which he separated white light into colored light with a prism and then "reconstituted" it through a second prism. Newton had shown that colored, not white, light is homogenous and primary. This meant that white light is heterogeneous and secondary. This exercise gave both theoretical validity and visual demonstration to the concept of mixing colors to form new end-products.

Newton also demonstrated an early version of a color wheel. This tool showed the relations among and between colors. It consisted of a range of colors in the form of a paper circle beginning with red, running through the other spectral colors, and circling back to red. On the wheel, each color was directly across from its opposite color. Newton thought that each color plus its opposite ended up with the color white. (Of course, in a true light spectrum model, colors are not so neatly defined, as in "wheel" form.) Newton also thought that certain of the colors were "primary." These were colors that could be mixed in different combinations to produce nearly any other color. However, Newton could not say how many primaries were necessary in a color combination or even what colors they should be.

In 1710, French printer and inventor Jakob Christof Le Blon (1667–1741) began a series of printing experiments that took Newton's corpuscular theory as their genesis. Le Blon set out to prove that three primary colors were sufficient to produce any color of the newly understood light spectrum. After many trials, Le Blon became convinced that cyan, magenta and yellow—or even blue, red and yellow—were the necessary primary colors for printing all other colors. He went on to publish the first theory of the fundamental nature of these three colors.

Le Blon created the first visual outcome of his theory when he began producing three-color mezzotints, the first mass-produced product of color printing. Mezzotints are created from a method of copper or steel engraving that was often used to reproduce paintings. Le Blon's process was a form of color separating in which he prepared a separate engraved plate for each of a painting's colors—blue, then yellow, then red. These plates were so graduated in texture that they accurately reproduced the required proportions of each color. Le Blon inked each plate with its appropriate color and pressed each one onto the paper, one impression on top of the other. After time and trials, Le Blon learned that his three primary colors didn't really produce a true black, but a dark brown. For these reasons, he customarily prepared a black plate separation, as well.

Many people recognize Le Blon's process as essentially the basis for modern color printing, using the familiar "C, M, Y, K" (cyan, magenta, yellow, black) method. By the middle of the eighteenth century, printers using Le Blon's process of color separation routinely produced a wide variety of sophisticated colors. In this process, a printer engraved a design on the surface of blocks— one for each of the three colors (plus black). The density of ink in the final piece was regulated by the depth of each block's varying incision. After inking each block, a printer pressed each block onto the paper, while taking extraordinary care as regards register and alignment.

In Le Blon's day, a printer had to mentally analyze the colors in a piece or painting to be copied, then prepare a color separation to match each color. For most of the twentieth century, a camera was used to photograph a copy four times to produce what printers call a *separation negative* of each color. For each shot, a different colored lens filter was used to block out all colors except the desired one. Today, color separations are usually produced by electronic scanners. These automated machines use laser beams to scan the copy for color separation. After separation, the process is not so vastly different than Le Blon's original one. On a rotary press, for example, a piece of paper passes between four sets of plates. Each plate prints a different color, until the paper emerges from the press as a fully printed piece.

Le Blon had invented modern color printing and conceived of the first three-color palate. However, he did not expand his conceptual framework into a complete, fully organized color system. In fact, this did not happen until a century after Newton's first trials with the experimentum cruces. In 1766, entomologist and engraver Moses Harris (1731–1785) published his book, *The Natural System of Colours*. Written in the tradition of English naturalism,

Harris's book examines the relationship between colors. It also examines how three primary colors can mechanically produce so many others. This book also contained a printed color wheel. For printers and other nonscientists, this was a first opportunity to study a visual representation of an abstract color theory.

Natural System also detailed Harris's primary scheme in specific detail for all to duplicate. Color printing techniques were a far cry from the closely guarded secrets of early movable type. For red, Harris used cinnabar, which he made from sulfur and mercury. For yellow, he used King's yellow, an orpiment of arsenic trisulfide. Finally, for blue, Harris used ultramarine, a pigment made by powdering the semi-precious stone lapis lazuli.

Harris organized two fundamental groupings of colors. The first group, "prismatic or primitive colors," consisted solely of the three pigments. When he mixed or compounded two or more of the "prismatic" colors, Harris created colors belonging to what he called the "mediate" group. This group contained the intermediate colors orange, green and purple. Harris arranged all six colors on his printed color wheel so that each of the three "prismatics" had the greatest, most equal distance between them. This reflected Harris's eighteenth-century belief in naturalism. He said that these colors—red, yellow and blue—were organically and fundamentally opposite in quality to one another. Therefore, Harris felt they should be similarly represented in an rational system of harmonious visual order.

SEE ALSO Corpuscular Model of Light and Color.

Selected Bibliography

Brock, William H. *The Norton History of Chemistry*. New York: W. W. Norton, 1992.

Farber, Eduard. *The Evolution of Chemistry: A History of Its Idea, Methods, and Materials*. New York: Ronald Press, 1952.

Krebs, Robert E. *Scientific Laws, Principles, and Theories: A Reference Guide*. Westport, CT: Greenwood Press, 2001.

U

Uranus Is a Planet
William Herschel, 1781

Seventh from the sun, **Uranus** is the farthest planet visible without aid of a telescope. Only large Neptune and tiny Pluto are further, as Uranus is about 2,875,000,000 kilometers (km) away. Along with Jupiter, Saturn and Neptune, Uranus is one of the four large planets in the solar system. Its diameter is over four times that of Earth. Unlike Earth, Uranus is gaseous, rather than rocky, with a surface of green clouds of methane crystals. It is therefore known as one of the "gas" planets. Beneath the methane clouds are clouds of liquid water and **ammonia** ice. Scientists believe that an ocean of ammonia dissolved in liquid water may lie beneath the clouds. Uranus's center may be a rocky mass similar in size to a small planet such as Earth. The mass (or actual quantity of matter) of Uranus is over fourteen times that of Earth and the force of gravity is about 90 percent of Earth's surface. An object weighing 100 kilograms (kg) would therefore weigh about 90 kg on Uranus. This large planet's atmospheric pressure is roughly 19 pounds per square inch, or about 1.3 times the pressure on the Earth's surface. Uranus's atmosphere is composed of around 83 percent **hydrogen**, 15 percent helium, 2 percent methane, and trace amounts of other gases. The temperature of Uranus's atmosphere is around $-214°$ **Celsius** (C, or $-353°$ **Fahrenheit** [F]). However, its ocean may be as warm as 2,300°C (4,200°F). Its core may be 7,000°C (12,600°F).

Until the eighteenth century, astronomers knew of only six of the nine planets: Mercury, **Venus**, Earth, Mars, Jupiter and Saturn. In fact, eighteenth-century astronomers were not searching the heavens for new planets. Very few dreamed that undiscovered planets even existed. The monarchical champion of enlightenment ideals, Frederick II of Prussia (also called

Frederick the Great, 1712–1786), famously remarked that everything of real importance had already been scientifically uncovered. Though today that comment seems quite shortsighted, it evidences just how many mysteries eighteenth-century science had already solved. It also betrays the mindset of many eighteenth-century astronomers. After all, the law of gravitation had helped scientists explain many mysteries of the universe—most especially planetary motion.

Another formulation, known as **Bode's law**, after the astronomer Johann E. Bode (1747–1826), allowed scientists to approximate distances of the planets from the sun. To use this 1772 law, a person begins with a series of numbers in which each one is two times the preceding one: 3, 6, 12, 24, 48, 96, 192, 384, 768, and so on. Then, a person adds 4 to every number and divides each sum by 10. Though there are some problems (such as a "missing" planet between Mars and Jupiter, where only large asteroids exist), the law approximates the relative distances of the planets from the sun in astronomical units (AU).

During the previous years, many astronomers had actually seen Uranus but (echoing Frederick) had never realized it was a new planet. Many believed it was a **comet** or a star. Beginning in 1690, Uranus was observed and documented at least twenty times by reliable astronomers. However, each observer failed to note Uranus's difference from ordinary stars in appearance or movement. One of these observers, James Bradley (1693–1762), was England's third Astronomer Royal. Bradley once puzzled over an object for four consecutive nights. Many scientists believe this object may have been Uranus. Nonetheless, even Bradley failed to realize he was looking at a planet. He believed the object was a puzzling star.

On March 13, 1781, the astronomer William Herschel (1738–1822) was examining a group of stars when he noticed one that he found peculiar for its size and possible movement. On most nights, this prolific amateur stargazer began watching the heavens at the appearance of the first stars and did not stop until daybreak washed the last one away. Herschel was not originally an astronomer by training, but a musician. Even today, many amateur astronomers follow Herschel's lead and continue to make important contributions to the field.

Herschel liked to observe the sky with the naked eye. Usually, he also employed telescopes of his own making—Herschel even ground his own lenses with meticulous accuracy. One of his telescopes was the largest ever built until 1845. Night after night, Herschel employed his own special observational technique, which he called "sweeping the sky." Herschel began by dividing the sky into strips bounded by arcs of longitude analogous to those on Earth. Patiently and assiduously, he surveyed each 2-degree strip for any objects that he had never before seen. Using this method, Herschel would eventually discover over 2,500 stellar objects and about 800 double stars. By 1779, however, Herschel had only finished his first comprehensive survey of the sky. He began a second, more detailed sweep.

On spotting Uranus on the March night in 1781, Herschel began to believe he had discovered a new comet. He could faintly make it out with his naked eye. Over the course of many nights, Herschel studied the object at magnifications of 227, 460 and 932. The object was exactly where Bode's law indicated a planet would be. However, Herschel became ever more convinced of his comet theory when he noted changes in the object's position among the stars. He also noted that its motion was nearly (but not completely) parabolic—a kind of extremely elongated ellipse. It was just as he would expect for a comet. (For more information, please see the entry on Comets Follow Predictable Orbits, Cycles and Returns.)

For several years, Herschel watched his comet and steadily charted its course. Other scientists, such as Simon de Laplace (1749–1827), also spent months trying to compute the object's orbit. These scientists, including Herschel, became puzzled at the supposed comet's irregular orbit. Their observations suggested that the comet was nearing its **perihelion**, its closest point to the sun. However, no comet had ever been observed with such a perihelion distance—greater than four times the earth's distance from the sun.

In the summer of 1781, Anders Johann Lexell (1740–1784), the Imperial Astronomer of St. Petersburg, who was visiting London, succeeded in calculating the orbit of the supposed comet. The mathematical methods he used were broadly acceptable, but the predicted orbit was not. It did not fit with any contemporary understanding of the orbit of known comets. After all, Edmond Halley (1656–1742) had shown that comets follow an altogether predictable elliptical orbit caused by gravitational pull from the sun. Herschel, Lexell, and Laplace began to believe that Herschel had become the first person in recorded history to discover a planet.

A new planetary orbit was calculated to be a circle nineteen times larger than the Earth's own orbit. This orbit was also far beyond that of Saturn, the previous outermost planet. The discovery of Uranus altered the entire understanding of the solar system. Instead of six, there were now seven planets. What's more, the known size of the planetary system roughly doubled. All at once, scientists realized that they had much to discover about the workings of the universe, and many began looking for other planets.

Herschel's discovery of Uranus was catastrophic in scope. Because no one had ever discovered a planet, scientists were not sure what to call it. There was simply no convention for such a discovery. Herschel named his discovery

Georgium Sidus, in honor of King George III of England. At the urging of Laplace, the planet was called *Herschelium* in France. Other countries adopted a variety of names. For almost 60 years, the planet had no single, agreed-on name. Eventually, Bode was allowed to name the planet *Uranus*, after the Greek god of the sky. Uranus was the father of Saturn and the grandfather of Jupiter—the namesakes of Uranus's two closest known planets in size and distance from the sun.

The discovery of a planet allowed scientists to accept Bode's law. It also ushered in a new age of planetary discovery. In 1787, Herschel discovered Uranus's two largest satellites, which he named *Titania* and *Oberon*. Three more satellites had been discovered by the middle of the twentieth century. The vast majority of today's scientific knowledge comes from the flight of the U.S. spacecraft *Voyager 2*, which flew very close to Uranus (80,000 km) in January 1986. This spacecraft sent back spectacular photographs of this outer planet. It also discovered ten additional satellites during its visit. Using powerful Earth-bound telescopes, scientists discovered two more satellites in 1997. This brought the current total to seventeen. One tiny satellite, *Cordelia*, is just 26 km across.

Uranus was the first of three planets discovered scientifically. As for Uranus, scientists predicted the existence of the solar system's eighth planet. Working independently, English astronomer John C. Adams (1819–1892) and French astronomer Urbain Leverrier (1811–1877) each predicted the planet's location. They found that the powerful gravitational pull of a large, undiscovered planet was affecting Uranus's orbit. In 1846, German astronomer Johann Galle (1812–1910) discovered the large, gaseous planet Neptune.

For a long time, scientists believed that another large planet was responsible for irregularities in the orbits of both Uranus and Neptune. After an exhaustive search, American astronomer Clyde Tombaugh (1906–1997) identified the tiny, rocky planet Pluto on a series of his photographic images in 1930. Much of today's knowledge of Pluto comes from images taken by the Hubble Space Telescope, beginning in 1996. Pluto may have once been a satellite of Neptune, as every 248 Earth-years, it enters Neptune's orbit for around 20 Earth-years. During this time, it comes closer to the sun than Neptune. Neptune and Pluto are the only planets that cannot be seen with the naked eye and Pluto is the only planet not yet visited by space probes. However, scientists agree that Pluto lacks the mass to have any considerable affect on the orbits of the two larger planets. Some scientists interpret this as evidence of another planet—a so-called Planet X—that may exist beyond Pluto. At this point, however, this prediction is based in speculation, rather than conclusive evidence.

SEE ALSO Comets Follow Predictable Orbits, Cycles, and Returns; Milky Way; Proper Motion of the Stars and Sun.

Selected Bibliography

Doig, Peter. *A Concise History of Astronomy*. London: Chapman & Hall, 1950.

Forbes, George. *History of Astronomy*. New York: G. P. Putnam's Sons, 1909.

Motz, Lloyd, and Jefferson Hane Weaver. *The Story of Astronomy*. New York: Plenum Press, 1995.

North, John. *The Fontana History of Astronomy and Cosmology*. London: HarperCollins, 1994.

Pannekoek, A. *A History of Astronomy*. New York: Interscience, 1961.

Turner, Herbert Hall. *Astronomical Discovery*. Berkeley: University of California Press, 1963.

V

Venus Has an Atmosphere
Mikhail Lomonosov, 1761
Johann Schröter, 1779

Venus is one of the six planets (including Earth) known since antiquity. The remaining three, **Uranus**, Neptune and Pluto, were discovered by scientists in the eighteenth, nineteenth and twentieth centuries, respectively. Through the ages, many cultures have popularized Venus as Earth's "twin," because of its similar mass, its relative closeness in the solar system and the fanciful belief that it might just be an Earthlike planet of water and life. The dazzling intensity of Venus against the early morning or evening sky has made it a favorite of astronomers for thousands of years. Because of its brilliance, it has long had an imagined association with goddesses and other traditional representations of beauty.

Around the year 3000 B.C., the Babylonians made detailed astronomical records of sightings of Venus. Ancient Chinese, Egyptian and Greek astronomers attempted to chart the planet's journey across the night sky. In what is now Central America, New World Indians devised sophisticated methods of observing Venus, noting seasonal and temporal changes in its appearance. The Roman astronomer Simplicius (ca. 490–560) puzzled over shadows that Venus appeared to cast. Though there is some debate, many scientists believe that the Syrian philosopher Al-Farabi (died 950) may have observed a transit of Venus across the Sun, from a point in Kazakhstan around the year 910. The first person to definitely witness a transit of Venus was English astronomer Jeremiah Horrocks in 1639. Solar transits of Venus would become vital events in the eighteenth-century scientific discovery and investigation of the planet's atmosphere. The Copernican sun-centered conception of the universe was vital to the understanding of solar transits.

In a solar transit, a planet passes between the Earth and the sun, providing a rare "silhouette" of the planet against the sun's disk. To eighteenth-century astronomers, solar transits offered the most accurate method then known for determining the distance from the Earth to the sun—the "astronomical unit." They also offered the possibility of determining the distance of other stellar objects through a practice known as **parallax**. Because Venus is closer to the sun than the Earth is, it appears during this time as a small, dark orb "crawling" across the disk of the sun. Transits of Venus are extremely rare, occurring in pairs that are 8 years apart and with more than 100 years between pairs of transits.

In 1610, Galileo (1564–1642) recorded the phases of Venus for the first time with a telescope. This was an important step, offering some evidence of a Copernican universe and greatly arousing scientific interest in Venus. In 1627, German astronomer Johannes Kepler (1571–1630) made a chart detailing when transits of Venus would occur. In the seventeenth century, French astronomer Giovanni Cassini (1625–1712) first observed dim marks on the planet. In the next century, the English astronomer Edmond Halley (1656–1742) observed similar markings and also suggested using a transit of Venus to measure the length of the astronomical unit. When the transits of 1761 and 1769 came (as widely predicted), astronomers throughout Europe turned their telescopes on the brilliant planet.

To measure parallax, astronomers knew that they would have to observe the 1761 transit from beginning to end in order to time its points of optical "contact" with the sun's disk. Britain, France, and other countries sent scientists to Asia, the Arctic, Newfoundland and the Atlantic island of St. Helena, where they could clearly study the transit. Their observations were not fruitful. The planet's exact times of "contact" with the solar sphere could not be distinguished because its outline was fuzzy through even the best lenses.

The Russian astronomer Mikhail Lomonosov (1711–1765) had also observed the transit. Lomonosov suggested that the infuriating ring around the planet was caused by intensified sunlight, created by something akin to a lens. This "something," Lomonosov correctly suggested, was an atmosphere. Lomonosov had become the first person to discover that Venus indeed had an atmosphere. This theory quickly fell into place with Lomonosov. An atmosphere would account for the planet's fuzzy outline and the difficulty with observation. An atmosphere with a permanent cloud layer beneath it could also explain Venus's dazzling appearance. This phenomenon is known as a planet's **albedo**, or reflecting power. Lomonosov correctly figured that Venus's atmosphere reflected far more sunlight than it absorbed. Finally, it occurred to Lomonosov, a cloudy atmosphere also explained the frustrating blankness of the planet. Venus was not featureless at all, however. Its surface was simply masked by clouds.

Lomonosov published his discovery in Russian. Unfortunately, Russian science was not yet well developed and few scientists outside of Russia read the language. During the next transit of Venus in 1769, colonial (and later American) astronomer David Rittenhouse (1732–1796) observed Venus's atmosphere through the first telescope built in the American colonies. Rittenhouse did not offer scientific proof of the atmosphere, but he did draw the same correct conclusion about its existence. When Rittenhouse published his finding in English, it was well received and earned him credit for the discovery. The greater scientific world did not learn of Lomonosov's original discovery until 1910—141 years later. Today, Lomonosov generally gets credit for this discovery.

The first hints toward scientific proof (or, at the very least, evidentiary confirmation) of Venus's atmosphere came only later in the century. In 1779, German astronomer Johann Schröter (1745–1816) began a series of Cytherian observations (that is, observations of Venus) that would last through 1788. This elusive and shifting planet was not an easy object of study for Schröter. He was unable to observe any markings on the planet through his telescope until February of 1788. On this day, Schröter noted a pronounced "filmy" streak on the otherwise uniformly bright surface of the planet. Subsequent close examinations revealed several other markings. However, Schröter found that each of the markings were wandering and imprecise. He drew the correct conclusion—that the markings stemmed from Venus's atmosphere.

During this same period of observation, Schröter closely studied Venus's movements. Almost immediately, he noted that Venus's reflected light tended to plunge noticeably in the direction of the planet's **terminator**, the optical line separating its light and dark parts. This impressive evidence of an atmosphere was assured in Schröter's mind when he studied Venus in its crescent phase. He clearly recorded that he could see the cusps of Venus extending far beyond the semicircle. In fact, they extended all the way around the planet's dark hemisphere, giving it a ringlike appearance. This was the most visible manifestation of Venus's atmosphere absorbing light. A planet devoid of atmosphere would certainly not exhibit these features.

Venus becomes a half-circle during a time known as **dichotomy**. Eighteenth-century astronomers were able to predict the dichotomy of many celestial bodies. However, the predicted time for this planet's dichotomy was not generally in agreement with the time of observed dichotomy. Here was a great disparity between the scientific (theoretical) and the observed dichotomy. Schröter made several studies of Venus's strangely shifting dichotomy and suggested that the effect was purely optical. He also thought, correctly, that the planet's atmosphere was responsible for this optical sleight of hand.

Schröter had provided more evidence (if not full-blown proof) of Venus's atmosphere.

For nearly 200 years, scientists could only wonder what lay beneath Venus's atmospheric blanket of clouds. For many years of the late nineteenth and early twentieth century, leading astronomers, such as Percival Lowell (1855–1916), believed they had photographed "canals" on Venus. Then, in the late 1920s, American astronomers William Wright (1871–1959) and Frank Ross (1874–1960) confirmed the clouded atmosphere of Venus by producing photographic images in the ultraviolet spectrum. Off and on for the next 30 years, many people (aided by science fiction writers) imagined Venus was a "swampy" planet—possibly even full of water, jungles, and life. However, beginning in the 1930s, scientists began gathering evidence suggesting Venus was much more than a little bit humid. In 1956, microwave measurements showed that the surface temperature was in the neighborhood of hundreds of degrees **Celsius**. Most scientists now credit the high temperatures to the greenhouse effect, caused by the thick cloud cover acting as solar **insulation**.

Then, following flybys by both Soviet and American space probes, the Soviet space probe *Venera 4* dropped an instrument capsule into Venus's dense atmosphere. Scientists found the data remarkable and surprising. As the probe pierced deeper into the atmosphere, the temperature and pressure continued to rise. Finally, around 24 kilometers (km) above the planet's surface, the instruments were crushed by the enormous pressure. Since *Venera 4*, many more probes have explored the planet. Beginning with *Venera 7* in 1970, a succession of probes have orbited the planet and successfully landed on the surface. In 1990, the American craft *Magellan* began orbiting Venus. It sends back radar images in unprecedented detail. It records objects as small as 100 meters (m) across.

As for Earth-based observations, the next solar transits of Venus will occur in 2004 and 2012. For the first time since 1874 and 1882, astronomers will watch the brightest object in the night sky pass in front of the sun.

SEE ALSO Comets Follow Predictable Orbits, Cycles, and Returns; Stellar Aberration; Uranus.

Selected Bibliography

Asimov, Isaac. *Asimov's Chronology of Science and Discovery*. New York: Harper & Row, 1989.

Asimov, Isaac. *Venus, Near Neighbor of the Sun*. New York: Lothrop, Lee & Shepard Books, 1981.

Hunt, Garry E., and Patrick Moore. *The Planet Venus*. London: Faber and Faber, 1982.

Moore, Patrick. *The Planets*. New York: W. W. Norton, 1962.

Motz, Lloyd, and Jefferson Hane Weaver. *The Story of Astronomy*. New York: Plenum Press, 1995.

North, John. *The Fontana History of Astronomy and Cosmology*. London: Fontana Press, 1994.

W

Water Is a Compound of Hydrogen and Oxygen
Henry Cavendish, 1784

Water is everywhere: in the huge oceans, tiny creeks and mighty rivers. It is in the Earth's atmosphere and in the cells of humans, whose bodies are about two-thirds water. It is locked in the thick polar ice caps and in the mammoth glaciers of Alaska. In fact, more than 70 percent of the Earth's surface is covered with water. It is curious to note, however, that there is exactly the same amount of water on the earth at this moment than there ever was in the past or ever will be in the future. The same water a person bathed in last night might have lapped the shore of Lake Michigan in 1887 or been drunk by Socrates thousands of years ago. This is because water is used and reused and never "used up." Indeed, many of water's properties are absolutely unique—so unique, in fact, that they violate many generic "rules" they might be expected to follow. Actually, for thousands of years, people believed that water was an **element**. Only in the eighteenth century did scientists realize that this time-honored classification was incorrect.

During the late eighteenth century, inventors were fundamentally redefining the world with water, through widespread industrial applications of the **steam engine**. At the same time, scientists were fundamentally redefining their understanding of this substance that sustains many of the basic processes of life on Earth.

Henry Cavendish (1731–1810), an eccentric English scientist, had discovered **hydrogen** in 1766. A staunch phlogistonist, Cavendish had collected hydrogen in a **pneumatic trough** by applying sulfuric or hydrochloric acid to iron, zinc or tin. Cavendish named his discovery **inflammable air**. Five years later, the famous pneumatic chemist Joseph Priestly (1733–1804) discovered that an electric spark applied to a mixture of hydrogen and ordinary air produced a "dew" inside his glass vessel. When Cavendish heard about the mysterious liquid formation, he began a series of key inquiries and experiments. Cavendish began these experiments in 1781 and published the results in a 1784 paper, "Experiments on Air."

Throughout the paper, Cavendish described his impeccable quantitative conclusions about the reactant gases. By repeatedly igniting hydrogen and ordinary air, Cavendish calculated that the volume of **oxygen** in the atmosphere was about 20.8 percent. Just as remarkably, his experiment showed the combining ratio of hydrogen to oxygen was approximately 2.01:1.

Equally important, Cavendish carefully described the moisture formed in these experiments. In his apparatus, he could burn 500,000 grain measures of inflammable air with around 2.5 times that quantity of common air. Through a great many trials, Cavendish found that he could reduce the volume of air by a maximum of one-fifth. When he conveyed the gases separately into a long glass chamber and burned them with a candle, Cavendish found that nearly 135 grains of liquid were formed in the cylinder.

Significantly, he also found that nearly all the inflammable air along with roughly one-fifth of the ordinary air lost its elasticity and was "condensed" into "dew." Utilizing a common (and dangerous) method of testing in his day, Cavendish reported that this resultant liquid had neither taste nor smell. In short, the "dew" behaved a lot like water. But why did the change take place in these proportions? Cavendish suspected that it was due to a part of the ordinary air—a part that, apparently, constituted roughly one-fifth of the whole.

Next, Cavendish took his inquiry a step further, replacing the ordinary air with highly combustible oxygen. As was his custom, Cavendish tried to eliminate all possible errors by preparing his gases from many alternate sources. As he expected, Cavendish found that water formed on the glass of his cylinder in every scenario. Additionally, he found that oxygen burned until it was completely gone, just as Priestly had found in his trials with the gas. After this experiment, Cavendish was finally ready for an announcement. He bluntly proclaimed, "[B]y this experiment it appears that this dew is plain water" (quoted in Riedman 1967, 120).

Cavendish only partly understood his discovery. He never fully grasped that water is the **compound** product of the gases hydrogen and oxygen. Cavendish was one of the last scientists to subscribe to the **phlogiston** theory. Nonetheless, he then held to it his entire life. This theory stated that, during **combustion**, the hypothetical substance phlogiston is released. **Dephlogisticated air** (or oxygen) supported combustion so strongly because it heartily grabbed up the phlogiston being released. In this context, Cavendish thought that air was "in reality nothing but dephlogisticated water, or water deprived of its phlogiston" (quoted in Jungnickel 1999, 364). Cavendish thus thought that, when he combined hydrogen and oxygen with the spark of combustion, each gave up its water, separately, in the form of condensation. (For more information, please see the entry on Phlogiston.)

Cavendish therefore never shared the modern interpretation of his experiment, hydrogen + oxygen → water. Likewise, he never fully grasped the more basic concept that water is a chemical compound of the two gases. Nonetheless, his stated conclusion is implicit that water is more or less "made" from these two gases. Therefore, by definition, it is a compound.

After this discovery, Cavendish kept experimenting with water. When he found the water he was producing was distinctly acidic, he correctly identified the presence of nitric acid. He reasoned that the electric spark had actually caused the gases to combine and that, from the resultant liquid, he could obtain nitrates and nitrites. Over a century later, Cavendish's discovery formed the basis for industrial production of nitrates from **nitrogen** in the air.

By studying the salts and gases dissolved in water, Cavendish went on to found the basic techniques of water analysis. These techniques are foundational to the public health of all communities. Cavendish tried lime as a water softener and also created a system to measure organic contamination by estimating the amount of **ammonia** present. These were major contributions from his work on water. But the discovery of water's compound nature formed a crowning achievement for Cavendish and even for English **pneumatic chemistry**. It was one of the most important discoveries in history.

As with most of pneumatic chemistry's eighteenth-century discoveries, however, it would fall to Antoine Lavoisier (1743–1794), the founder of modern chemistry, to correctly interpret the findings. Lavoisier most thoroughly accomplished this reinterpretation in 1785. In a public display in that year, Lavoisier actually "made" 5.5 ounces of water from hydrogen and oxygen. (For more on this topic, please see the entry on Oxygen.) On this day, he boldly claimed that "water is not an element, that it is on the contrary composed of two very distinct principles, the base of vital air [oxygen] and that of hydrogen gas" (quoted in Poirier 1996, 151).

Despite their fierce disagreements, Lavoisier's oxygen chemistry and the pneumatic chemistry of Priestley and Cavendish were mutually dependent. Where the latter discovered, the former interpreted. After Cavendish broke water into its constituent elements, Lavoisier put it back together, just 1 year later.

SEE ALSO Heat Is a Form of Energy; Hydrogen; Oxygen; Substances: Elements or Compounds.

Selected Bibliography

Berry, A. J. *Henry Cavendish: His Life and Scientific Work*. New York: Hutchinson, 1960.

Crowther, J. G. *Scientists of the Industrial Revolution: Joseph Black, James Watt, Joseph Priestly, Henry Cavendish*. London: Cresset Press, 1962.

Holmyard, Eric John. *Makers of Chemistry*. Oxford: Clarendon Press, 1931.

Jungnickel, Christa, and Russel McCormmach. *Cavendish: The Experimental Life*. Lewiston, PA: Bucknell University Press, 1999.

Poirier, Jean-Pierre. *Lavoisier: Chemist, Biologist, Economist*. Translated by Rebecca Balinski. Philadelphia: University of Pennsylvania Press, 1996.

Riedman, Sarah R. *Antoine Lavoisier, Scientist and Citizen*. New York: Abelard-Schuman, 1967.

Weaving Machines
John Kay, Flying Shuttle Loom, 1733
Edmund Cartwright, Steam-powered Loom, 1785

Weaving is the ancient practice of crossing two pieces of material over and under each other. Throughout history, people of different civilizations have woven many different materials, such as grass, leaves, strips of wood, animal fiber (such as wool) and plant fiber (such as cotton). Recovered wall paintings from 5000 B.C. depict ancient Egyptian weavers and suggest a mastery of their occupation. By 2500 B.C., weaving was widespread throughout Europe, South Asia and the Middle East. Historical evidence also suggests that Chinese weavers mastered the practice sometime between 2500 and 1300 B.C. on exquisite silk fabrics. In several Islamic countries, weaving reached high development beginning in the eighth century. By the 1500s, Persian weavers wove intricate patters and geometric shapes into their fabrics. Some even created miniature scenes with detail generally reserved for paintings.

New World Indians such as the Pueblo and Navajo learned weaving by the middle of the eighth century. The Navajos wove wool into sturdy and ornamental blankets and floor coverings. In the Northwest, Indians wove blankets of soft wood fibers. On the East Coast, the weaving techniques of several tribes were so advanced that European explorers mistook woven plant fibers for spun cotton cloth. Much further south, the Incas wove cotton and llama hair with such skill that modern power looms have yet to duplicate many recovered Incan products.

Since antiquity, weaving has been done on machines called looms. All weaves that are known today have existed for thousands of years. The basic action of a loom is to weave filling yarn over and under warp yarn to create a finished (woven) product. In Europe during the 1200s, a loom known as the two-bar or two-harness loom came into use. This machine allowed weavers to raise alternate warp yarns to create a *shed*—a space between the yarns. This sped the process of weaving filing yarn, which was carried by a *shuttle*. Within two hundred years, intricate tapestries for cathedrals and castles were in high demand. Many of these delicate woven tapestries still exist.

Before the eighteenth century, England generated great wealth through the export of fine English wool. However, by the early part of the eighteenth century, its New World colonies were on the brink of exploding cotton production. Cotton was already prized for its many advantages over wool.

Yet a major limitation in producing cloth and other finished products was that all weaving was done on a loom by hand—just as it had always been. With existing looms, the weaver had to pass the shuttle (with the filler yarn) across the warp yarn by hand. After all, this is what formed the crosswise yarn of the fabric. The weaver almost continually had to change the position of his or her hands. Worse yet, for broad cloth, two weavers per loom were necessary because the width of material was far beyond a single weaver's reach. This was a time-consuming, expensive process. Also, quality varied greatly with the technical proficiency of the individual weaver. The English textile industry desperately needed a better way.

In 1733, English inventor and reed maker John Kay (1704–1764) invented the flying shuttle loom. Kay put his shuttle on wheels and determined a way to make it run in a straight line by mounting it along the loom's edge. Because wheels greatly lessened the amount of friction on the shuttle, it was much easier for the weaver to send it flying across the loom with a swift push. Kay also mounted little wooden boxes at either end of the shuttle run.

The loom previously required a weaver's hand to change direction of the shuttle. Now, the weaver could simply jerk the string connected to the wooden box to send the shuttle flying back across the loom. With a little practice, experienced weavers could send the shuttle racing to and fro across the loom at much greater speed than either one or two weavers could previously manage with their hands.

Nearly instantly, Kay's shuttle doubled the productivity of any given loom with very little cost to its owner. Once fitted with Kay's flying shuttle apparatus, the loom required only one weaver, even for wide cloth. Moreover, one weaver could produce much more cloth than before. For clothiers who had to pay weavers, Kay's machine was an economic windfall. In less than a year, many clothiers adopted Kay's flying shuttle. However, few were willing to pay Kay the royalties he was legally due for his innovation. The flying shuttle was widely pirated and, in cases in which the offenders were prosecuted, they happily paid fines over royalties—the flying shuttle was that profitable.

Historians disagree on whether the flying shuttle loom was adopted more widely among English cotton or wool weavers. However, it is clear that many weavers feared the invention would ultimately make their skills archaic. In manufacturing regions, machines were frequently smashed in spontaneous riots by weavers. In 1753, Kay himself became the recipient of violence, as his house and loom were destroyed by a mob. Kay actually fled the country and spent the remainder of his years in exile in France. But his invention was to stay in England, where it would provide colossal (though mostly pirated) growth of the textile industry.

Later in the century, the power of steam engines perfected by James Watt (1736–1819) began to replace all previous forms of power, such as animal, water and wind. England was at the forefront of this power shift, which historians generally consider the single greatest factor of the Industrial Revolution.

During the middle of the century, another enormous shift had taken place. Originally conceived in response to the yarn shortness caused by Kay's machine, spinning machines such as the **spinning jenny** of James Hargreaves (1772?–1778), the **water frame** of Sir Richard Arkwright (1732–1792), and the **spinning mule** of Samuel Crompton (1753–1827) appeared on the scene. These machines ironically created inundating, even overwhelming amounts of yarn in England. In response to the demand of spinners for cotton, Eli Whitney (1765–1825) had created the **cotton gin** to mechanically clean New World cotton very quickly. Even with Kay's flying shuttle, weavers could not keep pace with the seemingly inexhaustible supply of cotton yarn or the insatiable demand for finished cotton products. Just as weavers were being squeezed from both sides, the power of the Industrial Revolution came to weaving. (For more information, please see the entry on Spinning Machines and also the one on the Cotton Gin.)

In 1785, inventor Edmund Cartwright (1743–1823) designed the first usable steam-powered loom (see Figure 28). Previously, Cartwright had never even seen a loom, and his first machine was clumsy and prone to breakdowns. Later machines addressed the main problem of converting the circular motion of the steam engine's rotary flywheel to the three different forms of reciprocating motion necessary for the loom. As the engine's piston provided an up-and-down power stroke, its rotary flywheel converted it into a circular motion. Cartwright's task to was to once again convert this motion—this time from circular to reciprocating.

The most difficult reciprocating motion was the critical one required to propel the flying shuttle to and fro across the loom. Cartwright first used springs aided by cams (or noncircular wheels). The springs sent the shuttle across the loom without problem, but the shuttle often rebounded out of its box and back into the web of yarn, where it created havoc. Cartwright tried to address this problem with several methods. He designed several elaborate methods for quick-stopping the loom to prevent damage to the yarn, should the shuttle rebound. He tried sidestepping the problem altogether by employ-

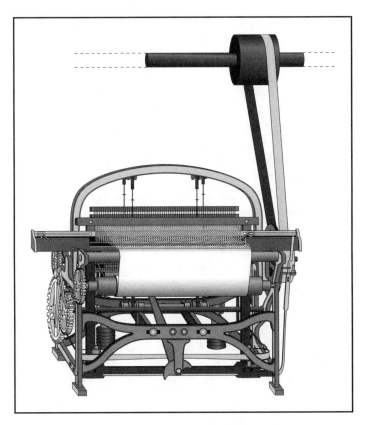

Figure 28. Weaving machine. Since antiquity, weaving has always been done on simple machines called looms. However, during the eighteenth century, demand for cotton products soared to levels that weavers could no longer meet. Whitney's cotton gin revolutionized the production of cleaning cotton and spinning machines revolutionized the spinning of yarn to meet this demand. The flying shuttle loom of John Kay and the steam-powered looms of Edmund Cartwright (similar to the one shown here) similarly revolutionized the ancient practice of weaving. They also modernized the booming English textile industry and defined the very means of modern factory-style production.

ing cranks for a sharper, swifter blow to the shuttle. Later, Cartwright employed a friction pulley with a hook to hold the peg tight against the loom's side in-between power strokes. Each of these improvements met with only limited success, and Cartwright's steam-powered loom never achieved total automation. Someone (often a weaver) always had to stand by, constantly watching it and making adjustments.

Cartwright met limited success with a system employing a spring that actuated a picker stick to drive the shuttle for the next shoot across the loom.

Several of these machines were used for several years at a factory in Gorton, where they are believed to have performed reasonably well. More importantly, Cartwright had shown that steam-powered looms were possible.

Beginning in 1796, John Austin (ca. 1820) installed improved steam-powered looms in Scotland. Austin's looms are generally credited with founding the modern Scottish textile industry, which is both vibrant and famous today. Both Cartwright's and Austin's steam-powered looms were important forerunners to mass produced, high-speed looms. These were employed with near universality in Europe during the early nineteenth century. Once perfected, the mechanical parts of steam-powered looms moved faster than the human eye could follow. Today, most power looms run on electricity and are completely automated. Some even employ sophisticated computer systems to develop and assign tables of "maximum weavability" for vastly different types of yarn, design patterns and fabric.

SEE ALSO Cotton Gin; Spinning Machines; Steam Engine.

Selected Bibliography

Burton, Anthony. *The Rise and Fall of King Cotton.* London: British Broadcasting Corporation, 1984.

Fitton, R. S. *The Arkwrights: Spinners of Fortune.* Manchester, UK: Manchester University Press, 1989.

Hills, Richard L. *Power in the Industrial Revolution.* Manchester, UK: Manchester University Press, 1970.

Hollen, Norma, and Jane Saddler, Anna Langford, and Sara Kadolph. *Textiles,* sixth edition. New York: Macmillan, 1988.

Mirsky, Jeannette, and Allan Nevins. *The World of Eli Whitney.* New York: Macmillan, 1952.

Rowland, K. T. *Eighteenth Century Inventions.* New York: David & Charles Books, 1974.

"Weighing" the World by Measuring Gravitational Constant
Henry Cavendish, 1797

Henry Cavendish (1731–1810) was indisputably one of the greatest scientists in history. He was also one of the least understood in his own time. The son of an English lord, Cavendish inherited a fortune, yet dressed in shabby clothes. He had a staff of servants, but communicated only with his housekeeper through written notes. He cared little for his finances, his health, his legacy or his reputation, but he is responsible for several of the most important scientific discoveries ever made.

Moreover, just as few people understood Cavendish, few scientists understood Cavendish's contributions. This was partly because of the complexity of his discoveries and partly because of Cavendish's stubborn indifference to publishing. In the end, it would take the better part of a century for the scientific community to fully appreciate his discoveries. This is true of the discoveries of gases with which Cavendish is most often associated. It is also true of less frequently cited but equally important experiments with meteorological and dynamical subjects.

The most famous of these latter experiments stemmed directly from Cavendish's education in Newtonian science. During the previous century, Sir Isaac Newton (1642–1727) proposed that all objects on Earth are attracted by gravitational force, which is proportional to the size of their masses and the distance between them. In other words, he hypothesized that the larger the mass of the objects, the greater the force between them. This became known as Newton's theory of gravitation, which included symbols for the gravitational constant, the masses of the given two bodies (in attraction), the distance between them, and their acceleration of movement toward one another. Thus, Newton's theory could be expressed as: $F = Gm_1m_2/d^2$, where F is the force of gravity, G is the gravitational constant, m_1 and m_2 are the respective masses of the bodies, and d^2 is the squared distance between them.

For Newton, the mass of two bodies was easy to measure, as was the squared distance between them. Additionally, since the Earth's gravity is much more powerful than that of any object, the Earth does much more attracting. Therefore, weight on Earth is an issue in the equation. (Because of this, one cannot simply use mass, which is constant regardless of gravity on Earth.) Thus, Newton's equation left two unknowns: the gravitational constant and the mass of the Earth. Finding one of the two variables would easily yield the other through simple calculation. However, neither of the two variables was ever found in Newton's day.

The gravitational constant was known to be the same for all objects. For this reason, the most logical method of solving a variable was to measure the gravitational attraction between two objects with established masses and at a measured distance from one another. Then, the gravitational constant could be calculated and, leaving only one variable, the mass of the Earth could be figured. In theory, this method was simple enough. Actually calculating the measurement was a different story. Objects of known mass were far too small for measurable gravitational attraction between them to exist.

Indeed, because the gravitational force of the Earth is so powerful, the mass of the objects to be measured would have to be enormous. In the 1770s, the Royal Society in England conducted experiments aimed at determining whether a mass the size of a mountain would cause detectable gravitational effect. An appointed committee chose the mountain Schiehallion in Scotland due to its observed uniform shape. The design and method of the observations fell to Cavendish.

Cavendish made some important contributions through a method in which he measured the vibration of a pendulum according to latitude and the attraction (which had been documented) of a plumb line toward the side of the mountain. However, his results were rough estimations due to a number of factors. The mountain's true composition was guesswork. Also, from the start, the project was limited because the mountain's mass could not be precisely determined. In other words, the problem was that an object whose mass was large enough to cause detectable gravitational effect contained too much mass to measure. At this point, it became clear to Cavendish that the experiment should be carried out with bodies of uniform, measurable shape as well as mass that could be determined exactly. Also, it became clear to him that such bodies would have to be manufactured, rather than found in nature.

Cavendish already owned an apparatus, the **torsion balance**, that he thought could greatly benefit the project. Cavendish made some drastic improvements to its design. The torsion balance consisted of a beam suspended from pulleys, which allowed its rotation, and a very fine wire suspended from the beam. At the ends of the wire, two small lead balls were suspended. Even a small force applied to these balls by two larger ones would produce a measurable twist. In fact, the entire apparatus worked on the premise that even the smallest movement could be measured under closely controlled conditions. From the start, Cavendish understood the challenge before him. He figured that the size of the force would be, by necessity, infinitesimal—not more than one-fifty-millionth the weight of the small lead balls.

In this context, Cavendish's extraordinary meticulousness was his biggest asset. He carefully eliminated outside variables such as temperature, wind and even light. He placed the apparatus in a small, enclosed, darkened room. From outside (in an adjacent room), he controlled the torsion balance by means of a system of pulleys. Often, Cavendish painstakingly waited a full two and a half hours for the machine's suspension wire to fully twist. Cavendish eventually published the results of seventeen experiments. Of these, three were made with a first wire and fourteen were made with a second, stiffer (and thereby more easily controlled) wire.

In 1798, Cavendish published one of his most important papers, "Experiments to determine the Density of the Earth." This paper details experiments in which he measured the thin twist of his apparatus's wire. From that, he calculated the gravitational force between the two small objects. With this information, it was a relatively simple matter to figure the gravitational constant. With this variable solved, Cavendish turned back to his equation to calculate the Earth's mass. Cavendish figured that the Earth's mass was approximately 6,600,000,000,000,000,000,000 tons. Based on the documented volume of the Earth, he also figured the average density of the Earth at roughly five and a half times that of water.

Cavendish thereby figured the mass of the Earth—not by weighing enormous objects like mountains, but by measuring the tiniest variations caused by the gravitational shift of small lead balls. If his method was absolutely unique, this idea of measuring the Earth through a small object was not. Two hundred years earlier, William Gilbert (1544–1603) had written in *De Magnete*, a classic work in early physics, that in a little stone the shape of the earth, he had "found the properties of the whole earth, in that little body" upon which he could experiment (quoted in Jungnickel 1999, 452). What was once a theoretical possibility had become a practical application of science in Cavendish's successful experiment.

Two hundred years after Newton, gravity is still at the very center of research into the physical world. Today's torsion balance also owes a great deal of its use and design to Cavendish. With a few significant improvements (such as a fine quartz strand or steel wire, as well as a tiny mirror whose reflection accurately measures movement), this instrument bears ready resemblance to its eighteenth-century predecessors. More than any other, this apparatus revolutionized precision measurements. Nearly all subsequent determinations of the gravitational constant after Cavendish's first one have been done with torsion balances. It has numerous other applications, from electrical calibration to seismological measurements.

In the nineteenth century, the density of the earth was measured at least six times using the method pioneered by Cavendish. It was also measured twice using the Royal Society's method of measuring the relatively enormous attraction of mountains. Finally, it was also measured by a method utilizing the seconds pendulum and the common balance. But the honor of originally determining the gravitational constant (and, from that, the Earth's mass) goes to the scientist who measured it first. Cavendish's inquiry was so important that, even today, scientists often refer to it simply as "The Cavendish Experiment." He never weighed the world, but Cavendish did determine the Earth's mass. And he did it so accurately that his figure turned out to be very close to today's estimate of 5.97×10^{24} kilograms.

SEE ALSO Newtonian Black Holes.

Selected Bibliography

Asimov, Isaac. *Asimov's Chronology of Science and Discovery*. New York: Harper & Row, 1989.

Berry, A. J. *Henry Cavendish: His Life and Scientific Work*. New York: Hutchinson of London, 1960.

Crowther, J. G. *Scientists of the Industrial Revolution: Joseph Black, James Watt, Joseph Priestly, Henry Cavendish*. London: Cresset Press, 1962.

Jungnickel, Christa and Russel McCormmach. *Cavendish: The Experimental Life*. Lewiston, PA: Bucknell University Press, 1999.

Krebs, Robert E. *Scientific Laws, Principles, and Theories: A Reference Guide*. Westport, CT: Greenwood Press, 2001.

Wet Transfer Copying Process (a.k.a. Copying Press)
James Watt, 1780

The first texts that are recognizable in function as books appeared on clay tablets several thousand years B.C. Since then, the physical form of books has gone through several dramatic metamorphoses. After clay tablets came wax boards, which were followed by papyrus scrolls and parchment codex. Paper was a relatively recent medium in Europe, as it appeared there only during the thirteenth century. (Paper was in China much earlier.) Durable, inexpensive linen or cotton paper swept across Europe, quickly becoming the medium of choice for most books. But despite these changes, the illuminated manuscripts of the Middle Ages shared one very basic constant with the first scrolls of antiquity—a total dependence on the scriptorium for production and dissemination of textual copies.

Between the fifth and sixteenth centuries in Europe, book production was the closely guarded province of the Church. Special classes of priests and monks called *scribes* practiced writing and bookmaking by hand. They mainly produced copies of the Bible and other religious books, though they also copied some important texts from ancient Greece and Rome.

Later, scribes often worked in royal courts, keeping tax records and recording laws. With the advent of mercantilism, scribes sold their services ad hoc for specific business transactions. Because common people were illiterate, scribes also wrote contracts, deeds, and wills—drawn up on the spot in a town square or public marketplace. Every book, every contract, and every business record had to be meticulously written by hand. Where more than one copy was desired, a scribe doubled the effort and produced what amounted to a second original.

Block printing had been practiced in China since the ninth century. This process involved carving or whittling each page of a book into a separate block of wood, smearing ink over the elevated surfaces, and transferring the ink by pressing onto paper. This form of printing arrived in Europe at the same time as paper. Now, printers could make several copies of a book with much greater rapidity than they could by hand. The present, recognizable form of the modern book resulted from the invention of *movable type*. Between the twelfth and fourteenth centuries, inventors in China and Korea perfected this technology. In this kind of printing, printers formed letters of the alphabet out of individual pieces of wood and later metal. Then they arranged the letters to form a page of print. With movable type, printers could produce high-quality, affordable books with great efficiency and regularity.

This technology appeared in Europe during the 1450s. German inventor Johannes Gutenberg (ca. 1395–1468) constructed his first printing press out of an old wine press. Along with several associates, Gutenberg printed a magnificent first book using movable type around the year 1454. This book came to be known as the *Mazarin Bible* or *Gutenberg Bible*.

By the time of the eighteenth century's scientific revolution, the printing of books had long been ubiquitous across Europe. An emergent, literate middle class that could afford time for reading meant a greater demand for a wider variety of books than ever before. Books had come a far way from the sole purview of clergy.

At the same time, the industrial revolution that swept across Europe created an ever-increasing need for business documents such as tax assessments and transaction records. Large-scale businesses and widespread enterprise required greatly expanded credit facilities and reliable institutions of banking. Early modern capitalism was heading into full swing. Investment in manufacturing facilities and products was really beginning to take off by the 1750s. In previous decades, investors routinely went to the continent, especially Amsterdam, for capitalization. Now, London was becoming a powerhouse financial center of the modern world, displacing all other cities. Banks were even opening up in smaller English cities and a few towns. Modern business organization was taking place, requiring accuracy, efficiency, and immediacy in all business practices. In the span of several decades, bookkeeping went from a piecemeal process of individual estimation to a routine, standardized, professional practice. As the term was on everybody's tongue, many businessmen even called bookkeeping a *science*.

At the very center of the industrial revolution (some even said its "cause") was the machinery powered by the newly harnessed energy of steam. And at the very center of this technology was the firm of James Watt & Co., founded by James Watt (1736–1819) and industrialist Matthew Boulton (1728–1809). Even today, Watt's name is nearly synonymous with the **steam engine**. During the final quarter of the century, orders for Watt's steam engine soared to unprecedented levels. It had engendered a social, scientific and economic transformation never before witnessed anywhere in the world.

During this time, it seemed that the business of virtually every company in England was—in one way or another—dependent on a Watt steam engine for its very prosperity. Indeed, around one hundred years ago, one of Watt's most famous biographers, Andrew Carnegie, wrote, "It would be difficult to name an invention more universally used in all offices where man labors in any field of activity." At first, a reader assumes Carnegie is talking about the steam engine. But then he continues, "In the list of modest inventions of greatest usefulness, the modern copying-press must take high rank, and this we owe entirely to Watt."

The "copying-press"? Well, it seems that Watt's steam engine had become so popular, he could barely keep up with the paperwork. During this period, captains of industry did not yet have specialized, organized staffs to handle every aspect of their companies. It was customary for a powerful man like Watt to personally handle much of the business—his most important customers expected nothing less. By the 1770s, it was more than apparent to Watt that he could not keep up with the staggering amount of paperwork. He began working on an invention to address his overwhelming need.

Early on, Watt realized that one of the most important aspects of the copying press was the ink. Its composition had to allow for transfer from one sheet of paper to another. The problem was, however, that the same ink had to become indelible when it dried. Ordinary ink would certainly not do. Earlier, while he had been perfecting the steam engine, Watt developed a lifetime friendship with one of Europe's preeminent chemists, Joseph Black (1728–1799). Black had discovered **carbon dioxide** and also **latent heat**—the latter of which explained how steam contained so much usable energy.

Watt realized the enormous business potential of his copying press. Black was one of only a handful of people he told about the project. He lent Watt his superior chemical knowledge in the form of advice about the ink's specific composition. By the end of 1779, Watt had perfected the composition of his top-secret ink and applied for a patent for the "letter-copying machine." Watt's ink was made of gum arabic—a sticky, water-soluble gum obtained from the acadia tree. This ink took a full 24 hours to dry, making press copies possible for a full day. Once it dried, the special ink was indelible, just like normal ink.

Watt also invented a special kind of tissue paper for the machine. This paper was strong enough not to tear even when wet, but thin enough to see through. When Watt wrote a letter or any business correspondence to be copied, he used his special ink on regular paper. If he needed to send multiple copies or keep a record for himself, Watt inserted a dampened sheet of tissue paper into his press. One design was actuated by a screw, and one was activated by a roller. In both cases, the original letter was pressed onto the dampened tissue paper. The pressure solubilized and transferred the dye component of the special ink in a process known as the *wet transfer copying* process. This yielded a mirror-image copy—a "negative" copy in *reverse* of the original. The document was therefore read from the back side. This is the reason that the sturdy tissue paper had to be so thin. And this is also another important property of the ink. It completely penetrated the paper so it was visible from both sides.

In addition to businessmen, eighteenth-century scientists rapidly adopted the copying press. The Scottish physician William Cullen (1710–1790) was such an authority that his students came to Edinburgh from all over Europe and even the New World. Many others wrote him letters—so many, in fact, that he could barely answer them all. During the late 1760s, Cullen began a "clinic by correspondence," in which he solicited specific medical questions from far-flung colleagues and answered them by mail. Later, when he had more than one letter with the same question, Cullen's assistant used the newly invented copying press to make multiple copies of his answer. Every one of his letters between April 1781 and December 1789 exist today in the form of letter-press copies that Cullen kept. They form much of the extant historical record of his career, as well as an important insight into the medical practices of the time.

Other important scientists benefited greatly from Watt's invention. As American ambassador to France, Benjamin Franklin (1706–1790) saw one in 1780 and brought it to the attention of his American colleagues. Both he and Thomas Jefferson (1743–1826) had been working on similar contraptions, and readily embraced Watt's invention for their personal use. In 1770, Jefferson lost his home, Shadwell, to fire. He was devastated to lose his books and other possessions. Most of all, he missed his many papers. Between 1770 and 1780, it is said that Jefferson meticulously copied every single piece of correspondence, official business, government document, philosophical tract—and even the daily minutiae. It seems that Jefferson wanted a backup copy for himself as well as a copy for posterity (after all, during this time he penned the Declaration of Independence). In the early 1780s, he bought a Watt copying press. For the next twenty years, Jefferson used the copying press for everything he wrote. He even designed a portable, spring-loaded version to use while he traveled. Like Cullen, much of the body of Jefferson's writing comes from copies of letters he kept. (Later, after 1800, Jefferson also used a *polygraph*, which consisted of five pens mounted together to move as one.)

The copying press became as ubiquitous in business offices as the steam engine was in factories. In this sense, it was analogous to the modern copying machine, an absolutely essential business tool. In 1780, Watt initially sold one thousand machines. By the century's end, he could no longer keep up with demand. During the next century, every office had at least one—and often several—copying presses. Letter-copying books became ubiquitous tools for business correspondence. These were bound volumes containing hundreds of leaves of high-quality paper for the purpose of copying outgoing correspondence and other business documents. Every office that could afford one also employed a copying clerk. This trustworthy person operated the copying press and archived each and every important document that the office sent out to clients, the government, or other businesses.

After a letter had been written in special ink, the clerk inserted a copy into the press, along with a sheet from a copying book. During the nineteenth century, improved copying presses could copy twenty or more one-page letters in one pressing of several minutes. Many historically important letters in museums and other archives are actually copies of letters made on a copying press.

During the late nineteenth and early twentieth centuries, portable, fold-out copying presses were widely used. In photographs of offices from this era, it is the most widely seen machine. As recently as 1905, the Watt biographer Andrew Carnegie referenced its universal use in offices. Eventually, the combination of typewriter and carbon paper began replacing the copying press in most offices. However, as recently as 1950—175 years after Watt's invention—advertisements for office copying presses routinely appeared in a variety of venues.

SEE ALSO Carbon Dioxide; Latent and Specific Heat; Steam Engine.

Selected Bibliography

Carnegie, Andrew. *James Watt*. New York: Page, 1905.

Constable, George (ed.). *TimeFrame AD 1700–1800, Winds of Revolution*. Alexandria, VA: Time-Life Books, 1990.

Rowland, K. T. *Eighteenth Century Inventions*. New York: David & Charles Books, 1974.

Rutland, Jonathan. *The Age of Steam*. New York: Random House, 1987.

Wheeler, Mark. "Declaration of Independence." *Inc Magazine*. September 15, 1996. Available online at: http://www2.inc.com/incmagazine/articles/2000.html.

APPENDIX: ENTRIES LISTED BY SCIENTIFIC FIELD

Astronomy

Comets Follow Predictable Orbits, Cycles, and Returns

Milky Way Is a Lens-Shaped Distribution of Stars

Navigational Quadrant (Octant and Sextant)

Newtonian Black Holes or "Dark Stars" Are Devoid of Light

Proper Motion of the Stars and Sun through Space

Stellar Aberration and Speed of Light

Uranus Is a Planet

Venus Has an Atmosphere

Biology

Animal Physiology and Experimental Pathology

Animal Respiration Obtains Oxygen and Eliminates Carbon Dioxide

Digestion Is Primarily a Chemical Process

Microscopic Organisms Reproduce, Rather Than Spontaneously Generate

Modern Entomology and the "History" of Insects

Plant Physiology Is Independent of Animal Physiology

Plants Use Sunlight to Absorb Gas, and They Reproduce through Insect Pollination

Scientific Morphology and Taxonomy of Animals and Plants

Chemistry

Atmospheric Composition Is Constant

Balloon Flight

Carbon Dioxide Is a Distinct Gas

Charles's Law of Gases

Chemical Cells Can Store Usable Electricity

Chlorine Is a Distinct Gas That Can Be Easily Produced in a Laboratory

Discovery of Nitrogen

Hydrogen Is an Element

Law of Conservation of Mass

Oxygen Is an Element That Supports Combustion

Phlogiston (or the First Reasonable, Rational Theory of Chemistry)

Priestley's Chemical Experiments, Discoveries, and Inventions

Proust's Law of Constant Proportions

Rational System of Chemical Nomenclature

Substances Fall into One of Two Distinct Chemical Categories: Elements or Compounds

Water Is a Compound of Hydrogen and Oxygen

General Inventions

Cotton Gin

Farmer's Almanac

Franklin's Practical Household Inventions

Metric System of Measurement

Modern Encyclopedia

Optical and Mechanical Telegraphs

Spinning Machines

Weaving Machines

Wet Transfer Copying Process (a.k.a. Copying Press)

Mathematics

Binary System

Calculus of Finite Differences and Differential Equations

Medicine

Citrus Prevents Scurvy

Immunization Prevents Smallpox

Modern Dentistry

Physics

Ballistic Pendulum and Physics of Spinning Projectiles

Corpuscular Model of Light and Color

Coulomb's Law

Electrical Conductors and Insulators

Electricity Can Produce Light

Electricity Is of Two Types: Positive and Negative

Franklin's Electrical Researches, Discoveries and Inventions

Heat Is a Form of Energy, Not the Element Caloric

Hydrodynamics

Latent and Specific Heat

Principle of Least Action

Steamboat

Steam Engine

Thermometry Instruments and Scales

Three "Primary" Colors Are Sufficient to Produce All Others

"Weighing" the World by Measuring Gravitational Constant

GLOSSARY OF TECHNICAL TERMS

Abiogenesis. *See* Spontaneous generation

Albedo. A planet or other body's ratio of light reflected to light received.

Alchemy. The ancestral practice of chemistry in China, North Africa, and the Middle East beginning in about 500 B.C. Versions of alchemy spread to Europe during the 1100s and 1200s. European alchemy was based in the ancient philosophy of Aristotle, who endorsed the idea of transmutation. Most alchemists sought two chemical results: the transmutation of base metals into gold and the creation of *elixir vitae*, a mythical substance believed to cure disease and lengthen life. In 1700, alchemy was first transformed into the phlogiston theory of chemistry by Georg Ernest Stahl. Beginning in the 1780s, alchemy was shattered by Antoine Lavoisier's theory of oxygen chemistry. *See also* Chemistry; Oxygen; Phlogiston; Transmutation.

Almanac. A book that lists a great deal of varying information, such as a calendar with important dates and events, as well as facts about government and history. Almanacs might also include important information about weather, movements of planets, and figures on population, agriculture, and industry. Though ancient civilizations published them, the quality and accuracy of information contained in almanacs was greatly increased with the application of eighteenth-century scientific methods. The almanac that had the greatest impact on this revolutionary improvement was the *Farmer's Almanac*, published by Benjamin Banneker between 1792 and 1796.

Amber. A hard, translucent, yellow-brown gum, formed from the resin of fossilized pine trees. When amber is rubbed with a cloth or similar object, it develops a negative charge of static electricity. This was important for early investigations of electricity. *See also* Lodestone; Static electricity.

Ammonia. A colorless gas comprised of nitrogen and hydrogen. This gas has a pungent, suffocating smell. It also has a strong alkaline reaction. Joseph Priestley first isolated ammonia gas in the eighteenth century.

Analogy. A comparison between two otherwise disparate objects or events. Until the 1727 publication of Stephen Hales *Vegetable Staticks*, most botanists tried to understand plant physiology in the methodological and factual framework of animal physiology. Hales replaced this analogist approach with an experimentalist one.

Animalcule. *See* Microorganism.

Aphelion. The position of a planet or comet when its distance to the sun is at its greatest.

Aqua regia. A mixture of nitric acid and hydrochloric acid capable of dissolving metals such as gold and platinum

Argon. A colorless, odorless inert gas that is used in producing electric light bulbs. Henry Cavendish first detected this gas in atmospheric composition tests. In addition to the vast majority of nitrogen and oxygen, Cavendish detected a "bubble" of gas of roughly 1/120 of the whole. Over one hundred years later, scientists working from his observation discovered argon gas as about 1 percent of the atmosphere's total composition.

Avogadro's law. Named after Amedeo Avogadro, this law states that when equal volumes of different gases all have the same pressure and temperature, they all contain the same number of particles. This law is independent of the properties of the specific gases themselves, but is true only for the "ideal" gases.

Ballistics. A branch of engineering that deals with the motion of projectiles. There are four main branches of ballistics: *Interior* deals with projectiles inside the barrel of a gun or weapon. *Exterior* deals with a projectile in flight. *Terminal* deals with a projectile's effect on its target and surroundings. *Forensic* deals with unique traits of specific firearms and projectiles for purposes of law enforcement. *See also* Ballistic pendulum.

Ballistic pendulum. A device consisting of a heavy wooden block suspended on a pivoting bar for the purpose of measuring projectile velocity. When a musket ball or bullet strikes the block, the block moves up and back in the arc of a pendulum. A scientist can mathematically determine the projectile's velocity by measuring the block's arc, amplitude, and period of oscillation.

Barlow's disease. This is the name for the disease scurvy, when it appears in infants. This disease is most common when a baby is being weaned from breast milk. Prevention involves administering orange or tomato juice after the first month. *See also* Scurvy.

Bernoulli's principle. Named for Daniel Bernoulli, this principle states that energy is conserved when a fluid (liquid or gas) is in motion. Bernoulli's principle explains that pressure decreases as the speed of a fluid increases (and vice versa). It also explains the principle of lift. For example, engineers give the top of an airplane's wing a curved shape. Because air travels faster over the wing's curved top than its flat bottom, a pressure differential results. This achieves vertical lift when the plane is "pushed up" by the greater air pressure on the bottom of the wing. *See also* Hydrodynamics.

Bifocal spectacles. First invented by Benjamin Franklin in 1784, these are glasses in which the top lens has a different focal point than the bottom. Typically, the top lens is for distance vision, and the bottom lens is for near vision. Modern bifocal spectacles no longer require two separate lenses, but one lens with two parts.

Binary system. A base 2 system of mathematics, which only uses the symbols 0 and 1 for the positional units twos, fours, eights, sixteens, and so on. In fact, this system takes its name from the Latin word for "two at a time." In the binary system, each number position has twice the value of the position to its right. For this reason, binary numbers tend to be much longer than their more familiar decimal counterparts. The binary system was originally conceived by Gottfried Leibniz in the early eighteenth century. However, scientists paid little attention, as the system offered no practical value. The binary system came into widespread use during the mid-twentieth century. During this period, scientists invented the first computers, which "think" in binary terms. *See also* Decimal system.

Binomial nomenclature. Literally from the Latin for "two-name naming." This is the modern system of naming organisms by two names—one for their genus and one for their species. In the early and mid-eighteenth century, Carolus Linnaeus first proposed this naming schema. Many of Linnaeus's groupings are still valid.

Biochemistry. The field of science that deals with chemical processes of living plants or animals. Through the study of respiration, Antoine Lavoisier and Simon de Laplace conducted early work in this field.

Biogenesis. The universally accepted theory that living things can only be produced by other living things, rather than nonliving materials. *See also* Spontaneous generation.

Black hole. An area of space in which the gravitational pull is so powerful that nothing—not even light—can achieve escape velocity. Because it traps light, a black hole is a "dark star," or invisible area. Due to the overwhelming gravitational pull, all matter in a black hole is located at a single point in its center, known as its *singularity*. John Michell first proposed the existence of black holes in purely Newtonian terms in 1783. Modern scientists have confirmed the existence of black holes. However, quantum theory conceives of them differently from Michell's original proposition. *See also* Corpuscular theory of light.

Bode's law. A formula for approximating planetary distances from the sun, first published in 1772 by Johann Bode. To use this law, begin with a series of numbers in which each one is twice the preceding one: 3, 6, 12, 24, 48, 96, 192, 384, 768, and so on. Then add 4 to every number and divide the sum by 10. Bode's law approximates the distance of each planet in the solar system, except for Neptune and Pluto. Additionally, Bode's law predicts a planet between Mars and Jupiter that astronomers have never located, but a number of large asteroids exist there. William Herschel's 1781 discovery of Uranus allowed contemporary astronomers to broadly accept and apply Bode's law.

Botany. The science that studies, classifies, and names plants. Though an ancient practice, modern biology was born from the work of many eighteenth-century scientists. *See also* Plant physiology; Taxonomy.

Boyle's law. Named for Robert Boyle, this law states that the volume (V) of a gas multiplied by the product of its pressure (P) remains constant only when there is no change in *both* temperature and the number of gas particles inside a container. This law is written as $VP = $ constant.

Calculus. An advanced mathematical system that uses algebraic symbols to solve questions of changing quantities. Originally, "the calculus" invented during the eighteenth century dealt especially with questions of moving objects. Today, this branch is known as differential calculus. *See also* Differential equation; Taylor's theorem.

Caloric. When Antoine Lavoisier shattered the phlogiston theory of substances with his modern theory of chemistry, he drew up a seminal table of elements and compounds. On the list of elements, he included a kind of invisible, weightless heat fluid—caloric. When a substance burned, Lavoisier thought it combined with oxygen and released caloric. He also thought caloric was transmitted from one object to another through conduction. Largely, caloric had inherited the hypothetical mantle of phlogiston. Count Rumford disproved the theory of caloric when he conclusively demonstrated that heat is a form of energy. *See also* Compound; Element; Ice calorimeter; Thermometer.

Calx. An ashy substance left after a chemical (especially a metal) has been roasted. The phlogistonists thought that this was the "essence" of a chemical, left after its phlogiston had been liberated. Antoine Lavoisier realized that this was actually a simple process of oxidation, which takes place during burning reactions. *See also* Phlogiston.

Carbon dioxide. A colorless, odorless gas present in ordinary air in the amount of approximately 0.033 percent. It occurs naturally in the atmosphere of many planets, including earth. Carbon dioxide is created by the burning of substances containing carbon (such as wood, coal, and gasoline). Carbon dioxide is especially important, because plants use it to manufacture tissue and food in the process of photosynthesis. Animals release this gas during respiration. Carbon dioxide is used in baking (wherein it causes products to rise), in fire extinguishers, and in its frozen form as dry ice. In 1756, Joseph Black discovered carbon dioxide long before any other gas was known. Noting its inability to support combustion or respiration and its frequent escape from solid chemicals, Black named this gas "fixed air."

Carnot cycle. A series of operations consisting of isothermal expansion, adiabatic expansion, isothermal compression, and adiabatic compression. These operations comprise the cycle of an "ideal" heat engine at maximum thermal efficiency. The Carnot cycle is named for Sadi Carnot, who demonstrated that such a "perfectly efficient heat engine" does not exist. During the early nineteenth century, the Carnot cycle provided important methods for improving the efficiency of steam engines. It did so by demonstrating that the efficiency of a reversible engine depends on the highest and lowest temperatures it reaches in a single cycle. As an equation, the Carnot cycle could be expressed as $E = (T1 - T2)/T1$, where E is efficiency and T is temperature. In other words, the Carnot cycle demonstrates that the difference between an engine's two temperature extremes forms a direct proportion of its thermal efficiency. *See also* Thermodynamics.

Celestial navigation. The ancient practice of finding one's own position (usually in a ship) in regard to the moon, planets, and especially stars of the night sky. This is done by precisely tracking the change in the paths of celestial bodies that *appears* to take place as a vessel journeys a far distance. Reliable navigational quadrants such as the sextant and octant were independently developed by John Hadley and Thomas Godfrey in 1731.

Celsius. A temperature scale first developed by Anders Celsius and calibrated to two temperatures: zero for the freezing point of water and 100 for its boiling point. In 1948, the General Conference of Weights and Measures officially named the scale Celsius (rather than centigrade). As the SI standard metric scale for temperature, the Celsius scale is used for everyday temperature measurement in all major countries except the United States. Scientists around the world also commonly use the Celsius scale, in addition to Kelvin. *See also* Metric system; SI.

Charles's law. Named after J.A.C. Charles, this law states that a gas expands by the same fraction of its original volume as its temperature rises. This law establishes that a given volume of gas is in direct proportion to the temperature of the gas (barring a pressure change). If the pressure remains constant, its volume (V) and temperature (T) will not change. Charles's law is written as V/T = constant. Scientists now generally accept this law for all "ideal" gases.

Charliere. *See* Gas balloon.

Chlorine. A green-yellow, poisonous gas with a pungent, unpleasant odor. Chlorine is a chemical element that mostly occurs in compound form with sodium, as common salt. This gas is extremely irritating to the nose, throat, and lungs. In large quantities, such as those used in chemical warfare, chlorine gas is deadly. Chlorine is widely used in bleaching and in the manufacture of many items, such as plastics, paints, and paper. It is also widely used as a disinfectant, especially in water for drinking and swimming. Carl Scheele first isolated chlorine in 1774 by treating manganese dioxide with hydrochloric acid. He named it "dephlogisticated acid of salt." In 1810, Humphry Davy called the gas *chlorine*, a name that he formed from the Greek word for "greenish yellow."

Chronometer. An instrument that keeps extremely accurate time, as needed by navigators making precise measurements of their positions at sea. Clockmakers John Harrison and Pierre LeRoy perfected the first reliable chronometers during the eighteenth century. The combination of chronometer and navigational quadrant led to extremely accurate celestial navigation for merchant, naval, and scientific ships. *See also* Navigational quadrants.

Churka. An early type of cotton gin used since antiquity in India. *See also* Cotton gin.

Chyme. The pulpy mass of partially digested food, which passes from the stomach into the small intestine. *See also* Digestion.

Combustion. The process of burning, generally through rapid oxidation, producing both heat and light. For many years, combustion was a source of mystery and confusion. It was also at the very center of eighteenth-century chemical inquires. In 1700, Georg Ernst Stahl proposed that combustion entailed mainly the release of the hypothetical substance phlogiston. This was a fruitful theory that led to many discoveries. At the century's end, however, phlogiston was demolished by the oxygen chemistry of Antoine Lavoisier. *See also* Calx; Phlogiston; Oxygen.

Comet. An icy celestial formation that consists of three basic parts: the *nucleus*, or center, made of ice and rock, often similar to a "dirty snowball"; the *coma*, or dust and gas, that collect around the nucleus; and the *tail*, or gases that spread out in a long trail from the coma. For many years, people feared comets as a sign of earthly demise. Then, astronomers reasoned that comets traveled in a straight or parabolic line—once through the solar system, never to return. However, in 1705, Edmond Halley concluded that he had seen the very same comet in 1682 as people had seen in 1607 and 1531. He correctly predicted that it would return in 1758. Halley demonstrated that most comets follow a highly elongated, elliptical orbit around the sun. *See also* Kuiper belt; Oort cloud.

Compound. Any substance that contains more than one kind of atom. According to Proust's law, every compound has a stable composition of two or more substances combined in definite proportions by weight. This composition can be described by a chemical formula. Antoine Lavoisier first categorized substances as either elements or compounds in 1789. Water is an example of a common chemical compound, as it consists of two kinds of atoms—hydrogen and oxygen. *See also* Element; Proust's law.

Conduction. *See* Electrical conduction.

Corpuscular theory. The ancient Greeks believed that all matter was composed of tiny, unbreakable particles called "corpuscles." In 1704, Isaac Newton proposed a corpuscular model of light. He said light is composed of tiny, discrete particles moving in straight lines at a finite velocity. Scientists generally accepted this model until the discovery of infrared light and later radio waves reinvigorated the wave model of light. Current quantum theory accepts that light behaves neither wholly like a wave nor wholly like a composite of particles.

Cotton gin. Invented by Eli Whitney, this was the machine that could clean the type of cotton most widely grown in the United States. It consisted of a square wooden frame around a wooden cylinder fitted with metal spikes. As the cylinder rotated, the spikes meshed with a wire grid. The cotton that became entangled in the spikes slid through the grid, but the seeds were too large to do so, and fell in a separate chamber. The cotton gin's impact on the United States was catastrophic—it helped build the powerful South around its "king" crop, cotton.

Coulomb's law. Named after Charles Coulomb, this law states that the force between two electric or magnetic charges varies inversely as the square of the distance between them. This law is also known as the rule of electrical forces and could be written as an equation:

$$F \propto \frac{qq'}{r^2}$$

Decimal system. A number system in which all numbers are written with only ten basic symbols (1, 2, 3, 4, 5, 6, 7, 8, 9, and 0). The relational placement of a number determines its value, as the decimal system is a place-value system. For example, the symbol 1 has different values in 10 and 100 because of the different place of the symbol in each number. In the eighteenth century, French scientists developed the metric system of weights and measures. This system was based on the base 10 decimal system. Today, all major countries except the United States and Burma have converted to the metric system. *See also* Binary system; Metric system; SI.

Dental drill. An instrument used to remove decaying parts of teeth, especially in preparation for filling them. The oldest description of a dental drill dates back to ancient Greece. However, most dental drills were too slow and clumsy for effective use. For this reason, extraction was the most common procedure. Pierre Fauchard, the founder of modern dentistry, invented a useable bow drill during the mid-eighteenth century. He also vastly improved drill-and-fill techniques, which dentists began to widely favor. This offered the enormous advantage of salvaging a damaged tooth.

Dephlogisticated air. *See* Oxygen.

Dichotomy. The phase of an inferior planet (or a moon) when one-half of its face is visible.

Differential equation. A specific kind of equation that involves differential coefficients and differentials, as expressed by Brooke Taylor in 1715. *See also* Calculus; Taylor's theorem.

Digestion. The process by which food is broken down for use in the body. Until the eighteenth century, scientists were divided over whether digestion was primarily a mechanical or a chemical process. The empirical trials of René Réaumur and Lazzaro Spallanzani proved that digestion is primarily chemical.

Electric arc. An early type of electric light, in which an electric current leaps between two electrodes. This dazzling invention was first demonstrated by Sir Humphry Davy. Today, particle accelerators employ modified electric arcs.

Electric machines. Vastly improved by Francis Hauksbee in the early eighteenth century, these machines were invaluable to early electrical inquiries. They consisted of a glass ball that was turned by means of a hand crank. When a person put his or her hand on the ball, electricity in the form of friction was created. Later models substituted leather cushions for a person's hand. *See also* Static electricity.

Electrical conduction. The actual flow of electricity. In other words, conduction is the transmission of electricity (or another form of energy, such as heat) from one particle to another. In common electrical conductors such as metal, electrons are weakly bound to their nuclei and therefore travel freely, rendering a strong flow of electricity. *See also* Electrical insulation.

Electrical insulation. The process by which a substance prevents the passage of electricity (or another form of energy, such as heat). In common electrical insulators such as glass, plastics, or rubber, electrons are tightly bound to their nuclei, so that they cannot move freely between atoms. Too few electrons can move through an insulator to produce a current. *See also* Electrical conduction.

Element. Any substance that contains only one kind of atom. By definition, an element cannot be separated into any simpler parts using ordinary chemical methods. Antoine Lavoisier first categorized substances as either elements or compounds in 1789. Hydrogen and oxygen are examples of common chemical elements. *See also* Compound.

Entomology. The science concerned with the study of insects, especially their physiology, development, anatomy, life history, ecology, behavior, and classification. Many entomologists also study arthropods, such as mites, ticks, and spiders. Entomology developed after the texts of Carolus Linnaeus established a system of taxonomic naming and classification. René Réaumur was the first strictly observational entomologist of the modern (post-Linnaean) tradition. *See also* Taxonomy.

Eudiometer. An instrument for measuring gas volume, formerly believed to be the "purity" of an air sample—or the amount of oxygen contained in it. After Henry Cavendish vastly improved its design, this instrument became crucial to his atmospheric composition trials.

Experimentum cruces. Literally, Isaac Newton's "crucial experiment." In a dark room, Newton directed a beam of sunlight through a prism and observed color separation. Then, he passed it through a second prism and observed the "reconstitution" of white light. Newton concluded that scientists beginning with the ancient Greeks had their understanding of light exactly backward. Colored light was primary and homogenous; white light was secondary and heterogeneous. After this discovery, scientists realized that sunlight is made up of multiple colored rays of light.

Fahrenheit. A temperature scale first developed by Daniel Fahrenheit and later calibrated to three temperatures: 0° for the freezing point of a slurry made of ice, water, and salt; 32° for the freezing point of water; and 212° for the boiling point of water. The United States is the only major country that still uses the Fahrenheit temperature scale for everyday temperature measurement.

Fiducial points. These are used as fixed bases or references for comparison. Thermometers are calibrated on the basis of universal fiducial points. For Celsius thermometers, these points are water's freezing point (zero) and its boiling point (100). *See also* Thermometer.

Fire air. *See* Oxygen.

Fixed air. *See* Carbon dioxide.

Franklin stove. Named after Benjamin Franklin, this stove was the first to achieve a high degree of thermal efficiency. The first heating stoves were little more than cast-iron fireplace inserts. Franklin saw that much heat was lost to the sides, back, and chimney. First, he extended the sides of the stove into the room. More importantly, Franklin applied scientific principles of heat exchange to create an air current. When cold air from the room entered his stove, red hot baffles heated it. From there, it was channeled back into the room. Franklin effectively made his room twice as warm with one-forth the fuel. The Franklin stove became a standard household fixture in England and the United States for well over a century.

Gas balloon. An airtight container filled with lighter-than-air gas (today, usually helium, but formerly hydrogen) so it will float in ordinary air. Most gas balloons carry either human passengers (for sport) or scientific instruments (for research). These balloons are sometimes called *charlieres*, after their eighteenth-century inventor, J.A.C. Charles. Charles experimented with hydrogen and discovered the law that bears his name. This law states that a gas expands by the same fraction of its original volume as its temperature rises. This law and the realization that hydrogen is much less dense than ordinary air convinced Charles that flight in a lighter-than-air gas balloon was possible. Charles also studied hot-air balloons, and probably understood more about how they function (the relationship between heat and density) than anyone else during the late eighteenth century. *See also* Charles's law; Hot-air balloon; Roziere.

Gastric juice. The digestive juice of the stomach, which consists of hydrochloric acid and the enzyme pepsin. Gastric juice begins the digestion of protein foods (such as meat, eggs, and dairy products). This term was coined by Lazzaro Spallanzani. *See also* Digestion.

Hot-air balloon. An airtight container filled with air heated by a burner so that the container will float in ordinary air. Most hot-air balloons carry human passengers. These balloons are sometimes called *montgolfiers* after their eighteenth-century inventors, Joseph and Étienne Montgolfier. The Montgolfier brothers experimented with hot air and found they could easily make increasingly larger, unmanned balloons rise high in the air. In 1783, the Montgolfiers launched the first full-scale lighter-than-air balloon flight. Later that year, they launched the first passengers—animals and, later, humans. Though their invention was wildly successful, the Montgolfiers never understood that lowering the density of air by heating was a physical process. They thought they had released the hypothetical chemical phlogiston, which (they reasoned) must be lighter than air. *See also* Charles's law; Gas balloon; Phlogiston; Roziere.

Humors. There existed a theory in early physiology (especially during the Middle Ages) that the body contained four principle fluids, or humors: black bile, blood, phlegm, and yellow bile. When any of these humors was out of balance and a person became sick, a physician performed a procedure—most often a blood letting—to restore the body's supposed balance. Though incorrect, this theory was enormously widespread and influential.

Hydrodynamics. The branch of physics that deals with the behavior of fluids in motion. This field includes studies of liquids or gases flowing either steadily or unsteadily. It is especially focused on the forces that such fluids exert while they are in motion. Bernoulli's principle was foundational to eighteenth-century work with hydrodynamics. *See also* Bernoulli's principle.

Hydrogen. An odorless, colorless gas and one of the most important chemical elements. Hydrogen is about fourteen times lighter than ordinary air. It also weighs less than any other known element. It is present in most all organic compounds. Hydrogen is also the most abundant element in the universe. This element has a variety of important industrial uses and was first used in lighter-than-air gas balloons. Henry Cavendish first recognized hydrogen as an element and called it "inflammable air" in 1766. Antoine Lavoisier named the gas *hydrogen*, from the Greek for "water former," because it combines with oxygen to form this compound.

Ice calorimeter. A scientific instrument invented by Antoine Lavoisier and Simon de Laplace to measure the production of caloric fluid (or heat) during combustion and animal respiration. This device consisted of three concentric chambers. The inside chamber held a live, respiring animal or a burning object. The middle chamber held a specific quantity of ice. The outside compartment held packed snow for insulation. By carefully collecting and measuring the water produced when the ice melted, Lavoisier and Laplace could measure heat produced in the innermost chamber. They invented a scale for this measurement. *See also* Caloric.

Incandescent light bulb. Invented by Sir Joseph Wilson Swan, this is the most common standard light bulb. In this kind of bulb, an electrical current encounters resistance in crossing a filament, which heats the filament, causing it to glow. Modern bulbs employ a tungsten filament and a mixture of gases to lengthen the life of the bulb.

Inflammable air. *See* Hydrogen.

Inoculation. The practice of purposefully infecting a healthy person with a weakened form of a disease. This then causes the body to produce antibodies, thereby granting a person immunity. *See also* Smallpox.

Insulation. *See* Electrical insulation.

Kelvin. An absolute temperature scale developed by Lord Kelvin, which calibrates zero at the theoretical temperature of absolute zero (−273.15 Celsius or −459.67 Fahrenheit). The Kelvin scale is the international standard for scientific temperature measurement. As such, it is widely used by scientists in a variety of fields. Physicists generally believe that attaining a temperature of precisely absolute zero is impossible. However, scientists have attained a temperature of less than one billionth of a degree Kelvin.

Kuiper belt. A region outside the orbit of Pluto that many scientists believe acts as a reservoir for short-period comets. *See also* Comet; Oort cloud.

Latent heat. This is the heat required to cause the state of matter to change from a solid to a liquid or from a liquid to a vapor. This heat is released in the reverse process. By paying close attention to temperature and noting that "additional" heat was required to make ice melt or water boil, Joseph Black first explained latent heat in 1761. Knowledge of this process was invaluable for the steam-powered industrial revolution. *See also* Specific heat; Watt steam engine.

Law of conservation of mass. The principle in chemistry that states that matter can neither be created nor destroyed. In a reaction, the mass of the products must equal the mass of reactants. For many years, scientists implicitly believed the conservation of mass, but none tested it or fully understood its implications. Antoine Lavoisier first paid close attention to the concept of mass in chemical reactions during the late eighteenth century. He found that oxygen was joining with metals, such as tin, in oxidation reactions: *chemical + oxygen → chemical oxide*. This discovery shattered the phlogiston theory of substances and became indispensable to modern chemistry. The discovery of nuclear reactions has caused scientists to state the law a little differently: mass-energy may not be created or destroyed, but each may be converted to the other. *See also* Phlogiston.

Leyden jar. Named for the University of Leyden, the Leyden jar was one of the first devices invented to successfully store an electric charge. This device acts as a capacitor. Simple Leyden jars consist of a glass jar sealed with a cork. Generally, sheets of thin metal foil (rather than water) cover roughly half of the inside and outside of the jar. The foil acts as an electrical conductor, in contrast to the insulating glass. Like the original Leyden jars, a brass rod is inserted through the cork and into the jar, so that it touches the foil inside the jar. When the brass rod is electrified, electrical current charges the inner foil. Because current cannot pass through the glass, the foil on the outside becomes charged by means of induction (assuming it is properly grounded). This charge is opposite to the charge held by the foil inside the jar. If the inner and outer layers of foil are connected by a conductor, their opposing charges discharge the jar of its electricity. *See also* Electrical conduction; Electrical insulation; Static electricity; Voltaic pile.

Lift. *See* Bernoulli's principle.

Light-year. In astronomy, the measure of the speed that light travels in one year—about 9.46 trillion kilometers.

Lightning rod. A large metal rod on or near a building which protects it by conducting lightning into the ground. The first lightning rods came directly out of Benjamin Franklin's study of electricity. Today's basic lightning rods differ very little from Franklin's original design 250 years ago.

Lodestone. A hard, black stone that attracts iron and steel as does a magnet. Lodestone is a kind of magnetite. Discovery of lodestone's attraction property was important for early delineations of electricity. *See also* Amber; Static electricity.

Magnus effect. The aerodynamic tendency of a spinning projectile, such as a bullet or baseball, to skew in one particular direction. This phenomenon is caused by improper (or uneven) spinning, in which the ball carries some of the surrounding air with it, due to the action of friction. This is roughly analogous to a whirlpool. When two airstreams traveling in opposite directions meet, they create an increased pressure on one side of the ball. This results in enhanced velocity. On the exact opposite side, the motion has been opposed and velocity has thereby decreased. The unequal effects of different air pressures thereby send the ball askew. *See also* Bernoulli's principle.

Mephitic air. *See* Nitrogen.

Method of fluxions. Isaac Newton's original name for "the calculus," which stemmed from the mathematical relationship between differentiation and integration of tiny changes in events. *See also* Calculus.

Metric system. The system of weights and measures, based on decimal division, adopted universally, except by the United States and the developing nation of Burma. First conceived in France in the 1790s, the metric system greatly simplified the older systems it replaced, especially the English inch–pound system. The metric system's name comes from its unit of length, the meter. Other units include the kilogram (for mass) and the Kelvin (for temperature). All international activities, such as trade and science, use the metric system. The United States government has made attempts to convert to the metric system, but these have been largely unsuccessful. *See also* Decimal system; SI.

Microorganism. An organism too small to be seen without the aid of a microscope. Bacteria are microorganisms.

Milky Way. The galaxy that contains the sun and the nine planets of its solar system. The portion of the Milky Way that is visible from earth appears as a milky swath of starlight in the clear night sky. This galaxy has a diameter of about 100,000 light-years, and the earth is located about 25,000 light-years from the galaxy's center. The galaxy's shape was first proposed by Thomas Wright in 1750. He correctly realized that it is a lens-shaped distribution of stars—thicker in the middle than at its thin edges. One of the most important achievements of twentieth-century astronomy was the classification of the Milky Way as a *spiral galaxy*. This classification came from the realization that dust, gases, and even stars unfold from the Milky Way's bulge in a pinwheel pattern.

Montgolfier. *See* Hot-air balloon.

Morphology. The science that deals with the form and structure of animals, plants, and nonliving matter (such as rocks). Generally, morphologists study form and structure rather than function. *See also* Taxonomy.

Navigational quadrants. These instruments were invented independently by John Hadley and Thomas Godfrey in 1731. They worked according to the optical rule, which set forth the principles of light reflection and angles of incidence as outlined by Edmond Halley and Isaac Newton. In use, these instruments—sextants or octants—measured the distance in angles between two points (such as the horizon and a star). Like older instruments, these consisted of a tri-shaped wooden frame and a swinging arm. In its so-called zero or starting position, the actual and reflected horizons were mutually aligned. To make measurements, the navigator held the instrument vertical to the horizon and moved the arm until the image of a certain celestial body appeared to "touch" the line of horizon. The double-reflecting instrument thereby "brought down" the celestial body to the level of the horizon, making it possible to record the vessel's relational position.

Newcomen steam engine. An early type of steam engine invented around 1712 by Thomas Newcomen. Although it was capable of pumping over 3.5 million gallons of water per day, this engine was extremely inefficient because it used only one chamber for both steam and condensation. This meant that, for every power stroke, it had to overcome the energy constraints of latent heat. *See also* Carnot cycle; Latent heat; Watt steam engine.

Nitrogen. A colorless, odorless gas that accounts for roughly 78 percent of the earth's atmospheric volume. Nitrogen is one of the most important chemical elements because it is necessary for the growth of all plants, as well as the tissues of animals. As a gas, nitrogen does not combine well with most other compounds. However, it is frequently condensed to useful liquid form, which boils at −195.8°C and freezes at −209.9°C. Daniel Rutherford first identified nitrogen in 1772.

Nitrous oxide. A colorless gas that dulls pain and induces fits of laughter in some people. Joseph Priestley first isolated nitrous oxide in 1772. Humphry Davy first experimented with this gas as a surgical anesthetic in 1798. In the 1860s, nitrous oxide became widely used as a dental anesthetic.

Ockham's razor. As stated by William of Ockham, the philosophical principle that a problem should be stated in its most basic and simplest terms. For enlightenment scientists, this meant that the simplest theory that fully explains a phenomenon or problem should be the one selected. *See also* Principle of least action.

Oort cloud. A collection of comets one thousand times farther away from earth than Pluto's orbit. Many scientists believe that the Oort cloud acts as a reservoir for long-period comets. In this cloud, comets probably move very slowly. However, they can be sent into the inner part of the solar system by a "passing" star's gravity. *See also* Comet; Kuiper belt.

The optic rule. This rule set forth the principles of light reflection and angles of incidence, as outlined by Edmond Halley and Isaac Newton. *See also* Navigational quadrants.

Oxygen. A colorless, odorless gas that forms about one-fifth of atmospheric air. Virtually all living things require this element to stay alive. Oxygen is present in plant and animal cells and is necessary for the energy of life processes. Most substances require oxygen in order to burn. Also, animals obtain oxygen during respiration. Oxygen was independently discovered by Carl Scheele and Joseph Priestley during the early 1770s. Scheele called it "fire air," and Priestley called it "dephlogisticated air." Antoine Lavoisier was first to understand oxygen's role and assign it its modern name. Lavoisier's realization that combustion involved oxygen *joining with* rather than phlogiston *separating from* a substance shattered the phlogiston theory of matter. Scientists have accepted Lavoisier's theory of oxygen chemistry with near universality since the late eighteenth century. *See also* Law of conservation of mass; Oxygen; Phlogiston; Pneumatic chemistry.

Parallax. The observed change in position of a heavenly object when viewed through a telescope from two different points. Modern surveyors depend heavily on parallax to measure distances of objects. James Bradley was researching parallax when he discovered stellar aberration in 1728. *See also* Stellar aberration.

Pathology. The scientific study of disease, with emphasis on its causes, symptoms, and treatments. Modern pathology developed quickly after the eighteenth century work of Stephen Hales, Albrecht von Haller, and, especially, John Hunter. Hunter showed early on that specific microorganisms cause specific diseases—a realization that was only fully understood many years later.

Perihelion. The position of a planet or comet when it is closest to the sun in its orbit.

Phlogiston. The first rational, reasonable theory of chemical substances, as proposed by George Ernst Stahl in 1700. Stahl said that when substances combust or calcify, or when animals respire, they release the substance he called phlogiston (from the Greek for "burning"). When a candle burns, Stahl believed it releases phlogiston. When the candle no longer will burn, he believed it had rendered its store of phlogiston. This theory also explained independent gases. For example, objects burn brightly in "dephlogisticated air" (or oxygen) because this air has lost all of its phlogiston, and therefore hungrily grabs it back during combustion. This powerful theory led to the eighteenth-century discoveries of the first independent gases, as well as many of the first applications of advanced chemical knowledge. Phlogiston was eventually replaced by a better theory, Antoine Lavoisier's theory of oxygen chemistry. Since the 1790s, the latter theory has been adopted with near universality. *See also* Alchemy; Carbon dioxide; Chlorine; Hydrogen; Law of conservation of mass; Oxygen; Nitrogen; Pneumatic chemistry; Respiration; *Terra pinguis*.

Photosynthesis. The term literally means "putting together with light." Photosynthesis is one of the most important chemical reactions in nature and the process by which plants make carbohydrates from carbon dioxide and water, given the presence of chlorophyll and light. Plants release oxygen as a by-product, which is essential for animal survival. Photosynthesis was first experimentally discovered by Jan Ingenhousz in the early 1770s.

Physiology. The scientific study of how living things function in the world. For thousands of years after antiquity, physiology was shackled with abstract theories about vital organ function. With the 1752 publication of Albrecht von Haller's work, animal physiology was opened for the first time to empirical observation. Another key event was proof of blood circulation, offered when Stephen Hales first measured blood pressure in a horse.

Planetary transit. This occurs when one celestial body passes optically across the disk of a larger one. Planetary transits of Venus across the sun were especially important to eighteenth-century astronomers trying to measure stellar parallax (the apparent change in an object's position due to two different observational points).

Pneumatic chemistry. The "chemistry of airs" founded by Stephen Hales and practiced by phlogiston-centered chemists in the eighteenth century. Pneumatic chemists explored the chemical and physical properties of the independent "airs," or gases, they discovered. Eventually, the discovery of one such gas, oxygen, led French chemists toward the theory of modern chemistry. *See also* Alchemy; Carbon dioxide; Chlorine; Hydrogen; Nitrogen; Oxygen; Phlogiston; Pneumatic trough.

Pneumatic trough. This apparatus was used by nearly all chemists of the eighteenth century. Collecting gas involved filling the trough with water and then forcing the gas into the trough until all of the water "bubbled" out. Joseph Priestley substituted dense mercury for water, making the trough practical for collecting water-soluble gases.

Pollination. The act of carrying pollen from floral anthers to stigmas in the process of fertilization. Common agents of fertilization are insects and the wind. Taken together, the independent work of Joseph Koelreuter and Christian Sprengel established overall conceptual understanding of this process during the second half of the eighteenth century.

Poxvirus. The virus that causes the disease smallpox in humans. *See also* Smallpox.

Principle of least action. Stated broadly, this is the idea that nature is thrifty in all of its actions. Mathematician Pierre-Louis de Maupertuis conceived of the principle of least action as the product of mass (m), velocity (v) and distance (s): m × v × s = action. In other words, any natural motion of a particle between two points tends to minimize the product (action). Examples of this principle include water running downhill, as well as the refraction of a ray of light in a body of water. The chief architect of quantum mechanics, Max Planck, often noted the importance of this principle to his theories. *See also* Ockham's razor.

Proust's law of constant proportions. In simple terms, this law states that compounds always contain fixed proportions of elements. In the late eighteenth century, scientists Claude Berthollet and Joseph Proust disagreed on whether the proportion of elements in a compound was definite. Proust demonstrated that every pure chemical compound consists of elements in definite, specific proportions. After endorsement by several influential chemists, Proust's law was broadly accepted as correct.

Respiration. In animals, the chemical process by which an organism secures oxygen, distributes it, combines it with food, produces energy, and releases carbon dioxide. By comparing respiration with combustion (which they previously had studied), Antoine Lavoisier and Simon de Laplace began the modern study of animal respiration as primarily a chemical process. *See also* Combustion.

Roziere. A hybrid gas and hot-air balloon, named for its inventor, Pilatre de Rozier. Today, rozieres are primarily flown as gas (helium) balloons. However, by firing burners at night, these balloons compensate for different temperatures between day and night flying. This method heats a "cone" of hot air surrounding the cache of trapped helium, making it easier for the balloon to maintain altitude. *See also* Gas balloon; Hot-air balloon.

Rule of electrical forces. *See* Coulomb's law.

SI. Since its inception in France in the 1790s, the metric system has gone through several refinements. *Systeme International d'Unites,* or International System of Units—SI for short—is the official name of the present version. *See also* Metric system.

Scurvy. A disease that results from the inability of humans to synthesize ascorbic acid or vitamin C. When a person does not obtain this substance dietetically for about 30 weeks, symptoms such as bleeding gums, sore joints, loose teeth, poor healing, excessive bruising, and anemia may occur. Eventually, death may ensue. Scurvy is an ancient disease. However, it showed up with ferocity in Europe only after the advent of long sea voyages in the fifteenth century. Beginning in 1747, James Lind successfully treated the scurvy-ridden sailors of several ships by administering citrus fruits and juices. Lind conducted controlled experiments on sick sailors to arrive at this treatment. However, Lind had no understanding of why these foods cured the disease, as vitamin C was not discovered until 1928.

Smallpox. The extremely contagious disease caused by the poxvirus. In humans, this disease causes high fever and a rash of small sores all over the body. The sores result in scars that range from mild to disfiguring. In severe cases, permanent blindness and even death can occur. The wife of the British ambassador to Turkey, Lady Mary Montagu, was one of the first Westerners to be inoculated against the disease in 1717. Inoculation had been practiced for many years in the East. Inoculation involved purposely infecting a healthy individual with a mild case of the disease. Later in the century, Edward Jenner discovered a much better prevention, known as *vaccination*. This involved the natural cross-immunity transferred to a human through infection with the relatively benign cowpox virus. Today, due to highly successful vaccination, smallpox exists only in rural "pockets" of developing countries.

Specific heat. This is the amount of heat (e.g., the number of calories) necessary to raise the temperature of 1 gram of a substance 1° Celsius. Specific heat is based on the fact that different substances of the same mass require different amounts of heat to reach the same temperature. Joseph Black first discovered that different bodies have different capacities for heat in 1761. *See also* Latent heat.

Spinning jenny. Invented by James Hargreaves, the spinning jenny allowed a spinner to (eventually) turn up to 120 spindles at once. A spinner turned a crank to spin the cotton, while he or she also slid a moveable frame to draw it out. In this manner, many threads could be drawn at once by a single spinner. This machine produced fine, delicate yarn. *See also* Spinning mule; Water frame spinning machine.

Spinning machines. A series of eighteenth-century machines that revolutionized the spinning of cotton into yarn. Previous machines allowed an individual to spin only a single thread of yarn. These newer machines allowed for spinning multiple threads at one time, thereby greatly increasing production. *See also* Spinning jenny; Spinning mule; Water frame spinning machine.

Spinning mule. Invented by Samuel Crompton, this machine (like its animal namesake) is a hybrid. It combined the best parts of the spinning jenny and the water frame spinning machine. The spinning mule used a movable carriage of forty-eight spindles to spin cotton. This innovation allowed a spinner to produce varying levels of coarse or fine yarn, as well as to achieve effective quality control. *See also* Spinning jenny; Water frame spinning machine.

Spiral galaxy. Along with elliptical and irregular types, this is one of the three classifications of galaxies. Spiral galaxies are shaped like a lens with a bulge in its center. Seen from its "side," the lens resembles a pinwheel, with bright arms of dust, gas, and stars spinning from its center. The galaxy of the sun and its solar system, the Milky Way, is a spiral galaxy. *See also* Milky Way.

Spontaneous generation. The ancient theory that living organisms come from non-living sources. For many centuries, people believed that snakes spontaneously generated in mud, when conditions were just right. They also thought that flies developed on the surface of rancid meat. This specific instance was disproved by Francesco Redi. In a controlled experiment, Redi segregated one piece of meat from houseflies and left one piece open to them. He found that maggots only "appeared" on the surface of the meat after he noticed adult flies on it. Redi concluded that complex organisms like flies come only from other complex organisms. Many scientists continued to believe that microorganisms spontaneous generate. Lazzaro Spallanzani demonstrated that, by boiling and vacuum sealing certain foods in glass containers, microorganisms did not develop. However, some scientists still insisted on the idea of spontaneous generation, claiming that air was a necessary component of the process. Finally, by segregating microorganisms, but not air, from his test foods, Louis Pasteur put the idea of spontaneous generation to rest in the nineteenth century. Some scientists believe that a much different form of spontaneous generation may have occurred in the primitive oceans of earth. They believe that atmospheric gases and energy (from lightning and the sun) may have compounded to form the first amino acids.

Static electricity. When a large number of atoms in an object gain or lose electrons, the object may take on an electric charge. Static electricity describes situations in which objects carry such an electric charge. When a person rubs a balloon on his or her shirt, friction causes electrons to transfer from shirt to balloon. This gives the shirt a positive charge because it has more protons than electrons. The balloon now has a negative charge because of its excess electrons. For this reason, it will stick to a person's shirt or similar surface. Ancient scientists noted that amber displays qualities of becoming charged with static electricity. This was an important observation for early electrical research. *See also* Amber; Electric machines.

Steamboat. A steam-driven vessel that journeys primarily on large rivers. The first steamboat to move under its own power was the *Pyroscaphe,* invented by Claude de Jouffroy, a French nobleman. American inventor John Fitch invented the first practical steamboat in 1787. Fitch also organized the first short-lived steamboat service on the Delaware River. Later, Robert Fulton's *Clermont* became the first financially stable steamboat for passenger service and ushered in a new era of transportation. *See also* Newcomen steam engine; Watt steam engine.

Steam engine. *See* Newcomen steam engine; Watt steam engine.

Stellar aberration. The observational displacement of a star or other heavenly body from its actual position in the sky. This effect is caused by the combined effect of the earth's motion and the length of time it takes a star's light ray to travel the length of a telescope. Because the earth is moving, an observer must tilt a telescope to properly "catch" the ray, which is analogous to a briskly walking person tipping an umbrella in order to shield vertically falling drops of rain. While searching for parallax, James Bradley discovered aberration in 1728. *See also* Parallax.

Stellar apex. A point in space toward which the sun is moving, as expressed in the theory proffered by William Herschel. Modern scientists have demonstrated that the sun is moving in the direction of the Hercules constellation at roughly 20 kilometers (12 miles) per second. Modern scientists have also shown that the sun makes a revolution as part of the Milky Way at about 250 kilometers (about 155 miles) per second. At this speed, the rotation period of the Milky Way and its stars is roughly 225 million years. *See also* Parallax; Stellar aberration.

Stratosphere. The layer of the atmosphere above the troposphere. This layer extends from about 10 kilometers (6 miles) above the earth's poles and about 16 kilometers (10 miles) above the equator. In the lower reaches of the stratosphere, temperature is nearly steady at about −55°C (−67°F). Because of absorption of sunlight by ozone, temperature can increase to −2°C (28°F) in the upper reaches of this layer. *See also* Troposphere.

Taxonomy. The scientific classification and naming of organisms. Modern taxonomy began with Carolus Linnaeus's first system of classifying plants and animals in the 1730s. *See also* Binomial nomenclature.

Taylor's theorem. Brook Taylor's proof of a well-known theorem:

$$f(x+h) = \frac{f(x) + hf^1(x) + h^2 f''(x) + \ldots}{2!}$$

In Taylor's theorem, a function of a single variable can be expanded in its own powers. It does not consider series convergence. The importance of this theorem was recognized as the foundational principle of differential calculus. *See also* Calculus; Differential equation.

Temperature. This measures how hot or cold something is, as indicated on a particular scale. Three widely used scales are Fahrenheit, Celsius, and Kelvin. Because heat transfer occurs only from a hotter object to a cooler one, temperature is closely related to the flow of heat.

Terminator. The optical line through a planet that separates its light and dark parts. This line is especially pronounced when viewed through a telescope.

Terra pinguis. Literally "fatty earth" in the alchemist Johann Joachim Becher's vague doctrine of chemicals. This concept served as the theoretical genesis for Georg Ernst Stahl's powerful theory of phlogiston. *See also* Phlogiston.

Thermometer. A device that measures temperature by employing materials that consistently change in ways that reflect changes in temperature. One common material is a liquid, such as alcohol or mercury, which expands as temperature rises. Most thermometers have numbered sides that correspond to one of several standardized temperature scales. *See also* Celsius; Fahrenheit; Kelvin.

Thermoscope. The earliest kind of thermometer. These instruments crudely indicated changes in temperature but did not measure their amounts in any meaningful or standardized way. *See also* Temperature; Thermometer.

Thermodynamics. The science that studies energy and the laws that govern its transfer. Thermodynamics often deals with the transfer of energy in the form of heat as well as other systems in which temperature is a factor. *See also* Caloric; Carnot cycle; Latent heat.

Torsion balance. An apparatus that measures a small, horizontal force by the amount of torsion (or twist) it causes in a wire. Henry Cavendish greatly improved its accuracy and design. Then, Cavendish used his torsion balance to measure the gravitational force of two small lead balls. He used this information to figure the gravitational constant—a long-elusive variable. Finally, he plugged this information into a Newtonian formula to accurately estimate the earth's mass.

Transmutation. European alchemists believed they could change the four ancient elements—earth, air, fire, and water—by varying levels of heat and moisture. This process was known as transmutation. They especially tried to change base metals into gold and never succeeded, although modern fission and fusion reactions transmute elements (albeit much differently than in alchemy). Fission splits a nucleus into two lighter nuclei; fusion joins two nuclei to form a heavier nucleus. *See also* Alchemy.

Transpiration. The act of passing or sending water vapor through a membrane, as takes place in the leaves of plants. This process was first understood as wholly independent from animal blood circulation by Stephen Hales in the 1720s.

Troposphere. The lowest layer of the atmosphere and the one in which terrestrial weather occurs. This layer extends about 10 miles over the equator and about 6 miles over the earth's poles. As altitude within this layer increases, temperature drops steadily. *See also* Stratosphere.

Universal gas law. This law combines Avogadro's law, Boyle's law and Charles's law into one single statement: $PV = nRT$ (where P is pressure, V is volume, n is the number of moles of a gas, R is a constant known as the universal gas constant, and T is temperature). This law states the three ways that the pressure of a gas can be doubled: by halving is original volume, doubling the amount of gas, and doubling the absolute temperature. *See also* Avogadro's law; Boyle's law; Charles's law.

Uranus. A large, gaseous planet about 2,875,000,000 kilometers from the sun. Neptune and Pluto are the only known planets that are farther away. In 1781, Uranus became the first planet discovered since prehistoric times.

Vaccination. The practice of purposefully infecting a healthy person with a weakened or dead form of a virus to confer immunity. Vaccination also may involve infecting a person with a similar, benign virus in order to achieve natural cross-immunity. In the case of the dread disease smallpox, Edward Jenner's vaccination technique involved infecting a person with relatively benign cowpox. This caused the body to produce antibodies, thereby granting a person immunity. *See also* Smallpox.

Venus. One of the five planets (aside from earth) known since antiquity and second in distance from the sun. In 1761, Mikhail Lomonosov discovered this planet's atmosphere. In the next decade, Johann Schröter offered evidence, through careful telescopic observations, of Venus's manipulation of light.

Volt. The unit, named after Alessandro Volta, in the modern metric system. A volt measures the potential difference between two points (where 1 joule of work is done in moving a charge of 1 coulomb between the points). In other words, the volt measures the force or pressure of a current: volts = amps × ohms.

Voltaic pile. Named for its inventor Alessandro Volta, the voltaic pile was the first successful electrolyte and an early type of chemical cell. To build his pile, Volta layered pairs of different metal disks. He found the best combination was a wide sheet of silver following a wide sheet of zinc. To increase conductivity, Volta carefully placed pieces of cloth saturated with a briny saltwater solution between each metal layer. He found that a chemical action occurred when his moist cloth was in contact with the two different metals. Volta surmised that there was a relationship between the "pile," or number of metal layers, and the strength of electrical charge rendered. When he increased the number of plates to sixty, he could effectively create a powerful charge. Though this was only partially correct, Volta said his results demonstrated that electricity is produced through the combination of different moist metals. This had the practical effect of creating a revolution in the use of small amounts of electricity for both scientific research and technological application. *See also* Electric conduction; Electric insulation; Volt.

Water frame spinning machine. Invented by Sir Richard Arkwright, this machine used two pairs of rollers moving at various speeds to draw yarn. One pair turned significantly faster than the other, thereby creating tension and drawing yarn. This machine made strong, coarse yarn. *See also* Spinning jenny; Spinning mule.

Watt steam engine. A later variety of steam engine invented by James Watt during roughly the final quarter of the eighteenth century. Though Watt made many improvements to older Newcomen steam engines, Watt steam engines generally share the characteristic of a separate condenser. Through his work with efficiency and the discovery of latent heat, Watt realized that a tremendous amount of heat was lost in heating and cooling the same cylinder. When he employed a separate condenser, the steam engine became much more efficient. Watt also invented the rotary flywheel, which used the principle of inertia to transfer the up-and-down stroke of a piston to a uniform circular motion. After these improvements, the steam engine became a nearly universal power source during the industrial revolution. *See also* Carnot cycle; Latent heat; Newcomen steam engine.

Weaving. The ancient practice of crossing two pieces of material over and under one another to create a useful product. Weaving has always been done on machines called *looms*. However, beginning in the eighteenth century, traditional weavers could no longer meet exploding demand for finished products. The flying shuttle loom of John Kay and the steam-powered loom of Edmund Cartwright revolutionized the entire English textile industry.

BIBLIOGRAPHY

Abetti, Giorgio. *The History of Astronomy.* Translated by Betty Burr Abetti. New York: Henry Schuman, 1952.

Allan, D. G. C., and R. E. Schofield. *Stephen Hales: Scientist and Philanthropist.* London: Scholar Press, 1980.

Allen, Will W. *Banneker, The Afro-American Astronomer.* Salem, NH: Ayer, 1971.

Asimov, Isaac. *Asimov's Chronology of Science and Discovery.* New York: Harper & Row, 1989.

Asimov, Issac. *Asimov's Guide to Halley's Comet.* New York: Walker, 1985.

Asimov, Isaac. *A Short History of Chemistry: An Introduction to the Ideas and Concepts of Chemistry.* New York: Anchor Books, 1965.

Asimov, Isaac. *Venus, Near Neighbor of the Sun.* New York: Lothrop, Lee & Shepard Books, 1981.

Baker, David. *Flight and Flying: A Chronology.* New York: Facts On File, 1994.

Baron, Margaret E. *The Origins of the Infinitesimal Calculus.* New York: Pergamon Press 1969.

Bedini, Silvio A. *The Life of Benjamin Banneker.* New York: Charles Scribner's Sons, 1972.

Beiser, Arthur. *Modern Physics: An Introductory Survey.* Reading, MA: Addison-Wesley, 1968.

Bensaude-Vincent, Bernadette, and Isabelle Stengers. *A History of Chemistry.* Translated by Deborah van Dam. Cambridge, MA: Harvard University Press, 1996.

Benumof, Reuben. *Concepts in Physics.* Englewood Cliffs, NJ: Prentice-Hall, 1965.

Berry, A. J. *Henry Cavendish: His Life and Scientific Work.* London: Hutchinson, 1960.

Bettman, Otto L. *A Pictorial History of Medicine.* Springfield, IL: Charles C. Thomas, 1956.

Blunt, Wilfrid. *The Compleat Naturalist: A Life of Linnaeus.* New York: Viking Press, 1971.

Bordeau, Sanford P. *Volts to Hertz: The Rise of Electricity.* Minneapolis: Burgess, 1982.

Bowen, Catherine Drinker. *The Most Dangerous Man in America: Scenes from the Life of Benjamin Franklin.* Boston: Little, Brown, 1974.

Bremner, M. D. K. *The Story of Dentistry, From the Dawn of Civilization to the Present.* Brooklyn: Dental Items of Interest, 1939.

Bridgett Travers (ed.). *World of Invention.* Washington, DC: Gale Research Group, 1994.

Briggs, Asa. *The Power of Steam: An Illustrated History of the World's Steam Age.* Chicago: University of Chicago Press, 1982.

Britton, Karen Gerhardt. *Bale o' Cotton: The Mechanical Art of Cotton Ginning.* College Station: Texas A&M Press, 1992.

Broberg, Gunnar (ed.). *Linnaeus: Progress and Prospects in Linnaean Research*. Pittsburgh: Hunt Institute for Botanical Documentation, 1980.

Brock, William H. *The Norton History of Chemistry*. New York: W. W. Norton, 1992.

Bruck, Richard Hubert. *A Survey of Binary Systems*. Berlin: Springer, 1958.

Buchmann, Stephen L., and Gary Paul Nabhan. *The Forgotten Pollinators*. Washington, DC: Island Press, 1996.

Burton, Anthony. *The Rise and Fall of King Cotton*. London: British Broadcasting Corporation, 1984.

Calder, Nigel. *The Comet is Coming! The Feverish Legacy of Mr. Halley*. New York, Viking Press, 1980.

Campbell, J. Menzies. *Dentistry Then and Now*. Glasgow, UK: privately printed (no publisher listed), 1963.

Canby, Edward Tatnall. *A History of Electricity*. New York: Hawthorn Books, 1968.

Carnegie, Andrew. *James Watt*. New York: Page, 1905.

Constable, George (ed.). *TimeFrame AD 1700–1800, Winds of Revolution*. Alexandria, VA: Time-Life Books, 1990.

Cook, Sir Alan. *Edmond Halley: Charting the Heavens and the Seas*. Oxford: Clarendon Press, 1998.

Cox, Bernard. *Paddle Steamers*. Dorset, UK: Blandford Press, 1979.

Crawley, Chetwode. *From Telegraphy to Television: The Story of Electrical Communications*. New York: Frederick Warne, 1931.

Crew, Henry. *The Rise of Modern Physics*. Baltimore: Williams & Wilkins, 1933.

Crouch, Tom D. *The Eagle Aloft: Two Centuries of the Balloon in America*. Washington, DC: Smithsonian Institution Press, 1983.

Crowther, J. G. *Famous American Men of Science*. New York: W. W. Norton, 1937.

Crowther, J. G. *Scientists of the Industrial Revolution: Joseph Black, James Watt, Joseph Priestly, Henry Cavendish*. London: Cresset Press, 1962.

Darrow, Floyd L. *Masters of Science and Invention*. New York: Harcourt, Brace, 1923.

Darrow, Floyd L. *The New World of Physical Discovery*. Indianapolis: Bobbs-Merrill, 1930.

Dayton, Fred Erving. *Steamboat Days*. New York: Frederick A. Stokes, 1925.

DeKruif, Paul. *Microbe Hunters*. New York: Harcourt, Brace, 1926.

Denniston, George. *The Joy of Ballooning*. Philadelphia: Courage Books, 1999.

Dickinson, H. W. *A Short History of the Steam Engine*. London: Cambridge University Press, 1938.

Dickinson, H. W., and H. P. Vowles. *James Watt and the Industrial Revolution*. New York: Longmans Green, 1943.

Doig, Peter. *A Concise History of Astronomy*. London: Chapman & Hall, 1950.

Drewry, Richard D. "What Man Devised That He Might See." Web article about the history of eyeglasses. At the time of publication, available at http://www.eye.utmem.edu/history/glass.html.

Edwards, C. H. *The Historical Development of the Calculus*. New York: Springer, 1979.

Efron, Alexander. *Direct Current Electricity: Franklinian Approach*. New York: Chapman & Hall, 1960.

Englebert, Phillis. *Astronomy & Space: From the Big Bang to the Big Crunch*. New York: UXL, 1997.

Engelbert, Phillis, and Diane L. Dupuis. *The Handy Space Answer Book*. Detroit: Visible Ink Press, 1998.

Etter, Roberta, and Stuart Schneider. *Halley's Comet: Memories of 1910*. New York: Abbeville Press, 1985.

Evans, Howard Ensign. *The Pleasures of Entomology: Portraits of Insects and the People Who Study Them*. Washington, DC: Smithsonian Institution Press, 1985.

Fahie, John J. *A History of Electric Telegraphy to the Year 1837*. New York: Arno Press, 1974.

Farber, Eduard. *The Evolution of Chemistry*. New York: Ronald Press, 1952.

Farrar, C. L., and D. W. Leeming. *Military Ballistics: A Basic Manual*. Elmsford, NY: Pergamon Press, 1983.

Fine, Leonard W. *Chemistry*, second edition. Baltimore: Williams & Wilkins, 1978.

Fitton, R. S. *The Arkwrights: Spinners of Fortune*. Manchester, UK: Manchester University Press, 1989.

Flexner, James Thomas. *Steamboats Come True: American Inventors in Action*. Boston: Little, Brown, 1978.

Flores, Ivan. *The Logic of Computer Arithmetic*. Englewood Cliffs, NJ: Prentice-Hall, 1963.

Forbes, George. *History of Astronomy*. New York: G. P. Putnam's Sons, 1909.

Foster, Mary Louise. *Life of Lavoisier*. Northampton, MA: Smith College, 1926.

Frängsmyr, Tore (ed.). *Linnaeus: The Man and His Work*. Los Angeles: University of California Press, 1983.

Garrison, Fielding H. *An Introduction to the History of Medicine*. Philadelphia: W. B. Saunders, 1917.

Gibbs, F. W. *Joseph Priestley: Revolutions of the Eighteenth Century*. Garden City, NY: Doubleday, 1967.

Gillispie, Charles Coulston (ed.). *Dictionary of Scientific Biography*, volume XI. New York: Charles Scribner's Sons, 1972.

Gillispie, Charles Coulston (ed.). *The Montgolfier Brothers and the Invention of Aviation: 1783–1784*. Princeton, NJ: Princeton University Press, 1983.

Glenner, Richard A. *The Dental Office, a Pictorial History*. Missoula, MT: Pictorial Histories, 1984.

Graham, Shirley. *Your Most Humble Servant*. New York: Julian Messner, 1952.

Green, Constance McLaughlin. *Eli Whitney and the Birth of American Technology*. Boston: Little, Brown, 1956.

Greenberg, Arthur. *A Chemical History Tour: Picturing Chemistry from Alchemy to Modern Molecular Science*. New York: John Wiley & Sons, 2000.

Gribbon, John. *Unveiling the Edge of Time: Black Holes, White Holes, Wormholes*. New York: Harmony Books, 1992.

Griffiths, Denis. *Steam at Sea: Two Centuries of Steam-Powered Ships*. London: Conway Maritime Press, 1997.

Guerlac, Henry. *Antoine-Laurent Lavoisier: Chemist and Revolutionary*. New York: Charles Scribner's Sons, 1975.

Haggard, Howard W. *Devils, Drugs and Doctors*. Garden City, NY: Blue Ribbon Books, 1929.

Hahn, Alexander. *Basic Calculus: From Archimedes to Newton to Its Role in Science*. New York: Springer, 1998.

Hall, Tomas S. *Ideas of Life and Matter: Studies in the History of General Physiology, 600 BC–1900 AD*, volume 2. Chicago: University of Chicago Press, 1969.

Halliday, David, and Robert Resnick. *Fundamentals of Physics*, third edition. New York: John Wiley & Sons, 1988.

Hankins, Thomas L. *Science and the Enlightenment*. New York: Cambridge University Press, 1985.

Hart, Ivor B. *James Watt and the History of Steam Power.* New York: Collier Books, 1961.

Hartley, Sir Harold. *Studies in the History of Chemistry.* New York: Oxford University Press, 1971.

Hays, H. R. *Birds, Beasts, and Men: A Humanist History of Zoology.* New York: G. P. Putnam's Sons, 1972.

Heilbron, J. L. *Electricity in the 17th and 18th Centuries: A Study of Early Modern Physics.* Los Angeles: University of California Press, 1979.

Herrmann, Ernest E. *Exterior Ballistics 1935.* Annapolis, MD: US Naval Institute, 1935.

Hill, Ralph Nading. *Sidewheeler Saga, A Chronicle of Steamboating.* New York: Rinehart, 1953.

Hills, Richard L. *Power from Steam: A History of the Stationary Steam Engine.* New York: Cambridge University Press, 1989.

Hills, Richard L. *Power in the Industrial Revolution.* Manchester, UK: Manchester University Press, 1970.

Hindle, Brooke. *Emulation and Invention.* New York: New York University Press, 1981.

Hoffmann-Axthelm, Walter. *History of Dentistry.* Chicago: Quintessence, 1981.

Hollen, Norma, and Jane Saddler, Anna Langford, and Sara Kadolph. *Textiles*, sixth edition. New York: Macmillan, 1988.

Holmyard, Eric John. *Makers of Chemistry.* Oxford: Clarendon Press, 1931.

Holt, Anne. *A Life of Joseph Priestley.* Westport, CT: Greenwood Press, 1970.

Hoyle, Fred. *Astronomy.* Garden City, NY: Doubleday, 1962.

Hudson, Robert P. *Disease and Its Control: The Shaping of Modern Thought.* Westport, CT: Greenwood Press, 1983.

Hunt, Garry E., and Patrick Moore. *The Planet Venus.* London: Faber and Faber, 1982.

Ichord, Loretta Frances. *Toothworms & Spider Juice: An Illustrated History of Dentistry.* Brookfield, CT: Millbrook Press, 2000.

Jackson, G. Gibbard. *The Book of the Ship.* New York: D. Appleton-Century, 1938.

Jungnickel, Christa, and Russel McCormmach. *Cavendish: The Experimental Life.* Lewiston, PA: Bucknell, 1999.

Karplus, Robert. *Physics and Man.* New York: W. A. Benjamin, 1970.

Ketcham, Ralph L. *Benjamin Franklin.* New York: Washington Square Press, 1966.

Knight, David M. *Natural Science Books in English: 1600–1900.* New York: Praeger, 1972.

Knight, David. *Ideas in Chemistry: A History of the Science.* New Brunswick, NJ: Rutgers University Press, 1992.

Krebs, Robert E. *Scientific Laws, Principles, and Theories: A Reference Guide.* Westport, CT: Greenwood Press, 2001.

Krebs, Robert E. *The History and Use of Our Earth's Chemical Elements: A Reference Guide.* Westport, CT: Greenwood Press, 1998.

Krider, E. P. "The Heritage of B. Franklin," in *Benjamin Franklin: Des Lumiéres a Nos Jours.* Edited by G. Hugues and D. Royot. Université Jean-Moulin-LYON III: Didier Erudition, 1991.

Lancaser-Brown, Peter. *Halley & His Comet.* New York: Sterline, 1985.

Larson, James L. *Reason and Experience: The Representation of Natural Order in the Work of Carl von Linné.* Los Angeles: University of California Press, 1971.

Lienhard, John H. "Diderot's Encyclopedia." *The Engines of Our Ingenuity* series. Web article available at: http://www.uh.edu/engines/epi122.htm. Copyright 1988–1997.

Lienhard, John H. "Encyclopaedia Britannica." *The Engines of Our Ingenuity* series. Web article available at: http://www.uh.edu/engines/epi203.htm. Copyright 1988–1997.

Lockemann, Georg. *The Story of Chemistry*. New York: Philosophical Library, 1959.

Lockett, Keith. *Physics in the Real World*. New York: Cambridge University Press, 1990.

Locy, William A. *The Growth of Biology*. New York: Henry Holt, 1925.

Loudon, Irvine (ed.). *Western Medicine: An Illustrated History*. New York: Oxford University Press, 1997.

Lufkin, Arthur Ward. *A History of Dentistry*. Philadelphia: Lea & Febiger, 1948.

McGraw-Hill Encyclopedia of Science and Technology, volume 1. Washington, DC: McGraw-Hill, 1997.

McGrew, Roderick E. *Encyclopedia of Medical History*. New York: McGraw-Hill, 1985.

McKie, Douglas, and Niels H. de V. Heathcote. *The Discovery of Specific and Latent Heats*. New York: Arno Press, 1975.

Meldrum, Andrew Norman. *The Eighteenth Century Revolution in Science—The First Phase*. New York: Longmans, Green, 1930.

Meyer, Herbert W. *A History of Electricity and Magnetism*. Cambridge, MA: MIT Press, 1971.

Middleton, W. E. Knowles. *A History of the Thermometer and Its Use in Meteorology*. Baltimore: Johns Hopkins Press, 1966.

Millar, David, Ian Millar, John Millar, and Margaret Millar. *The Cambridge Dictionary of Scientists*. New York: Cambridge University Press, 1996.

Mirsky, Jeannette, and Allan Nevins. *The World of Eli Whitney*. New York: Macmillan, 1952.

Moore, F. J. *A History of Chemistry*. New York: McGraw-Hill, 1931.

Moore, Patrick. *The Planets*. New York: W. W. Norton, 1962.

Motz, Lloyd, and Jefferson Hane Weaver. *The Story of Astronomy*. New York: Plenum Press, 1995.

North, John. *The Fontana History of Astronomy and Cosmology*. London: HarperCollins, 1994.

Olson, Richard (ed.). *Biographical Encyclopedia of Scientists*, volume 5. New York: Marshall Cavendish Corporation, 1998.

Panek, Richard. *Seeing and Believing: How the Telescope Opened Our Eyes and Minds to the Heavens*. New York: Viking Penguin, 1998.

Pannekoek, A. *A History of Astronomy*. New York: Interscience, 1961.

Parkinson, Claire L. *Breakthroughs: A Chronology of Great Achievements in Science and Mathematics, 1200–1930*. Boston: G. K. Hall, 1985.

Partington, J. R. *A Short History of Chemistry*. New York: St. Martin's Press, 1965.

Plambeck, James. "Chemical Sciences: Chemical Thermodynamics, From Heat to Enthalpy" (website). Copyright 1995 James A. Plambeck, updated July 4, 1996. http://www.chem.ualberta.ca/~plambeck/che/p101/p01073.htm.

Poirier, Jean-Pierre. *Lavoisier: Chemist, Biologist, Economist*. Translated by Rebecca Balinski. Philadelphia: University of Pennsylvania Press, 1996.

Porter, Roy (ed.). *The Biographical Dictionary of Scientists*. New York: Oxford University Press, 1994.

Porter, Roy (ed.). *The Cambridge Illustrated History of Medicine*. New York: Cambridge University Press, 1996.

Prinz, Hermann. *Dental Chronology, a Record of the More Important Historic Events in the Evolution of Dentistry*. Philadelphia: Lea & Febiger, 1945.

Proctor, Michael, Peter Yeo, and Andrew Lack. *The Natural History of Pollination*. Portland, OR: Timber Press, 1996.

Quinn, T. J. *Temperature*. New York: Academic Press, 1983.

Rhodes, Philip. *An Outline History of Medicine*. Boston: Butterworths, 1985.

Riedman, Sarah R. *Antoine Lavoisier, Scientist and Citizen*. New York: Abelard-Schuman, 1967.

Rolt, L. T. C. *The Aeronauts: A History of Ballooning, 1783–1903.* Gloucester, UK: Alan Sutton, 1985.

Rolt, L. T. C., and J. S. Allen. *The Steam Engine of Thomas Newcomen.* New York: Science History Publications, 1977.

Ronan, Colin. *The Astronomers.* New York: Hill and Wang, 1964.

Rowland, K. T. *Eighteenth Century Inventions.* New York: David & Charles Books, 1974.

Rowland, K. T. *Steam at Sea: A History of Steam Navigation.* New York: Praeger, 1970.

Rumford, Count Benjamin Thomson. "Heat Excited by Friction." Published in *Philosophical Transactions* (vol. 88), 1798. Reprinted editions of this paper are widely available in books or on the web, at sites such as: http://dbhs.wvusd.k12.ca.us/Chem-History/Rumford-1798.html and http://dev.nsta.org/ssc/pdf/v4-0937s.pdf.

Rutland, Jonathan. *The Age of Steam.* New York: Random House, 1987.

Sagendorph, Robb. *America and Her Almanacs: Wit, Wisdom & Weather, 1639–1970.* Boston: Little, Brown, 1970.

Schwarts, A. Truman, and John G. McEvoy (eds.). *Motion Toward Perfection: The Achievement of Joseph Priestley.* Boston: Skinner House Books, 1990.

Schwarz, W. M. *Intermediate Electromagnetic Theory.* New York: John Wiley & Sons, 1964.

Siegel, Beatrice. *The Steam Engine.* New York: Walker and Company, 1986.

Simmons, George Finlay. *Calculus Gems: Brief Lives and Memorable Mathematics.* New York: McGraw-Hill, 1992.

Singer, Charles. *A Short History of Medicine.* New York: Oxford University Press, 1928.

Smith, Betty. *A Tree Grows in Brooklyn.* New York: Harper, 1947.

Smith, Ray F. (ed.). *History of Entomology.* Palo Alto, CA: Annual Reviews, 1973.

Sootin, Harry. *12 Pioneers of Science.* New York: Vanguard Press, 1960.

Stillman, John Maxson. *The Story of Early Chemistry.* New York: D. Appleton, 1924.

Stowell, Marion Barber. *Early American Almanacs: The Colonial Weekday Bible.* New York: Burt Franklin, 1977.

Sullivan, Walter. *Black Holes: The Edge of Space, the End of Time.* Garden City, NY: Anchor Press/Doubleday, 1979.

Taylor, John G. *The New Physics.* New York: Basic Books, 1972.

Thiel, Rudolf. *The Discovery of the Universe.* Translated by Richard Winston and Clara Winston. New York: Alfred A. Knopf, 1967.

Travers, Bridget (ed.). *World of Invention.* Detroit: Gale Research, 1994.

Trefil, James S. *Space Time Infinity.* Washington, DC: Smithsonian Books, 1985.

Turner, Herbert Hall. *Astronomical Discovery.* Berkeley, CA: University of California Press, 1963.

Urdang, George. *The Apothecary Chemist: Carl Wilhelm Scheele.* Madison, WI: American Institute of the History of Pharmacy, 1958.

Virginskii, V. S. *Robert Fulton, 1765–1815.* Translated by Vijay Pandit. Washington, DC: Smithsonian Institution and National Science Foundation, 1976.

Waligunda, Bob, and Larry Sheehad. *The Great American Balloon Book: An Introduction to Hot Air Ballooning.* Englewood Cliffs, NJ: Prentice-Hall, 1981.

Weinberger, Bernhard W. *An Introduction to the History of Dentistry,* volume 1. St. Louis: C. V. Mosby, 1948.

Weinberger, Bernhard W. *Pierre Fauchard, Surgeon-Dentist.* Minneapolis: Lancet Press, 1941.

Wheeler, Mark. "Declaration of Independence. *Inc Magazine.* September 15, 1996. Available online at: http://www2.inc.com/incmagazine/articles/2000.html.

Whitmer, Robert M. *Electromagnetics*. New York: Prentice-Hall, 1952.

Wilson, Geoffrey. *The Old Telegraphs*. London: Phillimore, 1976.

World Book (encyclopedia) CD-ROM. San Diego: World Book, 1999.

Wynbrandt, James. *The Excruciating History of Dentistry: Toothsome Tales & Oral Oddities from Babylon to Braces*. New York: St. Martin's Press, 1998.

Yeomans, Donald K. *Comets: A Chronological History of Observation, Science, Myth, and Folklore*. New York: John Wiley & Sons, 1991.

INDEX